Managing Digital Transformation

This book provides practising executives and academics with the theories and best practices to plan and implement digital transformation successfully. Key benefits:

- an overview on how leading companies plan and implement digital transformation
- interviews with chief executive officers and chief digital officers of leading companies – Bulgari, Deutsche Bahn, Henkel, Lanxess, L'Oréal, Unilever, Thales and others – explore lessons learnt and roadmaps to successful implementation
- research and case studies on the digitalization of small and medium-sized companies
- cutting-edge academic research on business models, organizational capabilities and performance implications of the digital transformation
- tools and insights into how to overcome internal resistance, build digital capabilities, align the organization, develop the ecosystem and create customer value to implement digital strategies that increase profits

Managing Digital Transformation is unique in its approach, combining rigorous academic theory with practical insights and contributions from companies that are, according to leading academic thinkers, at the forefront of global best practice in the digital transformation. It is recommended reading both for practitioners looking to implement digital strategies within their own organizations, as well as for academics and postgraduate students studying digital transformation, strategy and marketing.

Andreas Hinterhuber is Associate Professor of Marketing, Ca' Foscari University of Venice, Italy.

Tiziano Vescovi is Full Professor of Strategic Marketing, Ca' Foscari University of Venice, Italy.

Francesca Checchinato is Associate Professor of Marketing, Ca' Foscari University of Venice, Italy.

Managing Digital Transformation

Understanding the Strategic Process

Edited by Andreas Hinterhuber,
Tiziano Vescovi and
Francesca Checchinato

LONDON AND NEW YORK

First published 2021
by Routledge
2 Park Square, Milton Park, Abingdon, Oxon OX14 4RN

and by Routledge
52 Vanderbilt Avenue, New York, NY 10017

Routledge is an imprint of the Taylor & Francis Group, an informa business

British Library Cataloguing-in-Publication Data
A catalogue record for this book is available from the British Library

Library of Congress Cataloging-in-Publication Data
Names: Hinterhuber, Andreas, editor. | Vescovi, Tiziano, editor. |
Checchinato, Francesca, editor.
Title: Managing digital transformation : understanding the strategic process /
edited by Andreas Hinterhuber, Tiziano Vescovi and Francesca Checchinato.
Description: Abingdon, Oxon; New York, NY: Routledge, 2021. |
Includes bibliographical references and index. |
Identifiers: LCCN 2020053460 | ISBN 9780367442682 (hardback) |
ISBN 9780367441975 (paperback) | ISBN 9781003008637 (ebook)
Subjects: LCSH: Information technology–Management. |
Technological innovations–Management. | Organizational change. | Strategic planning.
Classification: LCC HD30.2 .M36186 2021 | DDC 004.068–dc23
LC record available at https://lccn.loc.gov/2020053460

ISBN: 978-0-367-44268-2 (hbk)
ISBN: 978-0-367-44197-5 (pbk)
ISBN: 978-1-003-00863-7 (ebk)

Typeset in Bembo
by Newgen Publishing UK

Contents

Figures

Tables

Book editors

Andreas Hinterhuber is an Associate Professor at the Department of Management at Università Ca' Foscari Venezia, Italy. He has published articles in leading journals including *Journal of Business Research* and *MIT Sloan Management Review*, and has edited many books on pricing, including *Innovation in Pricing* (2012/2017), *The ROI of Pricing* (2014), *Pricing and the Sales Force* (2016), *Value First then Price* (2017) and *Pricing Strategy Implementation* (2020). He can be contacted at: andreas.hinterhuber@unive.it

Tiziano Vescovi is a Full Professor at the Department of Management at Università Ca' Foscari Venezia, Italy. His main research interests focus on cross-cultural and international marketing and on brand management. He has published books, articles in several academic journals, and conference proceeding. He was a Visiting Professor in several European, Chinese and American universities. He is the Director of International Management to Asia Lab at Università Ca' Foscari. He can be contacted at: tiziano.vescovi@unive.it

Francesca Checchinato is an Associate Professor at the Department of Management at Università Ca' Foscari Venezia, Italy. Her research focuses on the digital marketing investigating how social media affect the marketing strategy and on brand management, with a particular focus on the food sector. She has published several academic journals, books and conference proceedings. She can be contacted at: f.checchinato@unive.it

Contributors

Axel Adida is an advisor to the Boston Consulting Group. Previously, he was Digital Chief Operations Officer of L'Oréal Group (2014–2019) and was responsible for the advertising and business intelligence digitisation. He can be contacted at: axel.adida@ gmail.com

Marisa Agostini is a fixed-term Assistant Professor in accounting at the Department of Management (Ca' Foscari University, Venice, Italy). During her international PhD in Business, she was a visiting scholar at the McCombs School of Business (The University of Texas at Austin, USA). She is a member of several national and international academic associations. She can be contacted at: marisa.agostini@unive.it

Ajay Agrawal is the Geoffrey Taber Chair in Entrepreneurship and Innovation and Professor of Strategic Management at the Rotman School of Management, where he founded the Creative Destruction Lab, a seed-stage program for scalable, science-based companies. CDL has six locations: Toronto, Vancouver, Calgary, Montreal, Halifax and Oxford, UK. He can be contacted at: Ajay.Agrawal@rotman.utoronto.ca

Jean-Christophe Babin is the Chief Executive Officer of the Bulgari Group since 2013. He graduated from the HEC Business School in Paris. Babin began his career in 1983 in sales and marketing, later worked at Procter & Gamble in France and at Benckiser and Henkel in Italy. In 2000 he joined the LVMH group as CEO of the TAG Heuer watch brand, which he led for 13 years.

Thijs Broekhuizen is an Associate Professor of Innovation at the Department of Innovation Management and Strategy at the Faculty of Economics and Business, University of Groningen, the Netherlands. His research focuses on the intersection of innovation and marketing. He is a researcher and board member of the Groningen Digital Business Center. He can be contacted at: t.l.j.broekhuizen@rug.nl

Frank V. Cespedes is Senior Lecturer in the Entrepreneurial Management Unit at Harvard Business School, USA, where he has developed and taught a variety of MBA and executive courses, led the Strategic Marketing Management program for senior executives, and was co-lead of the Sustainable Market Leadership program for CEOs and their leadership teams. Before joining the faculty, he was a Research Associate at Harvard and worked at Bain & Company, USA. He can be contacted at: fcespedes@ hbs.edu

Francesca Checchinato is an Associate Professor at the Department of Management at Università Ca' Foscari Venezia, Italy. Her research focuses on the digital marketing investigating how social media affect the marketing strategy and on brand management, with a particular focus on the food sector. She has published several academic journals, books and conference proceedings. She can be contacted at: f.checchinato@unive.it

Nunzia Coco is a Post-Doctoral Research Fellow at Ca' Foscari University of Venice. She holds a PhD in Management from Ca' Foscari. Her main interest regards the role of design as a driver for innovation and co-creation. She can be contacted at: nunzia. coco@unive.it

Cinzia Colapinto is an Assistant Professor of Strategy and Entrepreneurship at the Department of Management of Ca' Foscari University of Venice. Her research is about the role played by digital technologies on business model transformations and on their impact on the achievement of sustainable development goals by SMEs. Her research has been published in reputable academic journals and in Italian and international books. She can be contacted at: cinzia.colapinto@unive.it

Nicolai E. Fabian is a PhD candidate at the Department of Innovation Management and Strategy / Department of Marketing, University of Groningen. His research focuses on the digital transformation of firms, integrating insights from information systems and marketing. He is also an active researcher within the Groningen Digital Business Center (GDBC), a knowledge centre that aims to bridge business practice and scientific research. He can be contacted at: n.e.fabian@rug.nl

Luca Ferrari holds a MSc in Electrical and Electronics Engineering from the University of Padua and one in Telecommunications Engineering from the Technical University of Denmark, both with honors. He worked at McKinsey & Company right after finishing his studies. He is a co-founder and the CEO of Bending Spoons – one of the leading mobile app developers globally.

Vladi Finotto is an Associate Professor of Strategy and Entrepreneurship at the Department of Management of Ca' Foscari University of Venice. His research focuses on the implications of digital technologies on SMEs' strategies, with a particular attention to low- and medium-tech sectors such as the food industry. He has published on business model innovation and on the processes of organizational mobilization; he investigates the conditions enabling the birth and growth of regional innovation and entrepreneurial ecosystems. His work has been published in international peer-reviewed journals, Italian and international books and it was presented at major international innovation and management conferences. He can be contacted at: vfinotto@unive.it

Giuseppe Folonari is the Executive Director Europe, Business Strategy, AKQA. Giuseppe joined AKQA at the beginning of 2014 and, from 2016, took a business development role in Hong Kong. He leads the European Business Strategy team, the unit responsible for strategic advisory and open innovation projects, delivering end-to-end services to investigate new (digital) products and services, from business modelling to design, development and delivery of prototypes. Previously, he was Investment Manager at H-Farm Ventures, leading the investment team in the screening and evaluation of a deal flow in excess of 1000 startups per year.

Jörg Hellwig is Chief Digital Officer of Lanxess, leading the company's digital transformation initiative and all related activities, reporting to the CEO. He additionally leads the "New Work" cultural transformation program to support the skill and talent necessary for the transition to a digital organization. As a champion of new technologies at the company, Jörg is also responsible for exploring new business models. As such, he founded the start-up company CheMondis and served as Managing Director and Supervisory Board member. CheMondis, now independent, has become the global leading online marketplace for the chemicals industry. He can be contacted at: joerg.hellwig@lanxess.com

Andreas Hinterhuber is an Associate Professor at the Department of Management at Università Ca' Foscari Venezia, Italy. He has published articles in leading journals including *Journal of Business Research* and *MIT Sloan Management Review* and has edited many books on pricing, including *Innovation in Pricing* (2012/2017), *The ROI of Pricing* (2014), *Pricing and the Sales Force* (2016), *Value First then Price* (2017) and *Pricing Strategy Implementation* (2020). He can be contacted at: andreas.hinterhuber@unive.it

Vinod Khosla is an Indian American venture capitalist. He is a co-founder of Sun Microsystems and the founder of Khosla Ventures, which invests in technology-based businesses both for profit and social impact, including in clean tech, microfinance and biomed tech. He can be contacted at: kv@khoslaventures.com

Dmitry Koroteev is a Professor at the Skolkovo Institute of Science and Technology (Skoltech) and the founder and CEO of Digital Petroleum LLC in Moscow, Russia. He holds a PhD in physics from Lomonosov Moscow State University. Previously, he worked in the oil and gas industry, for Schlumberger and Gazprom Neft. He can be contacted at: koroteev@petroleum.digital

Stephan M. Liozu is Chief Value Officer of the Thales Group (www.thalesgroup.com), the Founder of Value Innoruption Advisors (www.valueinnoruption.com) – a consulting boutique specializing in value-based pricing, data monetization and digital pricing – and an Adjunct Professor and Research Fellow at the Weatherhead School of Management at Case Western Research University, USA. Stephan sits on the Advisory Board of LeveragePoint Innovation and of the Professional Pricing Society. He is the author of multiple books about pricing, including *Pricing and Human Capital* (2015), and co-edited *Innovation in Pricing* (2012/2017), *The ROI of Pricing* (2014) and *Pricing and the Salesforce* (2015). He can be contacted at: sliozu@gmail.com

Dinh Khoi Nguyen is a PhD candidate at the Department of Innovation Management and Strategy, University of Groningen. His research focuses on the digital readiness of firms and employees, integrating insights from information systems, innovation and marketing. He is an active researcher of the Groningen Digital Business Center. He can be contacted at: d.k.nguyen@rug.nl

Michael Nilles is Chief Digital & Information Officer (CDIO) for Henkel and responsible for Henkel dx, in charge of all Digital, IT, Corporate Venture Capital activities and the Open Innovation Platform Henkel dx Ventures. Before joining Henkel, he was a member of the Group Executive Board and Chief Digital Officer (CDO) for Schindler as well as the CEO of Schindler Digital Business AG. Prior to Schindler, he held various executive management positions at Mannesmann and Bosch Rexroth. He also serves on the Supervisory Boards of Deutsche Lufthansa AG and Lufthansa Technik AG. He can be contacted at: michael.nilles@henkel.com.

Enrico Pieretto is a recent graduate student from Ca' Foscari University of Venice with a master thesis on the digital transformation of SMEs. He can be contacted at: pierettoenrico@gmail.com

Stefan Stroh is the Chief Digital Officer and Chairman of the Digital Executive Board at Deutsche Bahn Group. He is responsible for the design and implementation of DB's digital strategy across all Business Units and the development of DB's digital Ecosystem (tech companies, start-up relations, accelerator and incubator programs, company building, corporate venturing and digital factories). Before joining Deutsche Bahn, Stefan Stroh was a partner with PwC Strategy&. Stefan graduated in mechanical engineering at TU Darmstadt and holds a degree in environmental management from TU Berlin. He can be contacted at: Stefan.Stroh@deutschebahn.com

Zeljko Tekic is Assistant Professor at the Skolkovo Institute of Science and Technology (Skoltech) in Moscow. He is a member of Skoltech's Center for Entrepreneurship and Innovation. Zeljko received his PhD degree in Industrial Engineering and Engineering Management from the University of Novi Sad, Serbia. His research focuses on topics at the interplay of digital technology, open innovation and startups. Recently, Zeljko was a Visiting Professor and Researcher at Massachusetts Institute of Technology, Cambridge. He can be contacted at: z.tekic@skoltech.ru

Tiziano Vescovi is a Full Professor at the Department of Management at Università Ca' Foscari Venezia, Italy. His main research interests focus on cross-cultural and international marketing and on brand management. He has published books, articles in several academic journals and conference proceeding. He was a Visiting Professor in several European, Chinese and American universities. He is the Director of International Management to Asia Lab at Università Ca' Foscari. He can be contacted at: tiziano.vescovi@unive.it

Gregory Vial is Assistant Professor of Information Technology (IT) at HEC Montreal. His research interests are in the areas of digital transformation, systems development practices and methodologies, and data management. His work has been published in journals such as *Journal of Strategic Information Systems*, *European Journal of Information Systems*, *Decision Support Systems*, *IEEE Software* as well as in several IS conferences. He can be reached at: gregory.vial@hec.ca

Part 1

Introduction

Chapter 1

Digital transformation

An overview

Andreas Hinterhuber, Tiziano Vescovi
and Francesca Checchinato

Digital transformation is "a process that aims to improve an entity by triggering significant changes to its properties through combinations of information, computing, communication, and connectivity technologies" (Vial, 2019, p. 121). Digital transformation is multidisciplinary by nature, pervasive, and growth-oriented (Verhoef et al., 2020). Recently, the topic has attracted substantial interest and, accordingly, comprehensive literature reviews (Verhoef et al., 2020; Vial, 2019).

Current research suggests that for most companies the digital transformation is not a question of *if* or *when* (Hess, Matt, Benlian, & Wiesböck, 2016): the digital transformation is given; key questions relate to how to develop and implement a digital transformation strategy.

And, yet it is precisely on the question of how that managers lack guidance (Hess et al., 2016). This is because a large part of the managerially relevant studies on the digital transformation have been produced by consulting companies or solution providers that are not without self-interest in their recommendations (Peter, Kraft, & Lindeque, 2020).

Digital transformation thus represents a substantial problem of practice: the overwhelming majority of companies is not ready for a digital future (Kane, Palmer, Phillips, & Kiron, 2015). There are, to be sure, several cases of companies that have successfully mastered the digital transformation: Adobe, Netflix, *The New York Times* will probably stand the test of time. The textbox illustrates pertinent aspects of the digital transformation of Adobe (Hinterhuber & Liozu, 2020).

The digital transformation of Adobe

Adobe, the US$11 billion software company, is an illuminating example of a successful digital transformation: In 2011, more than 80% of the company's sales were product-based. In 2019, close to 90% of the company's sales are digital subscriptions, i.e., usage rights. The move from products to digital subscriptions is driven both by a strong customer orientation – subscription sales allow immediate product updates, as opposed to product sales that are driven by release cycles – as well as by the objective to increase performance. Mark Garrett, CFO of Adobe, comments: "We were driving revenue growth by raising our average selling price— either through straight price increases or through moving people up the product ladder. That wasn't a sustainable approach" (Sprague, 2015, pp. 1–2).

The digital transformation of Adobe is built on several elements that we high-light in this volume: (1) new organizational capabilities. Says Mark Garrett: "one of the most challenging aspects of the cloud transition [was] changing how we brought products to market. We had to educate and compensate the channel and our salesforce differently" (Sprague, 2015, p. 5). (2) business model innovation: digital transformation requires new business models, a transition from selling physical products to selling digital solutions. An important element in this process is internal and external analysis. Again, Mark Garett comments: "We spent hours knee-deep in Excel spreadsheets modeling this out. We literally covered the boardroom with pricing and unit models. Through this discussion, which took about a year, we saw that we could manage through it and that we, our customers, and our shareholders would come out on the other side much better off" (Sprague, 2015, p. 2). (3) customer value creation: the digital transformation is about creating meaningful customer experiences. Dan Cohen, VP of strategy of Adobe: the digital transformation allows "offering a broader and better value proposition," "attracting new customers," and a "better read of [customer] needs" enabled by closer relationships with customers (Sprague, 2015, p. 6).

In the context of the digital transformation, Adobe changed its capabilities, business model and improved the value delivered to customers (S. Gupta & Barley, 2015). The digital transformation is delivering tangible results: from 2011 to 2019 sales revenues more than doubled and operating margins increased from 26% to 29%.

Many other companies failed largely as a result of not getting the digital transformation right: Kodak and Blockbuster are trivial examples.

The point is: while many senior executives state that the digital transformation is critical for survival, only 3% of companies have actually completed organization-wide digital transformation efforts (Chanias, Myers, & Hess, 2019).

A key objective of this edited volume is to present scientifically grounded, practically relevant contributions on designing and implementing the digital transformation.

This book is thus intended as a guide that lays out how to transition from a largely pre-digital to a largely digital strategy, i.e., how to transition from marketing/selling physical products/services to conceiving and marketing digital products/services.

The structure of the book

"Part 1 Introduction" contains this introductory chapter, by *Andreas Hinterhuber, Tiziano Vescovi*, and *Francesca Checchinato*.

"Part 2 Digital transformation and organizational capabilities" contains several chapters that address the organizational capabilities needed in the context of the digital transformation.

In Chapter 2 "Understanding digital transformation: a review and a research agenda" *Gregory Vial* presents the definition of "digital transformation" outlined above. What makes the study of the digital transformation so fascinating and so important is the insight that it offers both game-changing opportunities as well as existential threats to businesses.

The literature analyzed converges on the following topics: digital transformation changes the value delivery process, from products to services/solutions and from single actors to value networks/ecosystems. The digital transformation requires structural changes within organizations, typically the establishment of separate organizational units under the leadership of the chief digital officer (CDO) – a question that this book explores in Part 3. Digital transformation then requires a change in organizational culture and in organizational capabilities – several interviews with CDOs explore this question in this volume. Typical obstacles to the successful digital transformation are inertia or active resistance. Finally, benefits of the digital transformation are improvements in efficiency, innovation, and organizational performance.

In Chapter 3 "The three pillars of the digital transformation: improving the core, building new business models and developing digital capabilities" by *Andreas Hinterhuber* and *Stefan Stroh* highlights pillars of the digital transformation of Deutsche Bahn, Europe's largest railway operator. Improving the core is about using digital technologies to improve value delivery. Building new business models is about building start-up businesses that able to compete with competitors that disrupt current business models: the willingness to cannibalize the core is a prerequisite (Kumar, Scheer, & Kotler, 2000). Developing digital capabilities is about data, IT infrastructure, innovation ecosystem, and new work styles. Stefan Stroh, CDO, also highlights obstacles to a successful digital transformation – with the difficulty in changing the culture of the organization featuring prominently. Metrics, according to this interview, are not different from metrics used to assess traditional initiatives – as the chapter by *Gregory Vial* also points out.

In Chapter 4 "Big data and analytics: opportunities and challenges for firm performance" *Marisa Agostini* analyzes challenges related to exploiting the opportunities offered by big data based on an extensive review of the literature. Three items three main items – people, culture, and technology – represent the main barriers (and the potential solutions) to required organizational changes. *Marisa Agostini* also highlights a growing gap between what is technologically possible and what is being implemented in organizations. This, of course, only increases the value of big data analytics capabilities (M. Gupta & George, 2016).

In Chapter 5 "Technology is just an enabler of digital transformation: The case of Unilever" *Francesca Checchinato* discusses the digital transformation process of Unilever, a leading company in the FMCG sector highlighting the main elements that drive to the success. A new business model based on a completely different organization, a clear vision and a strong support from CEO are some of the key elements of its digital transformation. Unilever offers training and learning programs to its employees to both improve their skills and coach them to better work on the new digital landscape. Without a new attitude and a cultural shift, digital technologies would just remain a tool without any value.

"Part 3 Digital transformation and business model innovation" contains several chapters highlighting the role of business models in the context of the digital transformation.

In Chapter 6 "Digital transformation and business models" *Tiziano Vescovi* analyzes the relationship between the digital transformation and the change of the business models, that is the influence of the digital technology adoption on the models of value creation of the company. In the chapter, four main aspects are highlighted: (a) different types of digital business models, (b) the main areas of business change induced by digitalization of the company, (c) difficulties and problems arose in changing the business model because of digital transformation, and (d) the pathway to a digital business model.

In Chapter 7 "From disruptively digital to proudly analog: a holistic typology of digital transformation strategies" *Zeljko Tekic* and *Dmitry Koroteev* provide a conceptual classification of digital transformation strategies in terms of two critical dimensions: usage of digital technologies and readiness of the business model for digital operation. The result is a typology of four generic digital transformation strategies that differ in target of transformation, leadership style, skills, and improvement tactics. The proposed framework – which is conceptual and thus needs empirical validation – defines digital transformation on a continuum with four typologies: (1) partial digitalization, (2) efficiency-driven digitalization, (3) business model-driven digitalization, and (4) disruptive digitalization. Digital transformation strategies are thus, in this view, driven by three distinct objectives: efficiency improvements, innovation, and disruption – with some companies largely unaffected by the digital transformation that will continue to operate mostly analogously.

In Chapter 8 "Digital transformation, the Holy Grail and the disruption of business models" *Andreas Hinterhuber* and *Michael Nilles* discuss the game-changing opportunities that the digital transformation opens to companies that embed a digital core into their business models. *Michael Nilles*, CDO of Henkel, one of the world's largest consumer good companies sees digital transformation as the search for the Holy Grail: a force that is not easy to find, not easy to capture, and that has the potential to dramatically improve the customer experience. In B2C, the Holy Grail is, in the specific case of Henkel, the beauty ID ecosystem – a series of connected devices harnessing big data and augmented reality – to create meaningful, personalized, and direct relationships with customers. In B2B, the Holy Grail is, for Henkel but also GE (Immelt, 2017), the digital twin along the digital thread. Henkel builds a digital twin along the entire value chain of the customer – starting with the initial customer request via deployment until and after sales service – in order to sell outcomes to customers. In B2B the digital twin enables outcome-based servitization (Bustinza, Vendrell-Herrero, & Baines, 2017). Digital technologies are thus more than enabling technologies: in horizon 1 – extend and defend core businesses (Baghai, Coley, & White, 2000) – digital technologies are, in fact, enablers. In horizon 2 – build emerging businesses – and in horizon 3 – create genuinely new businesses – digital technologies allow the creation of fundamentally new, disruptive business models. At a fundamental level, the digital transformation is, according to *Michael Nilles*, built on two shifts: a move from the product towards a software stack and a move from single products to open platforms. These two, fundamental forces are, in this view, the basic building blocks of every digital transformation: these are the pathways towards the Holy Grail. The interview then discusses the cultural and organizational aspects related to the digital transformation: it needs small, agile teams with end-to-end responsibility for project delivery, the mindset of a start-up company and customer obsession. The final aspect that this contribution discusses are metrics of the digital transformation. Key metrics are digital efficiency/productivity, digital revenues, and digital growth (net of cannibalization). These metrics are linked to the three horizons of the digital transformation: digital efficiency (horizon 1), digital revenues, and digital growth (horizon 2 and horizon 3). This is a great interview that every manager must read.

In Chapter 9 "How artificial intelligence and the digital transformation change business and society" *Vinod Khosla*, one of the world's most successful technology entrepreneurs, and *Ajay Agrawal* discuss the AI-generated opportunities and dangers that lie ahead, and how to prepare for them. This interview explores the role of academics in shaping the role of education in a world characterized by an increasing ability of machines to replace jobs formerly occupied by humans. This discussion anticipates the coming age of hyper-innovation. Technology, *Vinod Khosla*, reminds us, does not rule – it serves those masters that are able to use technology to better serve business and society.

In Chapter 10 "Digital transformations: the case of bending spoons" *Tiziano Vescovi* and *Luca Ferrari*, CEO of Bending Spoons, deepen the strategic way of thinking of a company born digital. It emerges that being digital is not about adopting software, which remains just a tool, but is a way of finding solutions to customer problems using, in large part, digital solutions. The business model of a digital company such as Bending Spoons is based on two key aspects: partnership, the network of valuable connections, and key resources, above all represented by people. In a digital company, people are more important than technology. The surprising changing trends of digital products are also highlighted.

In Chapter 11 "Consulting for digital transformation" *Tiziano Vescovi* asks *Giuseppe Folonari*, the European Head of Business Strategy of Akqa, a consulting company for digital transformation, discuss how to advise companies in adopting digital strategies and actions. It emerges that companies facing digital transformation must accept and know how to make decisions in conditions of uncertainty without waiting for everything to be clear, without expecting an immediate result, as evolution is continuous. The combination of digital and analog is a key to success, avoiding digital washing. The business model that emerges is an open model, capable of welcoming and understanding valuable partners. Companies are required to make a cultural change that is not always easy.

The subsequent chapter "L'Oréal digital consumer operating system" by *Axel Adida* highlights the digital transformation of one of Europe's largest consumer goods companies, L'Oréal. Traditional models of mass communication do not reach today's, mostly young, consumers. *Axel Adida* highlights the steps that the company took to digitalize communication and sales. Awareness of the opportunities and threats of the digital transformation, senior leadership commitment, new digital skills, true customer centricity, new rituals reflecting the importance of digital technologies, new KPIs, and a willingness to experiment, learn, and adjust are important elements of the digital transformation of L'Oréal.

In Chapter 13 "Internal start-ups as a driving force in the digitalization of traditional businesses" *Andreas Hinterhuber* and *Jörg Hellwig* discuss the role of internal, digital start-ups as a driving force for the digital transformation of traditional businesses. *Jörg Hellwig*, CDO of Lanxess, one of Europe's largest specialty chemical companies, played a key role in establishing CheMondis, a digital start-up that today is the largest marketplace for chemical products in the Western world. Digital, internal start-ups allow traditional, analogue companies to understand, absorb, develop, and diffuse the digital capabilities required to compete successfully in the digital future. Internal start-ups are a driving force allowing to pilot large-scale organizational transformation on a small scale. The culture of the start-up – fast communication and decision-making – is a force that potentially can positively change the culture of the established business. *Jörg Hellwig* stresses the importance of building a platform supporting the move from selling products to selling services or solutions. Platforms, by nature, are environments in which transactions are executed allowing the platform provider to add additional services which then can be profitably sold. The digital transformation has direct repercussions on the nature of work and required capabilities. Big data analytics capabilities are a key differentiator: Lanxess has hired an astrophysicist, an unusual qualification in a specialty chemical company, and other data scientists to internalize capabilities in data analytics. Digital technologies strongly influence innovation processes, both in terms of limiting the number of low-value targets as well as in terms of allowing an early parametrization of desirable innovation outcomes.

"Part 4 Digital transformation and customer value creation" contains several chapters that highlight how digital technologies can contribute to dramatically improve value creation and value delivery to customers.

In Chapter 14 "Digital transformation and consumer behavior: how the analysis of consumers data reshapes the marketing approach" *Francesca Checchinato* discusses the impact of big data and how it affects the marketing strategy development. Starting with a discussion about the impact on digital technologies on consumer behavior and on customer experience, she describes the data-driven approach and how companies can exploit this data to define their marketing strategies. Finally, she analyzes the impact of this new approach on the main marketing levers with a particular focus on communication and distribution.

In Chapter 15 "Digital transformation and the salesforce" *Frank V. Cespedes* offers warnings about current uses of the term "digital transformation" and guidelines concerning the managerial pre-requisites for productive use of digital technologies.

In Chapter 16 "Digital transformation and the role of customer-centric innovation" *Andreas Hinterhuber* and *Stephan Liozu* discuss several salient aspects of the digital transformation. *Stephan Liozu* is chief value officer (CVO) of Thales, one of the world's largest aerospace and defence companies that has invested heavily in digitalization. Digital transformation requires new business models that are frequently located in a separate organizational unit. The digital transformation requires a shift from single products to the ecosystem – also this is a consistent topic across interviews. In order to implement the digital transformation successfully, several issues are key: superior customer insights and true customer orientation; acquisition of new, digital analytics capabilities; partnerships; finally, a careful balance between broad experimentation and a few, narrow bets that deliver results and prove the viability of digitalized solutions.

In Chapter 17, *Tiziano Vescovi*'s interview "Digital transformation: the case of Bugari" with *Jean-Christophe Babin*, CEO of Bulgari Group, analyzes the particularities of digitization in the luxury industry. In this case, it is not a question of digitizing products, but of digitally integrating customer management and the relationship with them, through important effects on communication and branding. It is a question of maintaining the difficult balance between exclusivity and the use of social media, with innovation in languages and targets, but also of promoting a change internally in the skills of the staff in the approach with new generations of luxury customers. Once again, the key issue is the change in corporate culture.

In Chapter 18 "The Importance of data in transforming a traditional company to a digital thinking company: the case of Etro" *Francesca Checchinato* and *Fabrizio Viacava*, discuss the digital transformation of a family-run company operating in the fashion industry. The chapter highlights the challenges and success of integrating a data-driven approach in a traditional company. As in other chapters, a new culture and mindset emerges as a key pillar of the transformation. Data availability and data analysis capabilities enable data-driven decisions. This approach enhanced Etro's performance and helped it to define an effective strategy to become a modern company.

"Part 5 The digital transformation in SMEs: challenges and best practices" contains several chapters on the digitalization of small and medium-sized businesses (SMEs). SMEs do not have separate organizational units responsible for the digitalization – unlike large companies, as the chapters in this volume argue. Nevertheless, the articles in this section suggest that the digitalization increases the competitiveness and performance also for SMEs.

In Chapter 19 "Digital transformation of manufacturing firms: opportunities and challenges for SMEs" *Enrico Pieretto* and *Andreas Hinterhuber* provide a framework for the digital transformation for small and medium-sized companies, linking dynamic capabilities, organizational maturity, and external drivers to identify six key dimensions of the digital transformation (Gurbaxani & Dunkle, 2019). These six dimensions are: strategic vision, culture for innovation, know-how and intellectual property, digital capabilities, strategic alignment, and technology assets. The chapter describes core technologies of the digital transformation, namely: big data and analytics, autonomous robots, simulation, horizontal and vertical system integration, industrial IoT, cybersecurity, cloud computing, additive manufacturing, and augmented reality. *Enrico Pieretto* and *Andreas Hinterhuber* then report the findings of four semi-structured interviews with senior executives of early movers in Italy in the digitalization, reporting on the adoption of these key technologies in four companies. The authors use the six-dimensions framework (Gurbaxani & Dunkle, 2019) to benchmark the capabilities relevant for the digital transformation of the four companies examined. The chapter concludes by suggesting that this framework is indeed potentially useful in benchmarking critical capabilities relevant for the digital transformation of small and medium sized companies.

In Chapter 20 "Digital transformation and financial performance" *Nicolai E. Fabian, Thijs Broekhuizen*, and *Dinh Khoi Nguyen* aim to assess the impact of digital transformation on financial performance of SMEs. Using survey data of 189 Dutch SMEs, the study suggests that the presence of digital specialists does not directly improve financial performance, but that it helps to convert digital transformation more effectively: the positive association between digital transformation and financial performance only exists when firms have a digital specialist. It thus seems that the presence of a CDO improves performance also for SMEs.

In Chapter 21 "Supporting pervasive digitization in Italian SMEs through an open innovation process" *Cinzia Colapinto, Vladi Finotto*, and *Nunzia Coco* argue that open innovation practices may accelerate digital transformation and increase innovation development, and thus help firms to establish new sustainable strategies. The study also identifies barriers and enablers of digital transformation projects.

The final chapter, "Conclusion: Our roadmap to digital transformation" by *Francesca Checchinato, Andreas Hinterhuber, and Tiziano Vescovi*, summarizes some salient points of the interviews, case studies and academic research presented in this volume. The authors introduce a four-step framework – introducing digitalization, implementing digital strategies, being a digital company, and next challenges – highlighting necessary conditions, challenges, and capabilities of each stage of the digital transformation. We probably do not have definitive answers to many questions outlined, but we have examples and questions that illuminate salient aspects.

This book is thus a call to action highlighting how managers and leaders in organizations can successfully lead the digital transformation. We hope that readers will listen.

References

Baghai, M., Coley, S., & White, D. (2000). *The alchemy of growth*. New York, NY: Basic Books.

Bustinza, O. F., Vendrell-Herrero, F., & Baines, T. (2017). Service implementation in manufacturing: An organisational transformation perspective. *International Journal of Production Economics, 192*(5), 1–8.

Chanias, S., Myers, M. D., & Hess, T. (2019). Digital transformation strategy making in pre-digital organizations: The case of a financial services provider. *Journal of Strategic Information Systems, 28*(1), 17–33.

Gupta, M., & George, J. F. (2016). Toward the development of a big data analytics capability. *Information & Management, 53*(8), 1049–1064.

Gupta, S., & Barley, L. (2015). Reinventing Adobe. *Harvard Business School case study, 9-514-066*, 1–17.

Gurbaxani, V., & Dunkle, D. (2019). Gearing up for successful digital transformation. *MIS Quarterly Executive, 18*(3), 209–220.

Hess, T., Matt, C., Benlian, A., & Wiesböck, F. (2016). Options for formulating a digital transformation strategy. *MIS Quarterly Executive, 15*(2), 123–139.

Hinterhuber, A., & Liozu, S. (2020). Introduction: Implementing pricing strategies. In A. Hinterhuber & S. Liozu (Eds.), *Pricing strategy implementation*. Abingdon, UK: Routledge.

Immelt, J. (2017). How I remade GE. *Harvard Business Review, 95*(5), 42–51.

Kane, G. C., Palmer, D., Phillips, A. N., & Kiron, D. (2015). Is your business ready for a digital future? *MIT Sloan Management Review, 56*(4), 37.

Kumar, N., Scheer, L., & Kotler, P. (2000). From market driven to market driving. *European Management Journal, 18*(2), 129–142.

Peter, M. K., Kraft, C., & Lindeque, J. (2020). Strategic action fields of digital transformation. *Journal of Strategy and Management, 13*(1), 160–180.

Sprague, K. (2015). Interview: Reborn in the cloud *McKinsey & Company white paper*.

Verhoef, P. C., Broekhuizen, T., Bart, Y., Bhattacharya, A., Dong, J. Q., Fabian, N., & Haenlein, M. (2020). Digital transformation: A multidisciplinary reflection and research agenda. *Journal of Business Research, 121*(Jan), 889–901.

Vial, G. (2019). Understanding digital transformation: A review and a research agenda. *Journal of Strategic Information Systems, 28*(2), 188–144.

Part 2

Digital transformation and organizational capabilities

Chapter 2

Understanding digital transformation

A review and a research agenda

Gregory Vial

Introduction

In recent years, *digital transformation* (DT) has emerged as an important phenomenon in strategic information system (IS) research (Bharadwaj et al. 2013; Piccinini et al. 2015a) as well as for practitioners (Fitzgerald et al. 2014; Westerman et al. 2011). At a high level, DT encompasses the profound changes taking place in society and industries through the use of digital technologies (Agarwal et al. 2010; Majchrzak et al. 2016). At the organizational level, it has been argued that firms must find ways to innovate with these technologies by devising "strategies that embrace the implications of digital transformation and drive better operational performance" (Hess et al. 2016:123).

Recent research has contributed to increase our understanding of specific aspects of the DT phenomenon. In line with previous findings on IT-enabled transformation, research has shown that technology itself is only part of the complex puzzle that must be solved for organizations to remain competitive in a digital world. Strategy (Bharadwaj et al. 2013; Matt et al. 2015) as well as changes to an organization, including its structure (Selander and Jarvenpaa 2016), processes (Carlo et al. 2012), and culture (Karimi and Walter 2015) are required to yield the capability to generate new paths for value creation (Svahn et al. 2017a). Notwithstanding these contributions, we currently lack a comprehensive understanding of this phenomenon (Gray and Rumpe 2017; Kane 2017c; Matt et al. 2015) as well as its implications at multiple levels of analysis. This chapter therefore proposes to take stock of current knowledge on the topic by studying the research question: "*What do we know about digital transformation?*"

Consistent with the breadth of our research question, we adopt an inductive approach using techniques borrowed from grounded theory (Wolfswinkel et al. 2013) and review 282 works on DT culled from IS literature. Based on extant definitions, we develop a conceptual definition of DT as "*a process that aims to improve an entity by triggering significant changes to its properties through combinations of information, computing, communication, and connectivity technologies.*" We then present, based on our analysis of the literature, an inductive framework describing DT as a process wherein organizations respond to changes taking place in their environment by using digital technologies to alter their value creation processes. For this process to be successful and lead to positive outcomes, organizations must account for a number of factors that can hinder the execution of their transformation.

Based on these findings, we discuss the novelty of DT based on previous literature on IT-enabled transformation. We argue that the scale, the scope, as well as the speed associated with the DT phenomenon call for research to consider DT as an *evolution of* the IT-enabled transformation phenomenon. We then propose a research agenda for strategic IS research on DT articulated across two main avenues. The first avenue proposes to study how dynamic capabilities contribute to DT. The second avenue calls for research to study the strategic importance of ethics in the context of DT. Together, these two research avenues ask questions that are relevant for strategic IS research as well as practice.

Our work offers two contributions. First, we provide a review that integrates current knowledge on DT. Second, we identify avenues that future research may use as a guide to answer pressing questions on DT while contributing to expand the theoretical foundations we rely on to study IS. In the next sections, we present the methods of our review. We then detail our findings, including our definition of DT and our inductive framework. Finally, we present our agenda for future research on DT, the limitations of our work and provide concluding remarks.

Methods

In line with the breadth of our research question, we selected an inductive approach to reviewing the literature on DT. Our methods are informed by guidelines from Wolfswinkel et al. (2013) and their use of techniques borrowed from grounded theory for "rigorously reviewing literature" (p. 1). These guidelines comprise five steps to (1) define the scope of the review, (2) search the literature, (3) select the final sample, (4) analyze the corpus, and (5) present the findings. We outline the application of the first four steps in the following paragraphs (for a more detailed account, please refer to Appendix A) and present fifth step in the Findings section below.

We initially ran several queries against online databases to gain an initial understanding of the coverage offered by literature in various disciplines. To ensure that the size of our review sample would remain manageable, we decided to focus on peer-reviewed sources (both in research and in practice) pertinent to IS literature using three databases (AIS Library, Business Source Complete, ScienceDirect). Based on the reading of abstracts as well as a few highly cited articles, we designed our final search criteria using combinations of keywords containing the terms "digital" and "transform" or "disrupt". We also opted to exclude works in progress, research outlets not ranked in the Journal Citation Reports index, as well as teaching cases from our final sample.

We then proceeded to run our search query against our selected databases. For each database, we adapted our search query and performed several checks to ensure that works identified through our initial search query were included in our search results. For each search result, we downloaded the full paper in PDF format along with its associated references into a bibliographic software. To finalize our sample, we applied our criteria of inclusion and exclusion against our search results. Our initial sample size of 381 works was reduced to 248 works, which were subsequently augmented to 282 works through backward and forward search (226 from research outlets, 56 from practitioner-oriented outlets).

Our analysis consisted of four main steps that were performed iteratively. First, we collected, for each work, a number of data points such as the publication outlet, the type of publication outlet (research journal, conference proceedings, practitioner's journal), the type of paper (empirical, conceptual), the context of application (e.g., healthcare), the theoretical foundation used or developed, the methods, as well as any definitions of DT and other related concepts. Second, we performed open coding by annotating sources based on the arguments and findings relevant to our phenomenon of interest and tracked relationships between variables in each paper, whether those were hypothesized (in a conceptual paper) or validated (in an empirical paper). Third, we performed two rounds of axial coding to refine our coding scheme into a more manageable set of higher-order categories of relationships. Finally, we integrated these relationships using selective coding. We imported all our coding instances into a relational database to contrast and compare our emergent findings using SQL queries which we complemented with visualization techniques (see Figure A.2). The result of this process is a high-level framework that incorporates the findings from our analysis based on the coverage of the main relationships contained within our sample.

Findings

Defining digital transformation

The first step of our analysis consisted in studying extant definitions of DT. Within our sample, we found 28 sources offering 23 unique definitions (see Table 2.1). Although encouraging, this relatively small proportion (about 10%) reflects an overall enthusiasm toward the phenomenon of DT at the expense of its conceptual clarity. Studying these definitions, we make three observations. First, DT as it is defined in the reviewed studies, primarily relates to organizations. Second, important differences exist across definitions with regards to the types of technologies (Horlacher et al. 2016; Westerman et al. 2011) involved as well as the nature of the *transformation* taking place (Andriole 2017; Piccinini et al. 2015b). Third, in spite of differences, similarities exist across definitions, e.g., using common terms such as "digital technologies" (Matt et al. 2015; Singh and Hess 2017).

We then proceeded to analyze extant definitions based on recommendations for the creation of conceptual definitions. In particular, we referred to the rules offered by Wacker (2004) as well as guidelines from Suddaby (2010) (see Table 2.2) and evaluated existing definitions against these recommendations (see the third column of Table 2.1). Our analysis reveals that circularity, unclear terminology, and the conflation of the concept and its impacts, among other challenges, hinder the conceptual clarity of DT.

Based on these findings, we used semantic analysis to build a working definition of DT from extant definitions. We used *semantic decomposition* (Akmajian et al. 2017) to systematically decompose extant definitions into series of constituting primitives and compared those primitives across definitions to identify essential properties of DT (a detailed account of the semantic decomposition process is available in Appendix B). We identified four such properties: (1) *target entity*, i.e., the unit of analysis affected by DT; (2) *scope*, i.e., the extent of the changes taking place within the target entity's properties; (3) *means*, i.e., the technologies involved in creating the change within the target entity; and (4) *expected outcome*, i.e., the outcome of DT. Using these properties, we constructed a conceptual definition of DT as "*a process that aims to improve an entity by triggering significant changes to its properties through combinations of information, computing, communication, and connectivity technologies*".

Our definition warrants three important observations. First, it is not organization-centric. Although in most extant definitions the target entity primitive refers to an organization, two definitions refer to other forms of entities (society, industry) that exist in many studies where definitions of DT are absent (e.g., Agarwal et al. 2010; Hanelt et al. 2015b; Pagani 2013). Our definition is therefore consistent with the related concept of

Table 2.1 Extant definitions of DT

Definition	Source(s)	Conceptual clarity challenge(s)
The use of technology to radically improve performance or reach of enterprises.	Westerman et al. (2011) Westerman et al. (2014) Karagiannaki et al. (2017)	Conflation between the concept and its impacts.
The use of new digital technologies (social media, mobile, analytics, or embedded devices) to enable *major business improvements* (such as enhancing customer experience, streamlining operations, or creating new business models). [emphasis original]	Fitzgerald et al. (2014) Liere-Netheler et al. (2018)	Unclear term: "digital technologies" defined using examples. Conflation between the concept and its impacts.
Digital transformation strategy is a blueprint that supports companies in governing the transformations that arise owing to the integration of digital technologies, as well as in their operations after a transformation.	Matt et al. (2015)	Unclear term: "digital technologies". Circularity ("transformation").
Digital transformation involves leveraging digital technologies to enable major business improvements, such as enhancing customer experience or creating new business models.	Piccinini et al. (2015b)	Unclear term: "digital technologies". Conflation between the concept and its impacts.
Use of digital technologies to radically improve the company's performance.	Bekkhus (2016)	Unclear term: "digital technologies". Conflation between the concept and its impacts.
Digital transformation encompasses both process digitization with a focus on efficiency, and digital innovation with a focus on enhancing existing physical products with digital capabilities.	Berghaus and Back (2016)	Unclear terms: "digitalization", "digital capabilities".
Digital transformation is the profound and accelerating transformation of business activities, processes, competencies, and models to fully leverage the changes and opportunities brought by digital technologies and their impact across society in a strategic and prioritized way.	Demirkan et al. (2016)	Unclear term: "digital technologies". Circularity ("transformation"). Conflation between the concept and its impacts.
Digital transformation encompasses the digitization of sales and communication channels, which provide novel ways to interact and engage with customers, and the digitization of a firm's offerings (products and services), which replace or augment physical offerings. Digital transformation also describes the triggering of tactical or strategic business moves by data-driven insights and the launch of digital business models that allow new ways to capture value.	Haffke et al. (2016)	Unclear term: "digitalization". Conflation between the concept and its impacts. Lack of parsimony.

Table 2.1 Cont.

Definition	Source(s)	Conceptual clarity challenge(s)
Digital transformation is concerned with the changes digital technologies can bring about in a company's business model, which result in changed products or organizational structures or in the automation of processes. These changes can be observed in the rising demand for Internet-based media, which has led to changes of entire business models (for example, in the music industry).	Hess et al. (2016)	Unclear term: "digital technologies". Conflation between the concept and its impacts. Lack of parsimony.
Use of new digital technologies, such as social media, mobile, analytics or embedded devices, in order to enable major business improvements like enhancing customer experience, streamlining operations, or creating new business models.	Horlacher et al. (2016) Singh and Hess (2017)	Unclear term: "digital technologies" defined using examples. Conflation between the concept and its impacts.
Changes and transformations that are driven and built on a foundation of digital technologies. Within an enterprise, digital transformation is defined as an organizational shift to big data, analytics, cloud, mobile and social media platform. Whereas organizations are constantly transforming and evolving in response to changing business landscape, digital transformation are the changes built on the foundation of digital technologies, ushering unique changes in business operations, business processes and value creation.	Nwankpa and Roumani (2016)	Unclear term: "digital technologies" defined using examples. Circularity ("transformation"). Lack of parsimony.
Digital transformation is not a software upgrade or a supply chain improvement project. It's a planned digital shock to what may be a reasonably functioning system.	Andriole (2017)	Unclear term: "digital shock".
Extended use of advanced IT, such as analytics, mobile computing, social media, or smart embedded devices, and the improved use of traditional technologies, such as enterprise resource planning (ERP), to enable major business improvements.	Chanias (2017)	Unclear term: "advanced IT" defined using examples. Conflation between the concept and its impacts.
The changes digital technologies can bring about in a company's business model, which result in changed products or organizational structures or automation of processes.	Clohessy et al. (2017)	Unclear term: "digital technologies". Conflation between the concept and its impacts.
Distinguishes itself from previous IT-enabled business transformations in terms of velocity and its holistic nature.	Hartl and Hess (2017)	Circularity ("transformation"). Comparative definition ("previous IT-enabled business transformations")

(continued)

Table 2.1 Cont.

Definition	Source(s)	Conceptual clarity challenge(s)
Transformations in organizations that are driven by new enabling IT/IS solutions and trends.	Heilig et al. (2017)	Circularity ("transformation").
Digital transformation as encompassing the digitization of sales and communication channels and the digitization of a firm's offerings (products and services), which replace or augment physical offerings. Furthermore, digital transformation entails tactical and strategic business moves that are triggered by data-driven insights and the launch of digital business models that allow new ways of capturing value.	Horlach et al. (2017)	Unclear term: "digitalization". Conflation between the concept and its impacts. Lack of parsimony.
The best understanding of digital transformation is adopting business processes and practices to help the organization compete effectively in an increasingly digital world.	Kane (2017c) Kane et al. (2017)	Conflation between the concept and its impacts.
Digital transformation describes the changes imposed by information technologies (IT) as a means to (partly) automatize tasks.	Legner et al. (2017)	Conflation between the concept and its impacts.
Digital transformation highlights the impact of IT on organizational structure, routines, information flow, and organizational capabilities to accommodate and adapt to IT. In this sense, digital transformation emphasizes more the technological root of IT and the alignment between IT and businesses.	Li et al. (2017)	Conflation between the concept and its impacts. Lack of parsimony.
An evolutionary process that leverages digital capabilities and technologies to enable business models, operational processes, and customer experiences to create value.	Morakanyane et al. (2017)	Unclear term: "digital capabilities". Conflation between the concept and its impacts.
The use of new digital technologies, in order to enable major business improvements in operations and markets such as enhancing customer experience, streamlining operations, or creating new business models.	Paavola et al. (2017)	Unclear term: "digital technologies". Conflation between the concept and its impacts.
Fundamental alterations in existing and the creation of new business models [...] in response to the diffusion of digital technologies such as cloud computing, mobile Internet, social media, and big data.	Remane et al. (2017)	Unclear term: "digital technologies" defined using examples.

Note: Definitions are sorted chronologically and alphabetically.

Table 2.2 Guidelines for conceptual definitions

Rules for conceptual definitions (adapted from Wacker 2004: 384)

Rule 1: "Definitions should be formally defined using primitives and derived terms."

Rule 2: "Each concept should be uniquely defined."

Rule 3: "Definitions should include only unambiguous and clear terms."

Rule 4: "Definitions should have as few as possible terms."

Rule 5: "Definitions should be consistent within [their] field."

Rule 6: "Definitions should not make any term broader."

Rule 7: "New hypotheses cannot be introduced in the definitions."

Rule 8: "Statistical test for content validity must be performed after the terms are formally defined."

Guidelines for conceptual clarity (adapted from Suddaby 2010: 347)

"Offer definitions of key terms and constructs."

"The definition should capture the essential properties and characteristics of the concept or phenomenon under consideration."

"A good definition should avoid tautology or circularity."

"A good definition should be parsimonious."

digitalization, which includes the "broader individual, organizational, and societal contexts" (Legner et al. 2017:301). Second, our definition acknowledges *improvement* as an expected outcome of DT without guaranteeing its realization (see Wacker 2004:393). Finally, we purposefully do not define the means primitive using the term *digital technologies*. Rather, we use the definition of digital technologies provided by Bharadwaj et al. (2013) to reinforce the conceptual clarity of our definition as well as its applicability over time as technology changes.

Digital transformation: an inductive framework

We present in Figure 2.1 and in the sections below our inductive framework summarizing current knowledge on DT. This framework builds upon relationships that emerged through our analysis across eight overarching building blocks describing DT as a process where *digital technologies* play a central role in the creation as well as the reinforcement of *disruptions* taking place at the society and industry levels. These disruptions trigger *strategic responses* from the part of organizations, which occupy a central place in DT literature. Organizations use digital technologies to alter the *value creation paths* they have previously relied upon to remain competitive. To that end, they must implement *structural changes* and overcome *barriers* that hinder their transformation effort. These changes lead to *positive impacts* for organizations as well as, in some instances, for individuals and society, although they can also be associated with *undesirable outcomes*. Descriptive statistics as well as a complete list of the works reviewed are available in Appendices C and D.

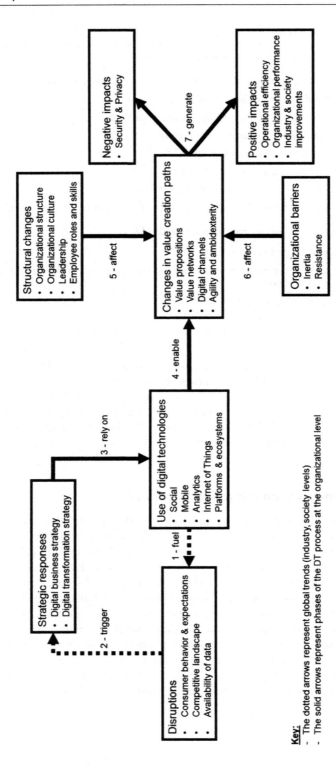

Figure 2.1 Building blocks of the DT process.

Note: The arrows do not represent a statistical relationship or a causality found in variance models. Rather, they detail an overarching sequence of relationships described by the literature on DT.

The nature of digital technologies

Most of the digital technologies mentioned within our sample fit with the popular SMACIT acronym (Sebastian et al. 2017), referring to technologies related to *social* (Li et al. 2017; Oestreicher-Singer and Zalmanson 2012), *mobile* (Hanelt et al. 2015a; Pousttchi et al. 2015), *analytics* (Duerr et al. 2017; Günther et al. 2017), *cloud* (Clohessy et al. 2017; Du et al. 2016), and the *internet of things* – IoT (Petrikina et al. 2017; Richter et al. 2017). We also found *platforms* as an important category, especially in research articles (Tan et al. 2015a; Tiwana et al. 2010) while other forms of digital technologies, including the internet (Lyytinen and Rose 2003b), software (e.g., Karimi et al. 2009; Setia et al. 2013), and blockchain (Glaser 2017) were seldom present. In line with Bharadwaj et al.'s definition of digital technologies, we observe that *combinations* of technologies are particularly relevant in the context of DT (Gray et al. 2013; Günther et al. 2017; Newell and Marabelli 2015; Westerman and Bonnet 2015). For example, the ability to implement algorithmic decision-making may be contingent upon a firm's ability to perform analytics on big data collected through individuals' use of social media on their mobile phones (Newell and Marabelli 2015).

Digital technologies as sources of disruption

The literature describes digital technologies as inherently disruptive (Karimi and Walter 2015). In this section, we report on the three types of disruptions revealed by our analysis (see Table 2.3): consumer behavior and expectations, competitive landscape, and the availability of data.

Altering consumer behavior and expectations

Digital technologies have a profound impact on the behavior (Chanias 2017; Hong and Lee 2017) of consumers who have ubiquitous (Yoo et al. 2010a) access to information and communication capabilities (e.g., using social media on a mobile device). Using these technologies, they become active participants in a dialogue that takes place between an organization and its stakeholders (e.g., Kane 2014; Yeow et al. 2017). An important implication of these changes is that customers no longer see themselves as captives of the firms with which they transact (Lucas Jr et al. 2013; Sia et al. 2016) and their expectations with regards to the services that should be provided to them are increasing. This is illustrated in the case study of DBS Bank (Sia et al. 2016) where Asian consumers expect to perform most of their banking operations using mobile digital banking solutions. In this case these expectations created pressure for DBS to offer new services to remain competitive as "the digital revolution has put banks under siege" (quote from DBS CEO, p. 107). As a result, *anticipating* rather than responding to changes in customer expectations has become a strategic imperative for firms.

Disrupting the competitive landscape

Digital technologies bring about disruption in the markets where firms operate (Mithas et al. 2013). They facilitate the (re)combination of existing products and services to generate new forms of digital offerings (Yoo et al. 2010b) favoring services over products

Table 2.3 Digital technologies as sources of disruption

Source type	Sources

Altering consumer behavior and expectations (n = 86)

Source type	Sources
Research paper – empirical (n = 50)	Agarwal et al. (2011); Bassano et al. (2017); Berghaus and Back (2017); Chanias and Hess (2016); Delmond et al. (2017); Duerr et al. (2018); Duerr et al. (2017); Fehér and Varga (2017); Felgenhauer et al. (2017); Gimpel (2015); Gimpel et al. (2018); Haffke et al. (2016); Hanelt et al. (2015a); Hanelt et al. (2015b); Henfridsson and Lind (2014); Hildebrandt et al. (2015); Hjalmarsson et al. (2014); Hong and Lee (2017); Karimi and Walter (2015); Lee and Lee (2013); Liere-Netheler et al. (2018); Mueller and Renken (2017); Oestreicher-Singer and Zalmanson (2012); Petrikina et al. (2017); Piccinini et al. (2015b); Ramasubbu et al. (2014); Rauch et al. (2016); Reinhold and Alt (2009); Richter et al. (2017); Roecker et al. (2017); Ross et al. (2016); Sachse et al. (2012); Saldanha et al. (2017); Schmidt et al. (2017); Selander and Jarvenpaa (2016); Setia et al. (2013); Smith and Webster (2006); Sørensen et al. (2015); Standaert and Jarvenpaa (2017); Tan et al. (2015a); Tan et al. (2017); Tanniru et al. (2016); Tiefenbacher and Olbrich (2016); Töytäri et al. (2017); Utesheva et al. (2012); Weinrich et al. (2016); Winkler et al. (2014); Wörner et al. (2016); Xie et al. (2014); Yeow et al. (2017); Zolnowski and Warg (2018)
Research paper – other (n = 20)	Agarwal et al. (2010); Bichler et al. (2016); Chowdhury and Åkesson (2011); Cziesla (2014); Fischer et al. (2018); Granados et al. (2008); Günther et al. (2017); Loebbecke and Picot (2015); Lucas Jr et al. (2013); Neumeier et al. (2017); Newell and Marabelli (2014); Newell and Marabelli (2015); Piccinini et al. (2015a); Pillet et al. (2017); Pousttchi et al. (2015); Prifti et al. (2017); Riedl et al. (2017); Seo (2017); Weiß et al. (2018)
Practitioner outlet (n = 16)	Dery et al. (2017); Earley (2014); Fitzgerald et al. (2014); Hansen and Sia (2015); Kane (2014); Kane (2015a); Kane (2016b); Kane et al. (2017); Nehme et al. (2015); NetworkWorld Asia (2015); Reinartz and Imschloß (2017); Sebastian et al. (2017); Soava (2015); Westerman and Bonnet (2015); Westerman et al. (2011); Wulf et al. (2017)

Disrupting the competitive landscape (n = 84)

Source type	Sources
Research paper – empirical (n = 46)	Agarwal et al. (2011); Barua et al. (2004); Berghaus and Back (2017); Chanias and Hess (2016); Delmond et al. (2017); Ebbesson (2015); Elbanna and Newman (2016); Gimpel (2015); Holotiuk and Beimborn (2017); Islam et al. (2017); Kamel (2015); Karagiannaki et al. (2017); Karimi et al. (2009); Karimi and Walter (2015); Kauffman et al. (2010); Klötzer and Pflaum (2017); Lee and Lee (2013); Leonhardt et al. (2017); Li et al. (2017); Li et al. (2016); Liere-Netheler et al. (2018); Mithas et al. (2013); Mohagheghzadeh and Svahn (2016); Nwankpa and Roumani (2016); Oestreicher-Singer and Zalmanson (2012); Oh (2009); Pagani (2013); Piccinini et al. (2015b); Ramasubbu et al. (2014); Reitz et al. (2018); Resca et al. (2013); Scott (2007); Selander et al. (2010); Setia et al. (2013); Staykova and Damsgaard (2015b); Tan et al. (2015a); Tan et al. (2017); Tan et al. (2015b); Tiefenbacher and Olbrich (2016); Wenzel et al. (2015); Woodard et al. (2012); Xie et al. (2014); Yeow et al. (2017); Zhu et al. (2006)

Table 2.3 Cont.

Disrupting the competitive landscape (n = 84)

Source type	Sources
Research paper – other (n = 24)	Barrett et al. (2015); Berghaus (2016); Bharadwaj et al. (2013); de Reuver et al. (2017); Dixon et al. (2017); Fichman et al. (2014); Glaser (2017); Granados et al. (2008); Günther et al. (2017); Heilig et al. (2017); Kahre et al. (2017); Krumeich et al. (2012); Krumeich et al. (2013); Lucas Jr et al. (2013); Myunsoo and Byungtae (2013); Nambisan et al. (2017); Neumeier et al. (2017); Newell and Marabelli (2014); Nischak et al. (2017); Rai and Sambamurthy (2006); Schmid et al. (2017); Seo (2017); Tanriverdi and Lim (2017); Tiwana et al. (2010); Yoo et al. (2010b)
Practitioner outlet (n = 14)	Du et al. (2016); Earley (2014); Fitzgerald et al. (2014); Hansen and Sia (2015); Henningsson and Hedman (2014); Hess et al. (2016); Kane (2017d); Kane et al. (2017); Kane et al. (2016); Kohli and Johnson (2011); Nehme et al. (2015); Porter and Heppelmann (2014); Sebastian et al. (2017); Sia et al. (2016); Wulf et al. (2017)

Increasing the availability of data (n = 37)

Source type	Sources
Research paper – empirical (n = 15)	Bravhar and Juric (2017); Chatfield et al. (2015); Constantiou et al. (2017); Ebbesson (2015); Gimpel (2015); Gimpel et al. (2018); Hjalmarsson et al. (2014); Hjalmarsson et al. (2015); Holotiuk and Beimborn (2017); Hong and Lee (2017); Lu and Swatman (2008); Pramanik et al. (2016); Saldanha et al. (2017); Tiefenbacher and Olbrich (2016); Trantopoulos et al. (2017)
Research paper – other (n = 13)	Agarwal et al. (2010); Bhimani (2015); Fichman et al. (2014); Günther et al. (2017); Heilig et al. (2017); Legner et al. (2017); Loebbecke and Picot (2015); Lucas Jr et al. (2013); Newell and Marabelli (2014); Newell and Marabelli (2015); Pousttchi et al. (2015); Rizk et al. (2018); Yoo et al. (2010b)
Practitioner outlet (n = 9)	Basole (2016); Dremel et al. (2017); Fitzgerald (2016a); Gust et al. (2017); Kane (2014); Kane (2016c); Sebastian et al. (2017); Westerman et al. (2011); Wulf et al. (2017)

(Barrett et al. 2015), lowering barriers to entry (Woodard et al. 2012), and hindering the sustainability of the competitive advantage of incumbent players (Kahre et al. 2017). For example, platforms enable the redefinition of existing markets (Tiwana et al. 2010) by facilitating exchanges of digital goods and services. As competition moves from a physical plane to a virtual plane where information flows more freely, previous forms of barriers to entry become less significant. For example, in the music industry (Lucas Jr et al. 2013), physical goods sold through intermediaries have been supplanted by music subscription services offered by firms that were not originally part of that industry (e.g., Apple, Spotify). More recently, it has been observed that Blockchain (Friedlmaier et al. 2018; Hayes 2016; Korpela et al. 2017), as a generic and extensible technology, enables the creation of decentralized, *digital infrastructures* (Tilson et al. 2010). These infrastructures can be applied to a variety of domains (e.g., banking, contract management) and act as complements or substitutes to more traditional, centralized institutions (e.g., to securely exchange funds from peer to peer rather than through authoritative intermediaries).

Increasing the availability of data

Beyond their immediate operational value, digital technologies also foster the generation of data (e.g., digital traces generated through the use of a mobile device). In the context of DT, firms strive to exploit the potential of data for their own benefit, or in some instances, to monetize those data by selling them to third parties (Loebbecke and Picot 2015). Using analytics, firms can offer services that better answer the needs of their customers or perform processes more efficiently (e.g., using data-driven algorithmic decision-making) for their competitive advantage (Günther et al. 2017). For instance, KLM (Kane 2014) uses social media such as Twitter and Facebook to perform customer service operations. They then use the data generated through those interactions to maintain and act upon their understanding of customers' sentiments in real time.

Strategic responses to digital disruption

In light of these disruptions, organizations must devise ways to remain competitive as digital technologies provide "both game-changing opportunities for – and existential threats to – companies" (Sebastian et al. 2017:197) (see Table 2.4). Indeed, although a majority of works within our sample treat DT as an endogenous phenomenon where initiatives are purposefully created to respond to *opportunities* afforded by digital technologies (Tan et al.

Table 2.4 Strategic responses to digital disruption

Digital business strategy (n = 31)	
Source type	Sources
Research paper – empirical (n = 18)	Berghaus and Back (2017); Fehér et al. (2017); Haffke et al. (2016); Holotiuk and Beimborn (2017); Islam et al. (2017); Karimi and Walter (2015); Leischnig et al. (2017); Li et al. (2016); Mithas et al. (2013); Nwankpa and Roumani (2016); Oestreicher-Singer and Zalmanson (2012); Pagani (2013); Ramasubbu et al. (2014); Richter et al. (2017); Ross et al. (2016); Setia et al. (2013); Woodard et al. (2012); Yeow et al. (2017)
Research paper – other (n = 7)	Bharadwaj et al. (2013); Dixon et al. (2017); Kahre et al. (2017); Matt et al. (2015); Morakanyane et al. (2017); Neumeier et al. (2017); Piccinini et al. (2015a)
Practitioner outlet (n = 6)	Dremel et al. (2017); Hess et al. (2016); Kane et al. (2017); Sebastian et al. (2017); Sia et al. (2016); Weill and Woerner (2018)

Digital transformation strategy (n = 19)	
Source type	Sources
Research paper – empirical (n = 10)	Berghaus and Back (2016); Berghaus and Back (2017); Chanias (2017); Chanias and Hess (2016); Fehér et al. (2017); Gimpel et al. (2018); Haffke et al. (2017); Hartl and Hess (2017); Lucas Jr and Goh (2009); Riasanow et al. (2017)
Research paper – other (n = 5)	Berghaus (2016); Matt et al. (2015); Morakanyane et al. (2017); Riedl et al. (2017); Weiß et al. (2018)
Practitioner outlet (n = 4)	Hess et al. (2016); Sia et al. (2016); Singh and Hess (2017); Westerman et al. (2011)

2015a), we found 49 sources where it is viewed as an exogenous *threat* for the focal organization (e.g., Li et al. 2016; Lucas Jr and Goh 2009; Sia et al. 2016). In the latter, DT is depicted as a higher-level phenomenon that disrupts the competitive environment and demands a response from the part of the organization. Although the generic concept of strategy is often invoked to explain these responses (e.g., Yoo et al. 2010b), the literature refers to two novel concepts in the context of DT: digital business strategy and digital transformation strategy.

Bharadwaj et al. (2013) argue that digital technologies call for researchers to study the *fusion* between organizational strategy and IS strategy (e.g., Kahre et al. 2017) rather than their *alignment*. They observe that competition among firms increasingly rests upon their ability to leverage digital technologies to accomplish their vision (Mithas et al. 2013) and that separating the two concepts may diminish their potential for synergies. To that end, they offer the concept of digital business strategy – DBS, defined as "organizational strategy formulated and executed by leveraging digital resources to create differential value" (p. 472). Since then, the concept of DBS has gained some traction in research and in practice (e.g., Holotiuk and Beimborn 2017; Leischnig et al. 2017; Mithas et al. 2013; Oestreicher-Singer and Zalmanson 2012; Sia et al. 2016). In three works, we also found evidence of DBS as an emergent concept (Chanias 2017; Henfridsson and Lind 2014; Yeow et al. 2017). For example, Yeow et al. (2017) studied a company incorporating a B2C model into its existing B2B model and found that tensions arising from the misalignment between a firm's existing resources and its emergent digital business strategy are continuously addressed via an *aligning* process, consistent with the view that DT is a "journey" (Kane 2017c) rather than a project (Gray et al. 2013).

Matt et al. (2015) propose the concept of *DT strategy* (DTS) to "focus on the transformation of products, processes and organizational aspects owing to new technologies" (p. 339). The authors argue that contrary to DBS, which focuses on "future states", DTS "is a blueprint that supports companies in governing the transformations that arise owing to the integration of digital technologies, as well as in their operations after a transformation" (p. 340). They view DTS as separate from "IT strategies and all other organizational and functional strategies" (p. 340) while *structural changes*, defined as "variations in a firm's organizational setup" (p. 341), must be carefully planned to leverage digital technologies for the benefit of the organization without forgoing financial constraints. Based on this concept, Hess et al. (2016) studied the DT of three German media companies and found that the specific financial constraints in place within each company had important implications on their respective ability to use digital technologies. Although less prominent (19 sources), this body of literature emphasizes the transformational process through which a firm leverages digital technologies to redefine its business model.

Leveraging digital technologies to uncover new paths for value creation

Digital technologies alone provide little value to an organization (Kane 2014). It is their use within a specific context that enables a firm to uncover new ways to create value, consistent with the enduring idea that organizational change is an *emergent* phenomenon (Markus and Robey 1988). In this section, we outline these new paths to value creation and present elements relevant to unlock the transformative potential of digital technologies (see Tables 2.5, 2.6, and 2.7).

Table 2.5 Transforming value creation and capture processes

Value propositions (n = 111)

Source type	Sources
Research paper – empirical (n = 63)	Antonopoulou et al. (2017); Asgarkhani (2005); Becker et al. (2018); Bravhar and Juric (2017); Clohessy et al. (2017); Delmond et al. (2017); Duerr et al. (2017); Fehér and Varga (2017); Friedlmaier et al. (2018); Gimpel (2015); Gimpel et al. (2018); Haas et al. (2014); Hanelt et al. (2015b); Hartl and Hess (2017); Henfridsson et al. (2014); Hildebrandt et al. (2015); Huang et al. (2017); Jha et al. (2016); Karimi et al. (2009); Karimi and Walter (2015); Kauffman et al. (2010); Kazan and Damsgaard (2014); Kleinschmidt and Peters (2017); Lee and Lee (2013); Leischnig et al. (2017); Li et al. (2016); Lucas Jr and Goh (2009); Lyytinen and Rose (2003b); Nwankpa and Roumani (2016); Oestreicher-Singer and Zalmanson (2012); Osmani et al. (2012); Pagani (2013); Petrikina et al. (2017); Piccinini et al. (2015b); Ramasubbu et al. (2014); Rauch et al. (2016); Remane et al. (2016a); Remane et al. (2016b); Resca et al. (2013); Riasanow et al. (2017); Richter et al. (2017); Roecker et al. (2017); Ross et al. (2016); Schmidt et al. (2017); Scott (2007); Selander et al. (2010); Shivendu and Zhang (2016); Srivastava and Shainesh (2015); Staykova and Damsgaard (2015a); Staykova and Damsgaard (2015b); Svahn et al. (2017a); Tan et al. (2015a); Tan et al. (2017); Tanniru et al. (2016); Terrenghi et al. (2017); Töytäri et al. (2017); Tumbas et al. (2015); Utesheva et al. (2012); Venkatesh et al. (2016); Woodard et al. (2012); Wörner et al. (2016); Yeow et al. (2017); Zolnowski and Warg (2018)
Research paper – other (n = 27)	Agarwal et al. (2010); Barrett et al. (2015); Bhimani (2015); Chowdhury and Åkesson (2011); Cziesla (2014); Fischer et al. (2018); Günther et al. (2017); Heilig et al. (2017); Jöhnk et al. (2017); Kahre et al. (2017); Krumeich et al. (2012); Krumeich et al. (2013); Lucas Jr et al. (2013); Lyytinen and Rose (2003a); Nambisan et al. (2017); Neumeier et al. (2017); Nischak et al. (2017); Pousttchi et al. (2015); Püschel et al. (2016); Rai and Sambamurthy (2006); Rizk et al. (2018); Seo (2017); Sørensen (2016); Tanriverdi and Lim (2017); Weissenfeld et al. (2017); Yoo (2013); Yoo et al. (2010b)
Practitioner outlet (n = 21)	Andal-Ancion et al. (2003); Basole (2016); Demirkan et al. (2016); Dremel et al. (2017); Earley (2014); Fitzgerald et al. (2014); Gray et al. (2013); Gust et al. (2017); Hansen and Sia (2015); Hess et al. (2016); Kane (2015a); Kane (2015b); Kane (2016b); Nehme et al. (2015); Porter and Heppelmann (2014); Reinartz and Imschloß (2017); Sebastian et al. (2017); Sia et al. (2016); Svahn et al. (2017b); Westerman et al. (2011); Wulf et al. (2017)

Value networks (n = 92)

Source type	Sources
Research paper – empirical (n = 54)	Asgarkhani (2005); Barua et al. (2004); Bazarhanova et al. (2018); Chanias (2017); Clohessy et al. (2017); Delmond et al. (2017); Dillon et al. (2015); Duerr et al. (2018); Duerr et al. (2017); Elbanna and Newman (2016); Friedlmaier et al. (2018); Gimpel (2015); Gimpel et al. (2018); Haffke et al. (2016); Hildebrandt et al. (2015); Hjalmarsson et al. (2015); Holotiuk and Beimborn (2017); Horlach et al. (2017); Huhtamäki et al. (2017); Islam et al. (2017); Karagiannaki et al. (2017); Karimi et al. (2009); Karimi and Walter (2015); Kiefer (2000); Korpela et al. (2017); Leonardi et al. (2016); Li et al. (2016); Medaglia et al. (2017); Mohagheghzadeh and Svahn (2016); Omar and Elhaddadeh (2016); Pagani (2013); Piccinini et al. (2015b); Ramasubbu et al. (2014); Reinhold and Alt (2009); Riasanow et al. (2017); Sachse et al. (2012); Saldanha et al. (2017); Schmidt et al. (2017); Selander et al. (2010); Setia et al. (2013); Smith and Webster (2006); Srivastava and Shainesh (2015);

Table 2.5 Cont.

Value networks (n = 92)

Source type	Sources
	Standaert and Jarvenpaa (2017); Svahn et al. (2017a); Tan et al. (2015a); Terrenghi et al. (2017); Töytäri et al. (2017); Utesheva et al. (2012); Winkler et al. (2014); Wörner et al. (2016); Xie et al. (2014); Yeow et al. (2017); Zhu et al. (2006); Zolnowski and Warg (2018)
Research paper – other (n = 17)	Barrett et al. (2015); Bharadwaj et al. (2013); Bhimani (2015); de Reuver et al. (2017); Fischer et al. (2018); Granados et al. (2008); Han et al. (2015); Hayes (2016); Legner et al. (2017); Lucas Jr et al. (2013); Nischak et al. (2017); Pousttchi et al. (2015); Prifti et al. (2017); Rizk et al. (2018); Seo (2017); Yoo (2013); Yoo et al. (2010b)
Practitioner outlet (n = 21)	Andal-Ancion et al. (2003); Basole (2016); Dremel et al. (2017); Fitzgerald (2014a); Gray et al. (2013); Hansen et al. (2011); Hansen and Sia (2015); Hess et al. (2016); Kane (2015a); Kane (2017b); Kane (2017d); Kane et al. (2016); Kohli and Johnson (2011); Nehme et al. (2015); NetworkWorld Asia (2015); Porter and Heppelmann (2014); Sia et al. (2016); Svahn et al. (2017b); Westerman and Bonnet (2015); Westerman et al. (2011); Wulf et al. (2017)

Digital channels (n = 72)

Source type	Sources
Research paper – empirical (n = 44)	Andrade and Doolin (2016); Barua et al. (2004); Berghaus and Back (2016); Berghaus and Back (2017); Bolton et al. (2017); Chanias (2017); Chanias and Hess (2016); Chatfield et al. (2015); Delmond et al. (2017); Duerr et al. (2017); Ebbesson and Bergquist (2016); Fehér and Varga (2017); Gimpel (2015); Gimpel et al. (2018); Haffke et al. (2017); Haffke et al. (2016); Holotiuk and Beimborn (2017); Horlacher et al. (2016); Karimi and Walter (2015); Kiefer (2000); Lee and Lee (2013); Li et al. (2017); Li et al. (2016); Oestreicher-Singer and Zalmanson (2012); Pagani (2013); Petrikina et al. (2017); Ramasubbu et al. (2014); Reinhold and Alt (2009); Ross et al. (2016); Sachse et al. (2012); Schmidt et al. (2017); Scott (2007); Selander et al. (2010); Shahlaei et al. (2017); Smith and Webster (2006); Standaert and Jarvenpaa (2017); Tan et al. (2015b); Terrenghi et al. (2017); Tiefenbacher and Olbrich (2016); Utesheva et al. (2012); Wenzel et al. (2015); Xie et al. (2014); Yeow et al. (2017); Zhu et al. (2006)
Research paper – other (n = 12)	Cziesla (2014); Fischer et al. (2018); Granados et al. (2008); Heilig et al. (2017); Krumeich et al. (2012); Krumeich et al. (2013); Legner et al. (2017); Lucas Jr et al. (2013); Morakanyane et al. (2017); Piccinini et al. (2015a); Rai and Sambamurthy (2006); Weissenfeld et al. (2017)
Practitioner outlet (n = 16)	Fitzgerald (2013); Gray et al. (2013); Hansen and Sia (2015); Hess et al. (2016); Johnson (2002); Kane (2014); Kane (2015a); Kane (2017d); Kane et al. (2016); Porter and Heppelmann (2014); Sebastian et al. (2017); Sia et al. (2016); Soava (2015); Westerman and Bonnet (2015); Westerman et al. (2011); Wulf et al. (2017)

Agility and ambidexterity (n = 79)

Source type	Sources
Research paper – empirical (n = 39)	Becker et al. (2018); Berghaus and Back (2017); Clohessy et al. (2017); Delmond et al. (2017); Duerr et al. (2017); Freitas Junior et al. (2017); Gimpel et al. (2018); Haffke et al. (2017); Haffke et al. (2016); Hartl and Hess (2017); Henfridsson and Lind (2014); Henfridsson et al. (2014); Holotiuk and Beimborn (2017); Horlach et al. (2017); Karagiannaki et al. (2017); Karimi et al. (2009);

(continued)

Table 2.5 Cont.

Agility and ambidexterity (n = 79)

Source type	Sources
	Karimi and Walter (2015); Leonhardt et al. (2017); Li et al. (2017); Li et al. (2016); Nwankpa and Datta (2017); Oh (2009); Osmani et al. (2012); Piccinini et al. (2015b); Ramasubbu et al. (2014); Reitz et al. (2018); Ross et al. (2016); Scott (2007); Shahlaei et al. (2017); Standaert and Jarvenpaa (2017); Svahn et al. (2017a); Tan et al. (2015a); Tanniru et al. (2016); Terrenghi et al. (2017); Tumbas et al. (2015); Woodard et al. (2012); Xie et al. (2014); Yeow et al. (2017); Zolnowski and Warg (2018)
Research paper – other (n = 18)	Bharadwaj et al. (2013); Dixon et al. (2017); Fichman et al. (2014); Fischer et al. (2018); Gerster (2017); Günther et al. (2017); Heilig et al. (2017); Jöhnk et al. (2017); Kahre et al. (2017); Le Dinh et al. (2016); Legner et al. (2017); Neumeier et al. (2017); Nischak et al. (2017); Piccinini et al. (2015a); Rai and Sambamurthy (2006); Schmid et al. (2017); Weiß et al. (2018); Yoo et al. (2010b)
Practitioner outlet (n = 22)	Demirkan et al. (2016); Dery et al. (2017); Dremel et al. (2017); Earley (2014); Fitzgerald (2014b); Gust et al. (2017); Hansen et al. (2011); Hansen and Sia (2015); Hess et al. (2016); Kane (2015b); Kane (2016b); Kane et al. (2017); Kane et al. (2016); Kohli and Johnson (2011); Maedche (2016); NetworkWorld Asia (2015); Sebastian et al. (2017); Sia et al. (2016); Weill and Woerner (2018); Westerman and Bonnet (2015); Woon (2016); Wulf et al. (2017)

Transforming the value creation process

The literature emphasizes the alteration as well as the redefinition of *business models* (Osterwalder and Pigneur 2010) in the context of DT (e.g., Morakanyane et al. 2017; Piccinini et al. 2015b). In this section, we detail four prominent changes related to (1) value propositions, (2) value networks, (3) digital channels, and (4) enabling agility and ambidexterity (see Table 2.5).

Value propositions

Digital technologies enable the creation of new value propositions that rely increasingly on the provision of *services* (Barrett et al. 2015). Organizations use digital technologies to transition from or augment the sales of physical products with the sales of services as an integral part of their value proposition to satisfy the needs of customers by offering innovative solutions as well as to gather data on their interactions with products and services (Porter and Heppelmann 2014; Wulf et al. 2017). A prime example of the creation of new value propositions through the use of digital technologies is Netflix, whose business model was originally based on the rental of movies stored on physical media. Over the years, Netflix has moved away from this value proposition to become the first large-scale provider of video streaming services. More recently, they have leveraged data collected from the use of their streaming service to better understand the content viewers enjoy as well as *how* content is consumed to help with the production of their own content (Günther et al. 2017). Overall, the literature highlights the potential for digital technologies to generate disruptive innovations that can significantly alter existing value propositions (Huang et al. 2017).

Value networks

Digital technologies also enable the redefinition of value networks (Delmond et al. 2017; Tan et al. 2015a). Andal-Ancion et al. (2003) argue that a firm can use digital technologies to implement one of three main mediation strategies. In a *disintermediation* strategy, digital technologies bypass intermediaries and enable direct exchanges among participants of a value network, e.g., customers (Hansen and Sia 2015). In a *remediation* strategy, the couplings between participants of a value network are reinforced as digital technologies enable close collaboration and coordination among participants, e.g., by using a platform to coordinate exchanges within a supply chain (Klötzer and Pflaum 2017). In *network-based* mediation, complex relationships among multiple stakeholders with potentially competing interests are created for the benefit of customers (Tan et al. 2015a). Digital technologies have also granted customers with the ability to become co-creators of value (*prosumers*) within a value network (Lucas Jr et al. 2013:379). For example, online communities (e.g., Oestreicher-Singer and Zalmanson 2012) and social media (e.g., Kane 2014) depend almost exclusively on the active contributions of users who have no obligation to use those technologies. Firms therefore have an imperative to incentivize customer engagement with digital technologies to drive the co-creation of value (Saldanha et al. 2017; Yeow et al. 2017).

Digital channels

In 72 sources, we found evidence that organizations use digital technologies to implement changes to their distribution and sales channels. This can be done in one of two ways. First, organizations can create new customer-facing channels, e.g., using social media, to reach and entertain a dialogue with consumers (Hansen and Sia 2015). For example, Hansen and Sia (2015) found that an organization can effectively use social media to bridge the gap between the physical and the digital world to support the creation of an omnichannel strategy, which the authors defined as "an integrated multichannel approach to sales and marketing" (p. 51). Second, the emergence of algorithmic decision-making afforded by digital technologies (Günther et al. 2017; Newell and Marabelli 2015) provides an unprecedented opportunity for organizations to effectively allow software to coordinate activities across organizations. In the manufacturing sector, sensors and other technologies associated with the IoT can improve supply change efficiency (Klötzer and Pflaum 2017) – e.g., through automated, *smart* procurement (Porter and Heppelmann 2014). Although IoT developments are still in their infancy when compared to other digital technologies (e.g., social media), we can expect that developments in smart products, digital goods, and the emergence of product upgrades "over the air" will drive further interest on this topic.

Agility and ambidexterity

Digital technologies can help firms rapidly adapt to changes in environmental conditions (Fitzgerald 2016b; Günther et al. 2017; Hong and Lee 2017; Huang et al. 2017; Kohli and Johnson 2011) by contributing to organizational agility, defined as a firm's "ability to detect opportunities for innovation and seize those competitive market opportunities by assembling requisite assets, knowledge, and relationships with speed and surprise" (Sambamurthy et al. 2003:245). Analytics and the IoT can be exploited to optimize existing business processes and reduce slack resources (Du et al. 2016). In other instances, these technologies can be implemented to provide insight into untapped market opportunities or to increase customer proximity (Hansen and Sia 2015; Setia et al. 2013). For example, a firm can offer innovative maintenance services based on the analysis of data generated by sensors embedded within its products (Porter and Heppelmann 2014). The literature also reports on the ability for firms to use of digital technologies to achieve ambidexterity – also referred to as bimodality in the practitioner literature (Haffke et al. 2017) and successfully combine the *exploration* of digital innovation with the *exploitation* of existing resources (Li et al. 2017; Svahn et al. 2017a). For example, in their study of 25 companies, Sebastian et al. (2017) found that ambidexterity is founded upon a firm's ability to maintain both an *operational backbone* as well as a *digital services platform*.

Structural changes required for changing the value creation process

Like any other initiative that has the potential to profoundly alter the fabric of an organization, DT is associated with a number of important structural changes (see Table 2.6).

Organizational structure

Consistent with the idea that agility and ambidexterity are necessary capabilities to compete in a digital world, the literature highlights cross-functional collaboration as an important element of DT (Earley 2014; Maedche 2016). Although the idea of fostering collaboration across business units and breaking functional silos is by no means new in IS research, the literature on DT highlights the reality that in many instances, a significant chasm must still be crossed for these forms of collaboration to emerge and to fuse organizational and IS strategy together (Duerr et al. 2017; Li et al. 2016; Seo 2017; Svahn et al. 2017a). One way to achieve this objective is through the creation of a separate unit that maintains a degree of independence from the rest of the organization (Maedche 2016; Sia et al. 2016). With this structure, the unit is granted with a relative degree of flexibility propitious to innovation while maintaining access to existing resources. Another way is to create cross-functional teams that remain within the current organization (Dremel et al. 2017; Svahn et al. 2017a). For instance, Dremel et al. (2017) studied the multi-year development of an analytics capability at Audi AG. They found that the formation of multidisciplinary *competence networks* that transcend Audi's traditional organizational structure helped the organization use analytics as an IT-driven initiative for the benefit of business units.

Organizational culture

The disruption spurred by DT also requires that the culture of the focal organization changes (Hartl and Hess 2017). In incumbent firms for instance, there is evidence that the traditional separation between IT and business functions is so ingrained into the fabric of the organization that they become part of the organization's values (Haffke et al. 2017). In the newspaper industry, Karimi and Walter (2015) found that the ability for a firm to build the capabilities required to alter their value proposition using digital platforms is

Table 2.6 Structural changes required to alter value creation and capture processes

Organizational structure (n = 59)

Source type	Sources
Research paper – empirical (n = 24)	Berghaus and Back (2016); Berghaus and Back (2017); Boland et al. (2003); Bolton et al. (2017); Driver and Gillespie (1992); Duerr et al. (2018); Duerr et al. (2017); Haffke et al. (2017); Holotiuk and Beimborn (2017); Horlach et al. (2017); Klötzer and Pflaum (2017); Leonardi and Bailey (2008); Lucas Jr and Goh (2009); McGrath et al. (2008); Mueller and Renken (2017); Piccinini et al. (2015b); Resca et al. (2013); Roecker et al. (2017); Ross et al. (2016); Selander and Jarvenpaa (2016); Svahn et al. (2017a); Tumbas et al. (2015); Yeow et al. (2017); Zhu et al. (2006)
Research paper – other (n = 17)	Dixon et al. (2017); Fischer et al. (2018); Günther et al. (2017); Jöhnk et al. (2017); Kahre et al. (2017); Krumeich et al. (2012); Krumeich et al. (2013); Legner et al. (2017); Loebbecke and Picot (2015); Lucas Jr et al. (2013); Lyytinen and Rose (2003a); Matt et al. (2015); Morakanyane et al. (2017); Neumeier et al. (2017); Schmid et al. (2017); Tilson et al. (2010); Weiß et al. (2018)
Practitioner outlet (n = 18)	Demirkan et al. (2016); Dremel et al. (2017); Du et al. (2016); Earley (2014); Fitzgerald (2014c); Fitzgerald (2016a); Hansen and Sia (2015); Hess et al. (2016); Kane (2017b); Kane et al. (2017); Kane et al. (2016); Kohli and Johnson (2011); Maedche (2016); Porter and Heppelmann (2014); Sebastian et al. (2017); Singh and Hess (2017); Svahn et al. (2017b); Wulf et al. (2017)

Organizational culture (n = 36)

Source type	Sources
Research paper – empirical (n = 22)	Berghaus and Back (2017); Bolton et al. (2017); Chatfield et al. (2015); Dasgupta and Gupta (2010); Duerr et al. (2018); Haffke et al. (2017); Hartl and Hess (2017); Holotiuk and Beimborn (2017); Kamel (2015); Karimi and Walter (2015); Klötzer and Pflaum (2017); Li et al. (2017); Li et al. (2016); Lucas Jr and Goh (2009); Mueller and Renken (2017); Piccinini et al. (2015b); Roecker et al. (2017); Schmidt et al. (2017); Scott (2007); Svahn et al. (2017a); Tan et al. (2017); Töytäri et al. (2017)
Research paper – other (n = 2)	Jöhnk et al. (2017); Morakanyane et al. (2017)
Practitioner outlet (n = 12)	Dremel et al. (2017); Fitzgerald (2014c); Gust et al. (2017); Hansen and Sia (2015); Kane (2016a); Kane et al. (2016); Sebastian et al. (2017); Svahn et al. (2017b); Watson (2017); Weill and Woerner (2018); Westerman et al. (2011); Wulf et al. (2017)

Leadership (n = 62)

Source type	Sources
Research paper – empirical (n = 27)	Agarwal et al. (2011); Becker et al. (2018); Bekkhus (2016); Benlian and Haffke (2016); Berghaus and Back (2016); Berghaus and Back (2017); Chanias (2017); Chanias and Hess (2016); Chatfield et al. (2015); Duerr et al. (2018); Fehér et al. (2017); Gimpel et al. (2018); Haffke et al. (2017); Haffke et al. (2016); Hesse (2018); Holotiuk and Beimborn (2017); Horlacher et al. (2016); Li et al. (2017); Li et al. (2016); Liere-Netheler et al. (2018); Oestreicher-Singer and Zalmanson (2012); Scott (2007); Tan et al. (2015a); Tanniru et al. (2016); Tiefenbacher and Olbrich (2016); Töytäri et al. (2017); Xie et al. (2014)

(continued)

Table 2.6 Cont.

Leadership (n = 62)

Source type	Sources
Research paper – other (n = 5)	Bharadwaj et al. (2013); Kahre et al. (2017); Legner et al. (2017); Matt et al. (2015); Riedl et al. (2017)
Practitioner outlet (n = 30)	Andriole (2017); Demirkan et al. (2016); Dery et al. (2017); Du et al. (2016); Earley (2014); Fitzgerald (2013); Fitzgerald (2014c); Fitzgerald (2016b); Fitzgerald et al. (2014); Hansen et al. (2011); Hansen and Sia (2015); Hess et al. (2016); Kane (2015a); Kane (2015b); Kane (2017c); Kane et al. (2017); Kane et al. (2016); Kohli and Johnson (2011); Maedche (2016); Nehme et al. (2015); Sebastian et al. (2017); Sia et al. (2016); Singh and Hess (2017); Weill and Woerner (2018); Westerman (2016); Westerman and Bonnet (2015); Westerman et al. (2014); Westerman et al. (2011); Wulf et al. (2017); Yee and Ng (2015)

Employee roles and skills (n = 69)

Source type	Sources
Research paper – empirical (n = 31)	Agarwal et al. (2011); Asgarkhani (2005); Chatfield et al. (2015); Delmond et al. (2017); Driver and Gillespie (1992); Duerr et al. (2018); Fehér et al. (2017); Gimpel et al. (2018); Hartl and Hess (2017); Hjalmarsson et al. (2014); Holotiuk and Beimborn (2017); Joshi et al. (2017); Klötzer and Pflaum (2017); Li et al. (2017); Li et al. (2016); Liere-Netheler et al. (2018); Lucas Jr and Goh (2009); Lyytinen and Rose (2003b); McGrath et al. (2008); Petrikina et al. (2017); Remane et al. (2017); Richter et al. (2017); Roecker et al. (2017); Ross et al. (2016); Shahlaei et al. (2017); Svahn et al. (2017a); Tan et al. (2015a); Tiefenbacher and Olbrich (2016); Utesheva et al. (2012); Xie et al. (2014); Zhu et al. (2006)
Research paper – other (n = 12)	Bharadwaj et al. (2013); Fichman et al. (2014); Günther et al. (2017); Jöhnk et al. (2017); Legner et al. (2017); Loebbecke and Picot (2015); Lyytinen and Rose (2003a); Matt et al. (2015); Morakanyane et al. (2017); Prifti et al. (2017); Sørensen (2016); Weiß et al. (2018)
Practitioner outlet (n = 26)	Demirkan et al. (2016); Dery et al. (2017); Dremel et al. (2017); Fitzgerald (2014b); Fitzgerald (2014c); Fitzgerald (2016a); Fitzgerald (2016b); Fitzgerald et al. (2014); Gust et al. (2017); Hess et al. (2016); Kane (2015a); Kane (2016a); Kane (2016b); Kane (2017a); Kane (2017d); Kane et al. (2017); Kane et al. (2016); Maedche (2016); Nehme et al. (2015); Porter and Heppelmann (2014); Singh and Hess (2017); Watson (2017); Weill and Woerner (2018); Westerman et al. (2014); Westerman et al. (2011); Wulf et al. (2017)

founded upon a combination of variables including *values* – which comprise an innovative culture, a common language, and a multimedia mindset. A question that arises from these findings therefore relates to our understanding of what a "digital culture looks like" (Kane et al. 2016:9). A common theme across studies points to the need for firms to cultivate a willingness to take risks and to experiment (Fehér and Varga 2017) with digital technologies on a small scale before scaling these successful experiments to the rest of the organization (Dremel et al. 2017). This theme highlights the necessity to align actions with the principles of agility (Horlach et al. 2017; Leonhardt et al. 2017) inspired by software development practices (Gust et al. 2017). In doing so, firms can foster learning through small, incremental, and iterative changes while maintaining their ability to adapt long-term plans based on the outcomes of such experiments as well as ongoing changes in their environment (Jöhnk et al. 2017).

Leadership

In the context of DT, organizational leaders must work to ensure that their organizations develop a digital mindset while being capable of responding to the disruptions associated with the use of digital technologies (Benlian and Haffke 2016; Hansen et al. 2011). To that end, the literature highlights the creation of new leadership roles (Haffke et al. 2016; Horlacher et al. 2016). For example, the creation of a chief digital officer (CDO) position signals the strategic nature of DT for the entire organization. CDOs are tasked to ensure that digital technologies are properly leveraged and aligned with the objectives of the organization (Horlacher et al. 2016; Singh and Hess 2017). They act as boundary spanners that can help to implement digital business strategy into series of concrete actions that influence a firm's *organizing logic* (Sambamurthy and Zmud 2000) and foster close collaboration between business and IT functions. In a few instances, the CDO position is also seen as an important but temporary role (Singh and Hess 2017:16), suggesting that there may be an end state to DT consistent with the notion of digital transformation strategy (Matt et al. 2015).

Employee roles and skills

In the context of DT, changes to the structure as well as the culture of an organization lead employees to assume roles that were traditionally outside of their functions. Specifically, the literature highlights the idea that DT fosters situations where employees who are not part of the IT function take the lead on technology-intensive projects (Yeow et al. 2017). Conversely, members of the IT function are expected to become active, business-savvy participants in the realization of those projects (Dremel et al. 2017). As digital technologies enable new forms of automation (Neumeier et al. 2017) and decision-making processes (Dremel et al. 2017; Hess et al. 2016), questions on the need to develop the skills of existing workers (Hess et al. 2016) as well as the skills required for future workers who will form the *digital workforce* (Colbert et al. 2016) are also becoming increasingly relevant (Watson 2017). Far from removing the need for organizations to depend on human capital, DT requires employees to depend more heavily on their analytical skills to solve increasingly complex business problems (Dremel et al. 2017), and accompanying employees through this transition poses significant challenges that extend beyond the domain of human resources (Karimi and Walter 2015; Singh and Hess 2017).

Barriers to changing the value creation process

Notwithstanding these changes, inertia and resistance can hinder the unfolding of an organization's DT, in line with the literature on IT-enabled organizational transformation (see Table 2.7).

Table 2.7 Barriers to changing value creation and capture processes

Inertia (n = 35)

Source type	Sources
Research paper – empirical (n = 21)	Bolton et al. (2017); Delmond et al. (2017); Hildebrandt et al. (2015); Karagiannaki et al. (2017); Kleinschmidt and Peters (2017); Li et al. (2017); Lucas Jr and Goh (2009); Mithas et al. (2013); Rauch et al. (2016); Remane et al. (2017); Schmidt et al. (2017); Scott (2007); Srivastava and Shainesh (2015); Töytäri et al. (2017); Tumbas et al. (2015); Wenzel et al. (2015); Woodard et al. (2012); Xie et al. (2014); Yang et al. (2012); Yeow et al. (2017); Zhu et al. (2006)
Research paper – other (n = 6)	Dixon et al. (2017); Granados et al. (2008); Legner et al. (2017); Nambisan et al. (2017); Schmid et al. (2017); Tilson et al. (2010)
Practitioner outlet (n = 8)	Andriole (2017); Fitzgerald (2014b); Fitzgerald et al. (2014); Kane (2016a); Kane (2016b); Kohli and Johnson (2011); Sia et al. (2016); Westerman et al. (2011)

Resistance (n = 40)

Source type	Sources
Research paper – empirical (n = 22)	Barua et al. (2004); Bazarhanova et al. (2018); Becker et al. (2018); Chatfield et al. (2015); Ciriello and Richter (2015); Duerr et al. (2018); Elbanna and Newman (2016); Hjalmarsson et al. (2014); Kleinschmidt and Peters (2017); Liere-Netheler et al. (2018); Lucas Jr and Goh (2009); Mohagheghzadeh and Svahn (2016); Omar and Elhaddadeh (2016); Paavola et al. (2017); Petrikina et al. (2017); Piccinini et al. (2015b); Selander and Jarvenpaa (2016); Serrano and Boudreau (2014); Svahn et al. (2017a); Töytäri et al. (2017); Yeow et al. (2017); Zhu et al. (2006)
Research paper – other (n = 4)	Bichler et al. (2016); Günther et al. (2017); Matt et al. (2015); Schmid et al. (2017)
Practitioner outlet (n = 14)	Andriole (2017); Dery et al. (2017); Du et al. (2016); Fitzgerald et al. (2014); Gust et al. (2017); Hansen et al. (2011); Hansen and Sia (2015); Kane (2016a); Kane et al. (2017); Kohli and Johnson (2011); Singh and Hess (2017); Svahn et al. (2017b); Westerman et al. (2011); Wulf et al. (2017)

Inertia

One of the most significant barriers to DT is inertia (35 sources). Inertia is relevant where existing resources and capabilities can act as barriers to disruption (Islam et al. 2017; Svahn et al. 2017a), highlighting the relevance of path dependence as a constraining force for innovation through digital technologies (Srivastava and Shainesh 2015; Wenzel et al. 2015). For example, incumbent firms are deeply embedded in existing relationships with customers and suppliers, have well-established production processes that are highly optimized, but often rigid (Andriole 2017) and rely on resources that cannot easily be reconfigured (Kohli and Johnson 2011; Woodard et al. 2012).

These issues have been identified in both research (Roecker et al. 2017; Töytäri et al. 2017) and practitioner (Westerman et al. 2011) literature. For example, the Kodak case (Lucas Jr and Goh 2009) illustrates how the core capabilities of an organization can become core rigidities that prevent the radical transformation afforded by digital technologies (in this instance, digital photography). Töytäri et al. (2017) found that organizational culture, identity, and legitimacy form strong institutional barriers that hinder the development of smart services. In all those instances, the issue is not that the organization's top management does not consider digital technologies as potentially beneficial to the organization. Rather, the structural components of the organization, both tangible (e.g., means of production) and intangible (e.g., organizational culture), are so embedded within everyday practices that they stifle the innovative and disruptive power of digital technologies.

Resistance

Another barrier to DT is the resistance that employees can demonstrate when disruptive technologies are introduced in the organization (40 sources) (Fitzgerald et al. 2014; Kane 2016a; Lucas Jr and Goh 2009; Singh and Hess 2017). The issue of resistance raises important questions with regards to the ways and the pace at which technologies are introduced into an organization and the practitioner literature highlights "innovation fatigue" (Fitzgerald et al. 2014: 9) as one of the causes of resistance. Singh and Hess (2017) found that the CDO position can be leveraged to ensure that digital technologies are used in a way that remains consistent with the organizational culture that employees are accustomed to and favor their acceptance. Conversely, Schmid et al. (2017) argued that resistance is a product of inertia rooted in everyday work that cannot be addressed by simply altering the behavior of employees. Rather, it requires that processes be altered to enable flexibility in the face of change. Svahn et al. (2017a:242) show that resistance can also be explained by a lack of visibility on the potential benefits of digital technologies. They found that workshops that involve organizational actors who will be affected by DT can help prevent resistance and improve cross-functional collaboration.

Assessing the impacts of digital transformation

It has been argued that DT has the potential to have wide ranging impacts (see Table 2.8), including at the society level (Agarwal et al. 2010; Majchrzak et al. 2016). Nevertheless, we find those impacts to be primarily assessed at the organization level, as illustrated in Appendix C.

Organizational level impacts

Operational efficiency

Although digital technologies have the potential to transform an organization, we found 36 studies highlighting operational efficiency – which includes the automation (Andriole 2017), the improvement of business processes (Gust et al. 2017) as well as costs savings (Pagani 2013), as a benefit of DT. For instance, cloud computing provides on-demand, elastic resources that do not need to be provisioned, managed, and maintained by IT staff (Kane 2015b). Big data and analytics are expected to speed up the decision-making process (Bharadwaj et al. 2013), enabling faster response time while smart products and services, through the embedding of artificial intelligence that leverages (big) data, can enable automated, algorithmic decision-making (Loebbecke and Picot 2015; Newell and Marabelli 2015).

Table 2.8 Impacts of digital transformation

	Organizational level impacts	
Operational efficiency (n = 36)		
Source type	Sources	
Research paper – empirical (n = 10)	Deliyannis et al. (2009); Holotiuk and Beimborn (2017); Liere-Netheler et al. (2018); Pagani (2013); Richter et al. (2017); Roecker et al. (2017); Ross et al. (2016); Schellhorn (2016); Scott (2007); Svahn et al. (2017a)	
Research paper – other (n = 7)	Agarwal et al. (2010); Bhimani (2015); Fischer et al. (2018); Heilig et al. (2017); Morakanyane et al. (2017); Neumeier et al. (2017); Weiß et al. (2018)	
Practitioner outlet (n = 19)	Andal-Ancion et al. (2003); Cummings (2012); Demirkan et al. (2016); Du et al. (2016); Fitzgerald (2013); Fitzgerald (2016a); Fitzgerald et al. (2014); Gray et al. (2013); Gust et al. (2017); Kane (2015b); Kane (2017a); Kohli and Johnson (2011); NetworkWorld Asia (2015); Porter and Heppelmann (2014); Sebastian et al. (2017); Westerman (2016); Westerman and Bonnet (2015); Westerman et al. (2014); Westerman et al. (2011)	
	Organizational performance (n = 49)	
Source type	Sources	
Research paper – empirical (n = 33)	Barua et al. (2004); Chanias (2017); Delmond et al. (2017); Felgenhauer et al. (2017); Freitas Junior et al. (2017); Gimpel (2015); Hildebrandt et al. (2015); Karimi and Walter (2015); Kauffman et al. (2010); Leischnig et al. (2017); Lienhard et al. (2017); Liere-Netheler et al. (2018); Mithas et al. (2013); Nwankpa and Datta (2017); Nwankpa and Roumani (2016); Oestreicher-Singer and Zalmanson (2012); Oh (2009); Pagani (2013); Piccinini et al. (2015b); Remane et al. (2017); Saldanha et al. (2017); Schellhorn (2016); Selander et al. (2010); Shivendu and Zhang (2016); Srivastava and Shainesh (2015); Srivastava et al. (2016); Staykova and Damsgaard (2015b); Svahn et al. (2017a); Tan et al. (2015a); Trantopoulos et al. (2017); Woodard et al. (2012); Yeow et al. (2017); Zhu et al. (2006)	
Research paper – other (n = 11)	Bharadwaj et al. (2013); Bhimani (2015); Dixon et al. (2017); Gerster (2017); Granados et al. (2008); Krumeich et al. (2013); Lucas Jr et al. (2013); Myunsoo and Byungtae (2013); Neumeier et al. (2017); Tanriverdi and Lim (2017); Yoo et al. (2010b)	
Practitioner outlet (n = 5)	Basole (2016); Du et al. (2016); Kane et al. (2017); Nambisan et al. (2017); Sia et al. (2016)	
	Higher-level impacts	
Societal impacts and well-being (n = 40)		
Source type	Sources	
Research paper – empirical (n = 15)	Andrade and Doolin (2016); Asgarkhani (2005); Chan et al. (2016); Chatfield et al. (2015); Deng et al. (2016); Ganju et al. (2016); Hanelt et al. (2015b); Jha et al. (2016); Leong et al. (2016); Miranda et al. (2016); Nastjuk et al. (2016); Oreglia and Srinivasan (2016); Selander and Jarvenpaa (2016); Srivastava and Shainesh (2015); Venkatesh et al. (2016)	

Table 2.8 Cont.

Societal impacts and well-being (n = 40)

Source type	Sources
Research paper – other (n = 19)	Agarwal et al. (2010); Bara-Slupski (2016); Barrett et al. (2015); Bryant (2010); de Reuver et al. (2017); Eymann et al. (2015); Fichman et al. (2014); Günther et al. (2017); Legner et al. (2017); Loebbecke and Picot (2015); Lucas Jr et al. (2013); Majchrzak et al. (2016); Newell and Marabelli (2014); Newell and Marabelli (2015); Rajan (2002); Riedl et al. (2017); Tilson et al. (2010); Urquhart and Vaast (2012); Yoo et al. (2010a)
Practitioner outlet (n = 6)	Earley (2014); Fitzgerald et al. (2014); Kane (2014); Soava (2015); Watson (2017); Westerman (2016)

Security and privacy issues (n = 44)

Source type	Sources
Research paper – empirical (n = 19)	Asgarkhani (2005); Bazarhanova et al. (2018); Chatfield et al. (2015); Dillon et al. (2015); Fehér et al. (2017); Gimpel et al. (2018); Islam et al. (2017); Kamel (2015); Kiefer (2000); Korpela et al. (2017); McGrath (2016); Medaglia et al. (2017); Paavola et al. (2017); Piccinini et al. (2015b); Roecker et al. (2017); Sachse et al. (2012); Tiefenbacher and Olbrich (2016); Töytäri et al. (2017); Zhu et al. (2006)
Research paper – other (n = 14)	Agarwal et al. (2010); Arens and Rosenbloom (2003); Arner et al. (2017); Eymann et al. (2015); Fichman et al. (2014); Fischer et al. (2018); Goes (2015); Günther et al. (2017); Legner et al. (2017); Newell and Marabelli (2014); Newell and Marabelli (2015); Pousttchi et al. (2015); Rai and Sambamurthy (2006); Tilson et al. (2010)
Practitioner outlet (n = 11)	Dremel et al. (2017); Gray et al. (2013); Kane (2015b); Kane et al. (2016); Nehme et al. (2015); Ng (2016); Porter and Heppelmann (2014); Singh and Hess (2017); Watson (2017); Woon (2016); Wulf et al. (2017)

Organizational performance

DT is also associated with increases in several dimensions of organizational performance, including innovativeness (Svahn et al. 2017a), financial performance (Karimi and Walter 2015), firm growth (Tumbas et al. 2015), reputation (Kane 2016c; Yang et al. 2012) as well as competitive advantage (Neumeier et al. 2017). For example, under the freemium model, a firm can use online communities to increase the sense of belonging of users and motivate them to purchase premium accounts (Oestreicher-Singer and Zalmanson 2012). In the context of entrepreneurial firms where the growth rate is nonlinear, Tumbas et al. (2015) found that successful firms put up a "digital façade" to enable connectivity with customers and business partners while later using this façade as an instrument to foster relationships with other customers and suppliers. This and other examples (e.g., Setia et al. 2013) show how digital technologies can, through higher customer engagement and participation, foster higher profits for firms. At a conceptual level, it has been proposed that digital technologies can support a firm's ability to sense the complexity of its environment in order to design a response that can help maximize its chances of survival through the adaptation or the redefinition of its core activities (Tanriverdi and Lim 2017).

Higher-level impacts

Positive impacts

Several articles also reflect on the impacts of DT at higher levels, including at the industry and the society levels. Research has argued that digital technologies afford a tremendous potential for the improvement of the quality of life of individuals (Agarwal et al. 2010; Pramanik et al. 2016). One such example is healthcare (Agarwal et al. 2010), where various types of technologies, including electronic health records (Kane 2015b), big data and analytics (Kane 2016c; Kane 2017a), as well as augmented physical products (Bravhar and Juric 2017) are perceived as valuable contributions to a sector that has traditionally been a laggard in technology adoption (Lucas Jr et al. 2013). Recent research has specifically highlighted those benefits in geographical areas that are impacted by poverty and resource disparities. For example, Srivastava and Shainesh (2015) studied the use of tele-ophthalmology in rural India and found that digital technologies enable healthcare organizations to increase access to care while simultaneously reducing costs for both the organization (e.g., by minimizing the physical space required to operate the clinic) and patients (e.g., by not having to travel long distances to reach a clinic). They likened this virtuous circle to a mechanism of "value reinforcement", where "the value created through one parameter can be leveraged to create value through another parameter" (p. 257).

Undesirable outcomes

Notwithstanding these positive outcomes, the literature also reflects on the potential issues associated with the pervasive use of digital technologies, primarily in the domain of security and privacy. For instance, Newell and Marabelli (2015) argue that algorithmic decision-making, for all its potential benefits, also carries significant risks for individuals and society in general and that security, privacy, and safety should remain important areas of consideration for researchers, government bodies as well as practitioners. In the automobile industry, Piccinini et al. (2015b) found that data security and privacy were also important issues. While our sample includes 19 empirical works containing references to security and privacy, our analysis reveals that these works do not address the crucial question as to *how* security and privacy can be effectively turned from a potential issue into a source of positive impacts for an organization as well as society. Rather, the focus is currently set on acknowledging these issues and their ramifications for organizations, society, as well as individuals (Newell and Marabelli 2015).

Digital transformation and IT-enabled transformation

Having reviewed the literature on DT, we now turn to the question of the novelty of the phenomenon. The notion that IT carries *transformative* potential is not new and has long been acknowledged in the literature (Zuboff 1988). In their study of IT investment announcements by public firms, Dehning et al. (2003) built on the seminal work of Zuboff (1988) and argued that the strategic role of IT falls into one of four main categories (see Table 2.9). Using these rules, Lucas Jr et al. (2013) assessed IT's transformative impact in three areas (financial markets, healthcare, customer experience). In two of those

Table 2.9 Strategic roles of IT (adapted from Dehning et al. 2003: 639; 653–654)

Automate	**Description**: Replacing human labor. **Goals**: Improve existing capabilities, efficiency, and effectiveness. **Outcomes**: Clearly identifiable and measurable.
Informate-up	**Description**: Providing information to top management. **Goals**: Improve decision-making, coordination, and collaboration. **Outcomes**: Difficult to anticipate because they may include intangible benefits.
Informate-down	**Description**: Providing information to employees across the firm. **Goals**: Improve decision-making, coordination, and collaboration. **Outcomes**: Difficult to anticipate because they may include intangible benefits.
Transform	**Description**: Redefining the business model, business processes and relationships of the firm. **Goals**: Alter existing capabilities, acquire new capabilities, both internally (through reconfiguration) and externally (through strategic partnerships). **Outcomes**: difficult to anticipate, include both tangible and intangible benefits; fundamentally alters the fabric of the firm.

areas – financial markets and customer experience – they argued that a profound transformation has already taken place as processes, market structures, and value networks have changed significantly. In the context of healthcare, Agarwal et al. (2010) observed that "an IT-enabled transformation of health care is just beginning, and it cannot happen too fast" (p. 377).

Although these criteria are referenced within our sample, they can be difficult to apply because studies often focus on innovation rather than change, while change is seen as a necessary step toward the achievement of organizational performance and operational efficiency. Nevertheless, if we consider that changes to one or more of the constituting dimensions of a firm's business model are reflective of a *transformation* (as per Dehning et al.'s criteria), then even the reconfiguration of key business activities using digital technologies is reflective of a transformation. For instance, algorithmic decision-making can be conceptualized as a form of automation. Yet as Newell and Marabelli (2015) argue, its implications are more far-reaching than that. As evidenced by our review, digital technologies (e.g., through the emergence of platforms and ecosystems) have significantly altered the way firms create value (Tan et al. 2015a). Even firms that build physical products are now faced with the pressing need to incorporate services and software as part their core offerings, turning their physical products into conduits for the generation, the collection, and the exchange of valuable data (Porter and Heppelmann 2014).

Still, the question remains: is DT different from other forms of IT-enabled transformation? To investigate this issue, we looked at prior literature on the topic of IT-enabled transformation. We analyzed evidence presented in studies on IT-enabled transformation against the evidence found in DT literature to compare both phenomena (see Table 2.10). We referred to the four properties of DT uncovered during the semantic decomposition process we used to build our definition of DT and augmented these properties with two other dimensions, which we labelled *impetus* and *locus of uncertainty* that emerged during our analysis. Together these observations lead us to view DT as an evolution of IT-enabled transformation. In our view, DT better reflects the complexity of the environment within

Table 2.10 Comparing IT-enabled transformation and digital transformation

Property	IT-enabled organizational transformation	Digital transformation
Impetus	Organizational decision.	Society and industry trends; organizational decision.
Target entity	Single organization or, less frequently, an organization along with its immediate value network.	Organization, platform, ecosystem, industry, society.
Scope	The transformation can, in some instances, be profound but is typically limited to an organization's processes and its immediate value network (e.g., suppliers).	The transformation can be profound and has implications beyond the organization's immediate value network (e.g., society, customers).
Means	Single IT artifact primarily focused on operations (e.g., ERP).	Combinations of digital technologies (e.g., analytics and mobile apps).
Expected outcome	Business processes are optimized and efficiency gains are realized; in some instances the business model of the focal organization is altered. Existing institutions remain unchanged.	Business processes are transformed and the business model of the focal organization is altered; in some instances business processes are optimized. Because of its ramifications at higher levels, the transformation raises important questions with regards to the relevance of current institutions (e.g., regulatory framework, ethics).
Locus of uncertainty	Internal: located inside the organization.	External (first): located outside of the organization. Internal (second): located inside the organization.
Illustrative example	A firm purchases an ERP and reengineers its business processes according to industry best practices as well as institutionalized accounting principles. The ERP implementation also enables increased coupling between the firm and its supply chain partners.	As consumers increasingly rely on mobile devices to purchase goods and services, a firm decides to capitalize on this trend by developing a mobile application to engage with customers. In doing so, it also captures and analyzes the data generated through customer interactions with their mobile application to increase customer proximity and enhance customer experience.

which firms operate and the disruptive impacts of digital technologies on individuals, organizations, and society. As a result, we concur with Bharadwaj et al.'s (2013) arguments on the relevance of the strategic role of digital technologies and their ability to impact the *scale* and the *scope* of the changes associated with their use along with the *speed* at which those changes take place.

Digital transformation: a research agenda

Our review highlights the significant contributions that research has made toward our understanding of DT. In this section, we extend these contributions through the outline of an ambitious agenda comprising two avenues for future strategic IS research and practice on DT. The first avenue is the contribution of dynamic capabilities as a theoretical foundation to study DT. The second avenue is the incorporation of ethics in strategic IS research on DT.

Avenue 1: how dynamic capabilities contribute to digital transformation

Our review and our inductive framework highlight the nature of DT as a process where digital technologies create an impetus for organizations to implement responses to gain or maintain their competitive advantage. Key questions related to the efficacy of these responses are the ability for firms to sense disruptions, seize them (e.g., through strategic responses), and to reconfigure elements of their business model accordingly. In this first avenue, we propose dynamic capabilities (DC) as a theoretical foundation to study those mechanisms that enable firms to engage with DT to enable strategic renewal. Specifically, we propose that research focus on three key areas. The first is the building of organizational dynamic capabilities to support the ongoing DT of a firm. The second is the role of integrative capabilities, an understudied form of dynamic capabilities, in the context of digital platforms and ecosystems. The third is the microfoundations that help us understand and explain how DT unfolds in practice.

Building organizational dynamic capabilities for digital transformation

The DC perspective contributes to explain how firms build and sustain competitive advantage (Helfat and Raubitschek 2018; Schilke et al. 2018; Teece 2007). DC extends the resource-based view of the firm (RBV) and focuses on the ability for firms to *purposefully alter* their resource base to increase their degree of fitness with their environment and ensure their survival (Jiang et al. 2015; Schilke et al. 2018). DC posits that firms possess both ordinary as well as dynamic capabilities. The former relate to "the performance of administrative, operational, and governance-related functions that are (technically) necessary to accomplish tasks"; the latter "involve higher-level activities that can enable an enterprise to direct its ordinary activities toward high-payoff endeavors" (Teece 2014: 328). Dynamic capabilities enable firms to innovate and adapt to changes in their environment through three main mechanisms (Teece 2007): *sensing*, i.e., the "identification, development, codevelopment, and assessment of technological opportunities in relationship to customer needs" (p. 332); *seizing*, i.e., the "mobilization of resources to address needs and opportunities, and to capture value from doing so" (p. 332); and *transforming*, that is, the "continued renewal" (p. 332) of the firm as its resources are reconfigured to strategically seize opportunities and respond to threats. Although DC has been found useful in IS research in general, and, as outlined in our review, in the context of DT (eight studies within our sample), we argue that research on DT could benefit from further engaging with this perspective.

There is an interesting fit between DC as a conceptual foundation and DT as a phenomenon of interest. The literature highlights the nature of DT as a source of *continuous* change and disruption in a firm's competitive environment. The ability for firms to design mechanisms that enable repeatable, continuous adaptation in spite of such rapid changes is therefore an important question. The contributions of DC have been found most useful in contexts fraught with environmental turbulence or hypercompetition as ordinary capabilities cannot explain – on their own – how firms build and sustain competitive advantage (Teece 2014:329). As physical resources – including products – become comparatively less relevant than services, as consumers contribute to influence trends related to the use of digital technologies, and as value networks become broader and more complex, firms experience higher levels of uncertainty. To manage this uncertainty, mechanisms to sense and adapt to changes that originate outside of the firm's competitive environment (e.g., Netflix entering the movie making business) and locus of control (e.g., online users moving away from Facebook and adopting Instagram) must be put in place. Although we have some evidence of firms managing to adapt to these changes (e.g., Yeow et al. 2017), our understanding of the ability for firms to design *repeatable* mechanisms for this purpose is limited. Recent developments in DT literature have proposed the concept of *digital maturity* (Kane 2017c) as a capacity to respond to change in an appropriate manner and we argue that DC may help us understand how firms achieve digital maturity as they design and maintain these higher-level mechanisms that enable adaptability through successive waves of digital innovation.

The role of integrative capabilities to support digital platforms and ecosystems

In the context of multi-sided platforms (MSP), research in organizational policy has argued that three types of dynamic capabilities are also relevant: innovation capabilities, environmental scanning and sensing capabilities, and integrative capabilities (Helfat and Raubitschek 2018). Research on DT has touched upon some elements of the first two (albeit without specifically referring to DC) (e.g., Tiefenbacher and Olbrich 2016), while the third has been largely left unexplored in DT literature as well as in IS research in general. Integrative capabilities (IC) "provide the capacity for reliable, repeatable communication and coordination activity directed toward the introduction and modification of: products; resources and capabilities; business models." (p. 1395) They can be internal (within the firm) or external (across firms, e.g., through alliances and partnerships).

In the context of DT, we argue that external IC are essential because the value networks that firms rely on to create and capture value are increasingly large and complex. For instance, firms have little choice but to depend on multiple parties to participate in platforms and ecosystems, whether they are leaders (Tan et al. 2015a), complementors (Ghazawneh and Henfridsson 2013), or customers (Li et al. 2017). The *integration* of digital technologies provided by multiple parties is a crucial piece of the puzzle enabling a firm to successfully participate in a digital platform or ecosystem. Most studies on DT acknowledge the need for firms to engage with other parties to generate digital innovation (e.g., Hansen and Sia 2015; Nehme et al. 2015). However, how these firms manage to remain abreast of the changes that take place within their value network remains unexplored. From the perspective of a platform leader for instance, the need to balance new functionality and technical debt, control and openness (Constantinides et al. 2018; Wessel et al. 2017) is paramount to recruit complementors and customers without running the risk of platform envelopment (Eisenmann et al. 2006) or desertion (Tiwana 2015). From the perspective of a complementor, sensing the extent of the changes implemented by platform leader(s) to reconfigure one's own processes is an equally important question.

Although research in strategy and organizational policy has indeed begun to turn toward digital platforms and ecosystems (Helfat and Raubitschek 2018), its coverage of the actual designs of the technological artifacts involved in these platforms and ecosystems and the impacts of those designs on the ability for firms to adapt to change in an appropriate and timely manner remains superficial. One notable exception is the case study of Alibaba by Li et al. (2017) who found that platform providers rely on *mentoring, facilitating and rule-making* as mechanisms to facilitate the building of cross-border e-commerce capabilities by SMEs. Nevertheless, there is a dire need to better understand how communication and coordination take place in the context of digital platforms and ecosystems.

Microfoundations of dynamic capabilities: how digital transformation unfolds in practice

At a micro level, the literature has called for research to further study the micro processes that support the building and the maintenance of dynamic capabilities (Schilke et al. 2018). Indeed, although they are conceptualized as organizational competencies, DC are founded on the performance of routines as repeated patterns of interdependent actions (Feldman and Pentland 2003), and are therefore anchored in the performances of individuals, including managers (Helfat and Martin 2015; Yeow et al. 2017). Notwithstanding the advances research in strategy has made to understand the contributions of individuals to the building and maintenance of DC (e.g., Abell et al. 2008), calls have been made to further engage with the nature of the work *performed* by actors that support these capabilities (Schilke et al. 2018; Teece 2007). In the context of DT, we view these calls as an opportunity for IS research to make a contribution to the literature on DC.

The literature on DT highlights changes to an organization's leadership structure as an important enabler of new business models. Specifically, it has been argued that DT calls for organizations to appoint chief digital officers (CDO) to help them undertake their transformation (e.g., Horlacher et al. 2016; Sia et al. 2016; Weill and Woerner 2018). However, we currently know very little with regards to the actual work CDOs do other than the fact that it is sometimes considered a temporary position (Singh and Hess 2017:16). One way to understand the implications of this new position would be to study the contributions of different leadership structures (e.g., firms with CEOs, CIOs and CDOs versus firms with CEOs and CIOs only) toward the decisions that enable the building of DC. This might involve, for example (1) *sensing* changes in markets by developing an analytics competency; (2) judiciously *seizing* upon these trends by augmenting products with services using mobile applications and social media; and (3) *transforming* the organization to become a platform provider enabling customers to act as complementors of the firm's digital services, thereby further contributing data that can be leveraged for future sensing.

Over the years, the need to build a better understanding of the performance of work underlying the alignment of business and IT strategies has become increasingly relevant. Specifically, it has been argued that a focus on the *practices* that undergird this process can help draw contributions that are closer to the actual work individuals perform and can better explicate the mechanisms that have been overlooked in the variance-based models that have been the hallmark of strategic alignment research since the early 1990s (Karpovsky and Galliers 2015; Peppard et al. 2014). This position is consistent with DC's account of the work of individuals (Schilke et al. 2018) and we posit that it can also help contribute to literature on DT, e.g., by unearthing those practices that are most effective depending on the "region of complexity" (Tanriverdi and Lim 2017) where a firm is situated at a given point in time based on the pace of digital innovation in a given competitive landscape.

In recent years, the strategy-as-practice (SaP) literature has applied and extended the contributions of the practice literature in strategy research (Jarzabkowski et al. 2007; Kaplan and Orlikowski 2013) to focus on IS strategy formulation and performance (Peppard et al. 2014). At a high level, the SaP literature seeks to understand what "managers and other organizational actors *do* in their day-to-day activities to achieve alignment" (Karpovsky and Galliers 2015: 137), thereby switching the focus from the study of alignment to that of *aligning* as a process accounting for "all activities that may contribute to tightening links between IT and business across an organization" (pp. 137–138). In their review of IS strategy articles on the topic, Karpovsky and Galliers (2015) identified eight core categories of practices pertaining to four overarching metaphors of the aligning process (see Table 2.11). In the DT literature, there is evidence of the importance of those practices in practitioner outlets (e.g., Sia et al. 2016; Singh and Hess 2017), but we could only find one research article within our sample that focused on the aligning process itself (Yeow et al. 2017). Gaining a better understanding of those practices and their relevance at different points in times during the digital transformation process would not only inform research, but also practitioners on this topic.

Table 2.11 Categories of aligning activities (adapted from Karpovsky and Galliers 2015: 141–142)

Aligning metaphor	Activity categories	Illustrations in the context of DT
Adaptation	Evaluating	Analyzing user-generated and third party data to *sense* shifts in consumer demand, possibly in real-time (Kane 2016c; Setia et al. 2013; Tiefenbacher and Olbrich 2016)
Translation	Developing	Writing a mobile application that enables direct business to consumer communication to increase customer proximity (Hansen and Sia 2015)
	Reconfiguring	Adapting an operational backbone to leverage digital services (Kohli and Johnson 2011; Sebastian et al. 2017)
Integration	Strengthening	Breaking functional silos to enable close collaboration between business and IT (Dremel et al. 2017; Maedche 2016)
	Signaling	Hiring a CDO to highlight the importance of a DT initiative in an organization (Horlacher et al. 2016; Singh and Hess 2017)
Experience	Negotiating	Creating forums to reconcile long-term strategic transformation objectives with short-term operational objectives (Svahn et al. 2017a; Yeow et al. 2017)
	Learning	Enacting a digital transformation strategy to enable knowledge sharing between business and IT units (Leonhardt et al. 2017; Matt et al. 2015)
	Decision-making	Using analytics to alter the decision-making process from one that is based largely on intuition to one that is based on evidence (Newell and Marabelli 2015; Watson 2017)

Additionally, it could help us study links between high level DC and the actual practices performed by organizational actors. For instance, it has been argued that research on DC currently has little knowledge on the decision-making processes that actors rely on, including heuristics (Schilke et al. 2018: 415). These processes could be mapped onto the *decision-making* aligning activity category unearthed by Karpovsky and Galliers (2015) and applied to the context of DT. Specifically, it has been argued that dynamic problem-solving and decision-making are paramount for successful digital innovation management (Nambisan et al. 2017). Yet it has been observed that "dynamic capability enables the repeated and reliable performance of an activity directed toward strategic change, as distinct from entirely *ad hoc* problem solving" (Schilke et al. 2018: 393). How do organizational actors design such dynamic problem-solving processes – e.g., using digital technologies, including algorithmic decision-making – that are not only effective and efficient, but also repeatable across waves of digital innovation is an important question that remains to be answered.

Avenue 2: the strategic relevance of ethics in digital transformation

The literature on DT focuses primarily on impacts located at the organizational level. Our review also points to a smaller but growing body of literature calling for a more comprehensive – and perhaps nuanced – understanding of the impacts of DT at different levels of analysis, including for individuals (Newell and Marabelli 2015) and society (Majchrzak et al. 2016). In this research avenue, we build on these calls to offer *ethics*, broadly defined as "abstract and theoretical reflection on moral statements" that "asks for the grounds on which moral statements are made" (Stahl 2012:641), as a reference discipline to tackle the question of the multifaceted nature of the impacts of DT. Specifically, we argue that theoretical approaches pertaining to ethics offer different vantage points through which we can better understand the long-term and higher-level impacts of DT. We do not purport that ethics provide a safeguard *against* DT. Rather, we view ethics as a complementary perspective that can help us peer into aspects of the phenomenon that are currently understudied. Central to those arguments is the notion that a firm must strike a *balance* between elements – an increasing number of which are outside of its locus of control – to ensure that it can generate and sustain performance.

We couch our arguments along three key areas for future research. The first is the role of ethics as a means to account for the multilevel implications of DT. The second is the growing need for firms to balance the tension between organizational performance and ethics. The third is the use of ethics as a means to address the concurrent, and often conflicting needs of value co-creators. Through this research avenue, our ambition is to foster strategic IS research on the role of ethics to study questions that are highly relevant for firms in the context of DT.

Ethics and the multilevel nature of digital transformation

Contrary to traditional views on IT-enabled organizational transformation, the ability for firms to innovate using digital technologies takes place within a context where environmental disruptions originate from the increased use of digital technologies at the industry and society levels. From a multilevel perspective, we can treat a firm's use of digital technologies as an outcome of a lower-level process, or as an antecedent of a higher-level outcome.

Focusing on the firm's use of digital technologies as the outcome of a lower level process, research on DT can study the alignment – or lack thereof – between decisions related to the firm's business model and the values and principles of their employees in the context of attracting and retaining a digital workforce (Baptista et al. 2017; Colbert et al. 2016). To illustrate the relevance of this question, we refer to the case of Google. As part of its initial public offering in 2004, the company that is at the source of many of the profound changes that continue to redefine the competitive landscape and influence our experiences as individuals stated: "Don't be evil. We believe strongly that in the long term, we will be better served – as shareholders and in all other ways – by a company that does good things for the world even if we forgo some short term gains" (The New York Times 2004). A somewhat similar consideration remains in the code of conduct of its parent company, Alphabet: "Employees of Alphabet and its subsidiaries and controlled affiliates ("Alphabet") should *do the right thing*" (Alphabet Inc. 2015, emphasis added). In spite of these claims, the company came under scrutiny from its employees and later the US government following news that it was working on the development of a censored version of its search engine for the Chinese market (Google employees against Dragonfly 2018). Considering the growing capabilities of digital technologies and the speed at which information spreads across networks, firms must now deal with the strategic imperative of ensuring that their objectives do not run against the moral views of their employees even if it may be counterintuitive from the perspective of financial performance.

Turning toward the impacts of DT at higher levels of analysis, theories of ethics can help us engage with broader objectives of strategic IS research to "stay abreast as well as anticipate the emerging organizational and societal problems around the world" (Galliers et al. 2012: 90). Indeed, it has been argued that the ethical implications of DT lie beyond the level of a firm's strategy and can impact society itself (Ganju et al. 2016; Majchrzak et al. 2016). Notwithstanding, research that has engaged with this aspect of DT has thus far remained scarce (e.g., Leonardi et al. 2016) and has focused primarily on DT in emerging environments. From this perspective, future research on DT may benefit from engaging with other streams of literature that focus on the higher-level implications of IT, including ICT4D as well as ICT4S. In other, non-emerging contexts, empirical studies on the higher-level impacts of DT remain scarce and discussions have primarily focused on issues of digital divide (Yoo et al. 2010a) as well as security and privacy (Newell and Marabelli 2015). However, we still know very little on the role of ethics in digital strategy formulation and execution as a means to ensure that the firm-level positive impacts of DT remain consistent at higher levels.

We argue that normative theories of ethics can help us better understand how the efforts undertaken by firms translate into higher-level impacts for industries and society. *Consequentialist* or *teleological* theories (e.g., utilitarianism) focus on evaluating the benefits of the ends that are sought through action. *Deontological* theories (e.g., Kantian ethics) are based on the existence of higher-level principles that provide a frame of reference against which actions are evaluated and generally focus on the means that are employed rather than their outcomes. Together, these theoretical foundations can help us gain a deeper understanding of the rationale behind firms' strategic decisions and the impacts of those decisions at higher levels of analysis. For example, we may find that some firms operate based on the existence of higher-level principles that guide their actions while others take a more consequentialist approach in reaching their objectives without much consideration for the means employed to achieve them (e.g., selling personal data to third parties). It is when these different approaches coexist and collide with one another that higher-level impacts may emerge through processes of *emergence* or *composition* (Klein and Kozlowski 2000) but our knowledge on these important questions that can guide IS and business strategy is limited.

In parallel, ethical theories can help us study the emergence of field-level disruptions. Indeed, in several instances, digital technologies have been under development for a period of time but their disruptive potential is only realized when individuals adopt them. For example, the failed introduction of Google Glass and the ongoing development of this project point to the notion that the timing of the introduction of digital innovations also accounts for the public's willingness to adopt digital technologies, beyond the technical or the financial feasibility of these technologies.

Sustaining organizational performance: the role of ethical performance

Consistent with the majority of strategic IS research, the literature on DT focuses on the ability for a firm to alter its business model and the short to medium-term impacts of those changes on organizational performance (Karimi and Walter 2015; Yeow et al. 2017). Although it may be attributable to the relative nascence of the phenomenon, we also find that the *sustaining* of organizational performance is much less discussed or studied in this body of literature. We argue that this question is highly relevant in the context of strategic research on DT as firms increasingly rely on digital technologies (e.g., mobile applications) used by multiple categories of stakeholders (e.g., individuals, other firms) to achieve their goals. As value networks become more complex (Andal-Ancion et al. 2003; Nehme et al. 2015) and involve more – and different – actors (Gray et al. 2013), exerting total control over a firm's ability to sustain organizational performance over time becomes more challenging. Within this context, we view ethics as particularly relevant because they can contribute to guide the design as well as use of digital technologies to ensure that the achievement of short-term goals does not compromise a firm's ability to sustain their performance over time, as illustrated in Figure 2.2.

For example, firms are able to increase customer proximity and tailor service offerings based on latent preferences that seldom need to be made explicit through the analysis of data collected from primary (e.g., social media, hardware) and/or secondary sources (e.g., data brokers). Although such tactics can be beneficial to a firm's performance, recent news has shown that there can be undesirable consequences associated with the very practices that make a firm successful in the first place (Cadwalladr and Graham-Harrison 2018), as illustrated with the case of Facebook (Neate 2018). In many instances, the undesirable outcomes of such tactics do not occur because they are illegal. Rather, it is because they are deemed morally reprehensible by some of the stakeholders who are co-creators of value for the firm.

Although ethical considerations have been conceptualized as an important element of a firm's strategy (Carroll 1999), their relationship to IT has thus far been largely ignored by IS research, save for a few notable exceptions (Smith 2002; Smith and Hasnas 1999; Stahl 2012). In a context where digital technologies occupy an increasing portion of

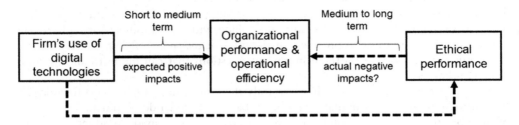

Note: Solid arrows represent the current focus of DT literature. Dashed arrows are unexplored aspects of DT.

Figure 2.2 The contribution of ethics to long-term organizational performance.

the value proposition of firms, traditional constructs that embed ethical considerations (e.g., corporate social responsibility) need to be revisited based on the building blocks of the DT process uncovered during our review. In light of the current manifestations of such considerations in the realm of practice (e.g., http://datafordemocracy.org, www.partnershiponai.org/, and, more recently, www.blog.google/topics/ai/aiprinciples/), we argue that it is time for strategic IS research to (re)engage with these topics by leveraging concepts that impact strategic research in related areas such as ethical performance, information ethics and data governance (Floridi 2018), which provide part of the scaffolding that can help us study these topics. In doing so, we will gain a more comprehensive, theoretically richer understanding of DT while addressing questions that are highly relevant for practitioners (Newton 2018; Westerman 2016).

Using ethics to account for the conflicting demands of value co-creators

Related to the previous argument is the notion that the increased complexity of value networks leads to a situation where firms must tend to the multiple, sometimes conflicting demands of value co-creators. This issue is particularly permeable for multi-sided platforms and digital ecosystems, which, by definition, rely on the contributions of multiple parties (Tan et al. 2015a; Wessel et al. 2017). In other instances, the business model of a firm may depend on the collection, the sharing, and the selling of data. These data are not a by-product of a firm's operations – as was often the case in the past – but are an integral part of the value proposition of the firm.

Within this context, firms must adequately balance the demands of multiples parties as well as the respective frames of reference that guide their perception of what is considered right and wrong. For a platform owner, the challenge resides in ensuring that tending to one party's needs does not happen at the expense of others. For example, granting more access to data to one party might be perceived as a breach of security and privacy by another (Newell and Marabelli 2015). To study this issue, we propose that strategic IS research turn toward normative theories of ethics. Specifically, theories of business ethics, such as stakeholder theory or social contract theory, underline the challenges associated with the evaluation of actions and goals in pluralistic business contexts in general and in the context of technology design and use by multiple parties in particular (Smith 2002; Stahl et al. 2014). In this instance again, regulatory frameworks and institutional pressures may prove insufficient to gain a deep understanding of the challenges firms face in a digital world because they provide a single frame of reference. In contrast, value networks associated with DT translate into contexts where multiple frames of reference exist and firms must devise ways to strike a balance between the needs of multiple parties, without compromising the performance of the firm itself or its ability to sustain competitive advantage. How to study these plural contexts is a complex issue that is currently under-researched or has primarily been studied from the perspective of the economics of digital platforms and ecosystems (Constantinides et al. 2018).

Together, these two research avenues highlight six key areas where research on DT is currently lacking. In our view, they represent exciting opportunities for strategic IS research to engage with topics that are interesting from a research standpoint as well as highly relevant in light of current debates on the phenomenon. Over the years, strategic IS research has established itself as a crucial link bridging the realms of technology and business. These two research avenues contribute to this tradition by calling for a better understanding of the theoretical linkages between the process of DT and its associated outcomes at different levels of analysis as well as over time.

Limitations

Our work has four limitations. Although care was taken to ensure that the review process was performed with rigor and *systematicity* (Rowe 2014: 243), the analysis was performed by a single researcher. To help overcome this limitation, we engaged with professionals and scholars working on DT throughout our analysis to assess the face validity of our findings. Second, our review is restricted to the IS discipline in spite of the relevance of DT in other domains (e.g., Preuveneers et al. 2016). While this limits the generalizability of our findings, we found that the size of IS literature on the topic was sufficient to adopt this focus and future research could use our findings to inform other disciplines where DT is relevant. Third, we acknowledge that DT is a very active topic in IS research, as evidenced by the conference papers in our sample. Although this means that some of the avenues or questions offered in our research agenda may already be under study, we welcome this possibility as a testament to the relevance of DT. Finally, our objective to take stock of current knowledge on DT called for an approach that favored breadth over depth. While this means that the intricacies of the relationships we have studied are not presented in the current paper, we deemed it necessary to provide this overall picture of DT. Future research on the topic may zoom in on specific relationships that have emerged from our analysis.

Conclusion

Our review of IS research on the digital transformation phenomenon highlights a rich body of literature that contributes to our understanding of the benefits as well as the challenges associated with digital transformation at multiple levels. Our findings underline the increasing complexity of the environment within which firms operate. As digital technologies afford more information, computing, communication, and connectivity, they enable new forms of collaboration among distributed networks of diversified actors. In doing so, they also create dependencies among actors whose interests may not fully be aligned. This new reality offers tremendous potential for innovation and performance in organizations, and extends beyond the boundaries of the firm to affect individuals, industries, and society. At the same time, it renders firms' ability to sustain their competitive advantage more fragile than ever as they control fewer elements of their operating environment.

Although they are by no means exhaustive, we believe that the two research avenues that we have proposed can help us better understand the strategic implications of DT and the dynamic interactions that take place between firms and their environment as digital technologies continue to impact these interactions. Beyond these avenues, future research may use our framework as a guide to zoom in and investigate specific relationships of our inductive framework such as to determine under which conditions an organizational design performs better than another (e.g., cross-functional teams versus digital ventures) or explore under-researched relationships (e.g., a potential feedback loop between firms' use of digital technologies and changes in consumer behavior and expectations). Overall, we hope that this review contributes to help future research further explore the nature and the implications of this highly relevant phenomenon for organizations as well as for society.

Acknowledgments

The author is grateful to the editors, especially Sirkka Jarvenpaa, for their guidance and to the three anonymous reviewers for their comments and suggestions. The author is also grateful to participants of the HEC Montréal writing workshop, Suzanne Rivard, and Camille Grange for their insightful comments and suggestions on earlier versions of this paper.

Appendices. Supplementary material

Appendices A, B, C, D, and E can be found online at https://doi.org/10.1016/j.jsis.2019.01.003.

References

Abell, P., Felin, T., and Foss, N. 2008. "Building microfoundations for the routines, capabilities, and performance links," *Managerial & Decision Economics* (29:6), pp. 489–502.

Agarwal, R., Guodong, G., DesRoches, C., and Jha, A. K. 2010. "The digital transformation of healthcare: Current status and the road ahead," *Information Systems Research* (21:4), pp. 796–809.

Agarwal, R., Johnson, S. L., and Lucas Jr, H. C. 2011. "Leadership in the face of technological discontinuities: The transformation of Earthcolor," *Communications of the Association for Information Systems* (29), pp. 628–644.

Akmajian, A., Farmer, A. K., Bickmore, L., Demers, R. A., and Harnish, R. M. 2017. *Linguistics: An introduction to language and communication.* Cambridge, MA: MIT Press.

Alphabet Inc. 2015. "Code of conduct." Retrieved October 12, 2017, from https://abc.xyz/investor/other/code-of-conduct.html

Andal-Ancion, A., Cartwright, P. A., and Yip, G. S. 2003. "The digital transformation of traditional businesses," *MIT Sloan Management Review* (44:4), pp. 34–41.

Andrade, A. D., and Doolin, B. 2016. "Information and communication technology and the social inclusion of refugees," *MIS Quarterly* (40:2), pp. 405–416.

Andriole, S. J. 2017. "Five myths about digital transformation," *MIT Sloan Management Review* (58:3), pp. 20–22.

Antonopoulou, K., Nandhakumar, J., and Begkos, C. 2017. "The emergence of business model for digital innovation projects without predetermined usage and market potential," *Hawaii International Conference on System Sciences*, Waikoloa Beach, HI, pp. 5153–5161.

Arens, Y., and Rosenbloom, P. S. 2003. "Responding to the unexpected," *Communications of the ACM* (46:9), pp. 33–35.

Arner, D. W., Barberis, J., and Buckley, R. P. 2017. "Fintech, regtech, and the reconceptualization of financial regulation," *Northwestern Journal of International Law & Business* (37:3), pp. 373–415.

Asgarkhani, M. 2005. "Digital government and its effectiveness in public management reform," *Public Management Review* (7:3), pp. 465–487.

Baptista, J., Stein, M.-K., Lee, J., Watson-Manheim, M. B., and Klein, S. 2017. "Call for papers: Strategic perspectives on digital work and organizational transformation." Retrieved January 2, 2019, from www.journals.elsevier.com/the-journal-of-strategic-information-systems/call-for-papers/digital-work-and-organizational-transformation

Bara-Slupski, T. K. 2016. "Holistic approach: Paradigm shift in the research agenda for digitalization of healthcare in sub-Saharan Africa," *African Journal of Information Systems* (8:4), pp. 1–21.

Barrett, M., Davidson, E., Prabhu, J., and Vargo, S. L. 2015. "Service innovation in the digital age: key contributions and future directions," *MIS Quarterly* (39:1), pp. 135–154.

Barua, A., Konana, P., Whinston, A. B., and Yin, F. 2004. "An empirical investigation of net-enabled business value," *MIS Quarterly* (28:4), pp. 585–620.

Basole, R. C. 2016. "Accelerating digital transformation: Visual insights from the API ecosystem," *IT Professional* (18:6), pp. 20–25.

Bassano, C., Gaeta, M., Piciocchi, P., and Spohrer, J. C. 2017. "Learning the models of customer behavior: From television advertising to online marketing," *International Journal of Electronic Commerce* (21:4), pp. 572–604.

Bazarhanova, A., Yli-Huumo, J., and Smolander, K. 2018. "Love and hate relationships in a platform ecosystem: A case of Finnish electronic identity management," *Hawaii International Conference on System Sciences*, Waikoloa Beach, HI, pp. 1493–1502.

Becker, W., Schmid, O., and Botzkowski, T. 2018. "Role of CDOs in the digital transformation of SMEs and LSEs: An empirical analysis," *Hawaii International Conference on System Sciences*, Waikoloa Beach, HI, pp. 4534–4543.

Bekkhus, R. 2016. "Do KPIs used by CIOs decelerate digital business transformation? The case of ITIL," *Digital Innovation, Technology, and Strategy Conference*, Dublin, Ireland.

Benlian, A., and Haffke, I. 2016. "Does mutuality matter? Examining the bilateral nature and effects of CEO-CIO mutual understanding," *Journal of Strategic Information Systems* (25:2), pp. 104–126.

Berghaus, S. 2016. "The fuzzy front-end of digital transformation: Three perspectives on the formulation of organizational change strategies," *Bled eConference*, Bled, Slovenia, pp. 129–144.

Berghaus, S., and Back, A. 2016. "Stages in digital business transformation: Results of an empirical maturity study," *Mediterranean Conference of Information Systems*, Cyprus.

Berghaus, S., and Back, A. 2017. "Disentangling the fuzzy front end of digital transformation: Activities and approaches," *International Conference of Information Systems*, Seoul, South Korea.

Bharadwaj, A., El Sawy, O., Pavlou, P., and Venkatraman, N. 2013. "Digital business strategy: Toward a next generation of insights," *MIS Quarterly* (37:2), pp. 471–482.

Bhimani, A. 2015. "Exploring big data's strategic consequences," *Journal of Information Technology* (30:1), pp. 66–69.

Bichler, M., Frank, U., Avison, D., Malaurent, J., Fettke, P., Hovorka, D., Krämer, J., Schnurr, D., Müller, B., and Suhl, L. 2016. "Theories in business and information systems engineering," *Business & Information Systems Engineering* (58:4), pp. 291–319.

Boland, R., Lyytinen, K., and Yoo, Y. 2003. "Path creation with digital 3D representations: Networks of innovation in architectural design and construction," *Digital Innovation, Technology, and Strategy Conference*, Seattle, WA, pp. 1–23.

Bolton, A., Murray, M., and Fluker, J. 2017. "Transforming the workplace: Unified communications & collaboration usage patterns in a large automotive manufacturer," *Hawaii International Conference on System Sciences*, Waikoloa Beach, HI, pp. 5470–5479.

Bravhar, K., and Juric, R. 2017. "Personalized drug administration to patients with Parkinson's disease: Manipulating sensor generated data in Android environments," *Hawaii International Conference on System Sciences*, Waikoloa Beach, HI, pp. 3489–3498.

Bryant, A. 2010. "The metropolis and digital life," *Communications of the Association for Information Systems* (27), pp. 665–676.

Cadwalladr, C., and Graham-Harrison, E. 2018. "Revealed: 50 million Facebook profiles harvested for Cambridge Analytica in major data breach." Retrieved March 18, 2018, from www.theguardian.com/news/2018/mar/17/cambridge-analytica-facebook-influence-us-election

Carlo, J. L., Lyytinen, K., and Boland Jr, R. J. 2012. "Dialectics of collective minding: Contradictory appropriations of information technology in a high-risk project," *MIS Quarterly* (36:4), pp. 1081–1108.

Carroll, A. B. 1999. "Corporate social responsibility: Evolution of a definitional construct," *Business & Society* (38:3), pp. 268–295.

Chan, J., Ghose, A., and Seamans, R. 2016. "The internet and racial hate crime: Offline spillovers from online access," *MIS Quarterly* (40:2), pp. 381–403.

Chanias, S. 2017. "Mastering digital transformation: The path of a financial services provider towards a digital transformation strategy," *European Conference of Information Systems*, Guimaraes, Portugal, pp. 16–31.

Chanias, S., and Hess, T. 2016. "Understanding digital transformation strategy formation: Insights from Europe's automotive industry," *Pacific Asia Conference on Information Systems*, Chiayi, Taiwan.

Chatfield, A., Reddick, C., and Al-Zubaidi, W. 2015. "Capability challenges in transforming government through open and big data: Tales of two cities," *International Conference of Information Systems*, Fort Worth, TX.

Chowdhury, S., and Åkesson, M. 2011. "A proposed conceptual framework for identifying the logic of digital services," *Pacific Asia Conference on Information Systems*, Brisbane, Australia: Association for Information Systems.

Ciriello, R., and Richter, A. 2015. "Idea hubs as nexus of collective creativity in digital innovation," *International Conference of Information Systems*, Fort Worth, TX.

Clohessy, T., Acton, T., and Morgan, L. 2017. "The impact of cloud-based digital transformation on ICT service providers' strategies," *Bled eConference*, Bled, Slovenia, pp. 111–126.

Colbert, A., Yee, N., and George, G. 2016. "The digital workforce and the workplace of the future," *Academy of Management Journal* (59:3), pp. 731–739.

Constantinides, P., Henfridsson, O., and Parker, G. G. 2018. "Introduction – platforms and infrastructures in the digital age," *Information Systems Research* (29:2), pp. 381–400.

Constantiou, I., Shollo, A., Kreiner, K., and Vendelø, M. 2017. "Digitization in maritime industry: Coping with a vessel's engine failure," *European Conference of Information Systems*, Guimaraes, Portugal.

Cummings, R. B. 2012. "Are sales & SSE taxes getting cloudier with the advent of cloud computing?," *Journal of State Taxation* (30:2), pp. 7–56.

Cziesla, T. 2014. "A literature review on digital transformation in the financial service industry," *Bled eConference*, Bled, Slovenia, pp. 1–13.

Dasgupta, S., and Gupta, B. 2010. "Organizational culture and technology use in a developing country: An empirical study," *Special Interest Group on ICT and Global Development*, St. Louis, MN.

de Reuver, M., Sørensen, C., and Basole, R. C. 2017. "The digital platform: A research agenda," *Journal of Information Technology*, (33:2), pp. 1–12.

Dehning, B., Richardson, V. J., and Zmud, R. W. 2003. "The value relevance of announcements of transformational information technology investments," *MIS Quarterly*, (27:4), pp. 637–656.

Deliyannis, I., Antoniou, A., and Pandis, P. 2009. "Design and development of an experimental low-cost internet-based interactive TV station," *Mediterranean Conference of Information Systems*, Athens, Greece.

Delmond, M.-H., Coelho, F., Keravel, A., and Mahl, R. 2017. "How information systems enable digital transformation: A focus on business models and value co-production," *IUP Journal of Business Strategy* (14:3), pp. 7–40.

Demirkan, H., Spohrer, J. C., and Welser, J. J. 2016. "Digital innovation and strategic transformation," *IT Professional* (18:6), pp. 14–18.

Deng, X., Joshi, K., and Galliers, R. D. 2016. "The duality of empowerment and marginalization in microtask crowdsourcing: Giving voice to the less powerful through value sensitive design," *MIS Quarterly* (40:2), pp. 279–302.

Dery, K., Sebastian, I. M., and Meulen, N. v. d. 2017. "The digital workplace is key to digital innovation," *MIS Quarterly Executive* (16:2), pp. 135–152.

Dillon, S., Deakins, E., Hofmann, S., Räckers, M., and Kohlborn, T. 2015. "A longitudinal study of local e-government development: The policy maker perspective," *European Conference of Information Systems*, Munster, Germany.

Dixon, J. A., Brohman, K., and Chan, Y. E. 2017. "Dynamic ambidexterity: Exploiting exploration for business success in the digital age," *International Conference of Information Systems*, Seoul, South Korea.

Dremel, C., Wulf, J., Herterich, M. M., Waizmann, J.-C., and Brenner, W. 2017. "How AUDI AG established big data analytics in its digital transformation," *MIS Quarterly Executive* (16:2), pp. 81–100.

Driver, S., and Gillespie, A. 1992. "The diffusion of digital technologies in magazine print publishing: Organizational change and strategic choices," *Journal of Information Technology (Routledge, Ltd.)* (7:3), p. 149.

Du, W. Y., Pan, S. L., and Huang, J. S. 2016. "How a latecomer company used IT to redeploy slack resources," *MIS Quarterly Executive* (15:3), pp. 195–213.

Duerr, S., Holotiuk, F., Beimborn, D., Wagner, H.-T., and Weitzel, T. 2018. "What is digital organizational culture? Insights from exploratory case studies," *Hawaii International Conference on System Sciences*, Waikoloa Beach, HI, pp. 5126–5135.

Duerr, S., Wagner, H.-T., Weitzel, T., and Beimborn, D. 2017. "Navigating digital innovation – The complementary effect of organizational and knowledge recombination," *Wirtschaftsinformatik Conference*, St. Gallen, Switzerland: AIS Electronic Library, pp. 1363–1377.

Earley, S. 2014. "The digital transformation: Staying competitive," *IT Professional* (16:2), pp. 58–60.

Ebbesson, E. 2015. "Fragmented digital infrastructures – The case of social (news) media," *Americas Conference of Information Systems*, Puerto Rico

Ebbesson, E., and Bergquist, M. 2016. "Dancing in the dark-social media tactics in the news industry," *Mediterranean Conference of Information Systems*, Cyprus.

Eisenmann, T., Parker, G., and Van Alstyne, M. W. 2006. "Strategies for two-sided markets," *Harvard Business Review* (84:10), pp. 92–101.

Elbanna, A., and Newman, M. 2016. "Disrupt the disruptor: Rethinking 'disruption' in digital innovation," *Mediterranean Conference of Information Systems*, Cyprus.

Eymann, T., Legner, C., Prenzel, M., Krcmar, H., Müller, G., and Liggesmeyer, P. 2015. "Addressing grand challenges," *Business & Information Systems Engineering* (57:6), pp. 409–416.

Fehér, P., Szabó, Z., and Varga, K. 2017. "Analysing digital transformation among Hungarian organizations," *Bled eConference*, Bled, Slovenia, pp. 139–150.

Fehér, P., and Varga, K. 2017. "Using design thinking to identify banking digitization opportunities – Snapshot of the Hungarian banking system," *Bled eConference*, Bled, Slovenia, pp. 151–168.

Feldman, M. S., and Pentland, B. T. 2003. "Reconceptualizing organizational routines as a source of flexibility and change," *Administrative Science Quarterly* (48:1), pp. 94–118.

Felgenhauer, A., Klier, J., Klier, M., and Lindner, G. 2017. "The impact of social engagement on customer profitability – Insights from a direct banking institution's online customer network," *European Conference of Information Systems*, Guimaraes, Portugal, pp. 2101–2118.

Fichman, R. G., Dos Santos, B. L., and Zheng, Z. E. 2014. "Digital innovation as a fundamental and powerful concept in the information systems curriculum," *MIS Quarterly* (38:2), 329–354.

Fischer, M., Imgrund, F., Friedrich-Baasner, G., Winkelmann, A., and Janiesch, C. 2018. "Connected enterprise meets connected customer – A design approach," *Hawaii International Conference on System Sciences*, Waikoloa Beach, HI, pp. 4641–4650.

Fitzgerald, M. 2013. "How Starbucks has gone digital," *MIT Sloan Management Review* (54:5), pp. 1–8.

Fitzgerald, M. 2014a. "Audi puts its future into high(tech) gear," *MIT Sloan Management Review* (55:4), pp. 1–4.

Fitzgerald, M. 2014b. "How digital acceleration teams are influencing Nestle's 2000 brands," *MIT Sloan Management Review* (55:2), pp. 1–5.

Fitzgerald, M. 2014c. "Inside Renault's digital factory," *MIT Sloan Management Review* (55:3), pp. 1–4.

Fitzgerald, M. 2016a. "Building a better car company with analytics," *MIT Sloan Management Review* (57:4), pp. 1–8.

Fitzgerald, M. 2016b. "General Motors relies on IoT to anticipate customers' needs," *MIT Sloan Management Review* (57:4), pp. 1–9.

Fitzgerald, M., Kruschwitz, N., Bonnet, D., and Welch, M. 2014. "Embracing digital technology: A new strategic imperative," *MIT Sloan Management Review* (55:2), pp. 1–12.

Floridi, L. 2018. "Soft ethics and the governance of the digital," *Philosophy & Technology* (31:1), pp. 1–8.

Freitas Junior, J. C. d. S., Maçada, A. C. G., and Brinkhues, R. A. 2017. "Digital capabilities as key to digital business performance," *Americas Conference of Information Systems*, Boston, MA.

Friedlmaier, M., Tumasjan, A., and Welpe, I. M. 2018. "Disrupting industries with Blockchain: The industry, venture capital funding, and regional distribution of Blockchain ventures," *Hawaii International Conference on System Sciences*, Waikoloa Beach, HI, pp. 3517–3526.

Galliers, R. D., Jarvenpaa, S. L., Chan, Y. E., and Lyytinen, K. 2012. "Strategic information systems: Reflections and prospectives," *Journal of Strategic Information Systems* (21:2), pp. 85–90.

Ganju, K. K., Pavlou, P. A., and Banker, R. D. 2016. "Does information and communication technology lead to the well-being of nations? A country-level empirical investigation," *MIS Quarterly* (40:2), pp. 417–430.

Gerster, D. 2017. "Digital transformation and IT: Current state of research," *Pacific Asia Conference on Information Systems*, Langkawi, Malaysia.

Ghazawneh, A., and Henfridsson, O. 2013. "Balancing platform control and external contribution in third-party development: The boundary resources model," *Information Systems Journal* (23:2), pp. 173–192.

Gimpel, G. 2015. "Alternative views of ICT & time: An application of scenario analysis and platform theory," *International Conference of Information Systems*, Fort Worth, TX.

Gimpel, H., Hosseini, S., Huber, R. X. R., Probst, L., Röglinger, M., and Faisst, U. 2018. "Structuring digital transformation: A framework of action fields and its application at ZEISS," *Journal of Information Technology Theory and Application* (19:1), pp. 31–54.

Glaser, F. 2017. "Pervasive decentralisation of digital infrastructures: A framework for blockchain enabled system and use case analysis," *Hawaii International Conference on System Sciences*, Waikoloa Beach, HI, pp. 1543–1552.

Goes, P. 2015. "Big data-analytics engine for digital transformation: Where is IS?," *Americas Conference of Information Systems*, Puerto Rico.

Google employees against Dragonfly. 2018. "We are Google employees. Google must drop Dragonfly." Retrieved December 23, 2018, from https://medium.com/@googlersagainstdragonfly/we-are-google-employees-google-must-drop-dragonfly-4c8a30c5e5eb

Granados, N. F., Kauffman, R. J., and King, B. 2008. "How has electronic travel distribution been transformed? A test of the theory of newly vulnerable markets," *Journal of Management Information Systems* (25:2), pp. 73–95.

Gray, J., and Rumpe, B. 2017. "Models for the digital transformation," *Software and Systems Modeling* (16:2), pp. 307–308.

Gray, P., El Sawy, O. A., Asper, G., and Thordarson, M. 2013. "Realizing strategic value through center-edge digital transformation in consumer-centric industries," *MIS Quarterly Executive* (12:1), pp. 1–17.

Günther, W. A., Mehrizi, M. H. R., Huysman, M., and Feldberg, F. 2017. "Debating big data: A literature review on realizing value from big data," *The Journal of Strategic Information Systems* (26:3), pp. 191–209.

Gust, G., Flath, C. M., Brandt, T., Ströhle, P., and Neumann, D. 2017. "How a traditional company seeded new analytics capabilities," *MIS Quarterly Executive* (16:3), pp. 215–230.

Haas, P., Blohm, I., and Leimeister, J. M. 2014. "An empirical taxonomy of crowdfunding intermediaries," *International Conference of Information Systems*, Auckland, New Zealand.

Haffke, I., Kalgovas, B., and Benlian, A. 2017. "The transformative role of bimodal IT in an era of digital business," *Hawaii International Conference on System Sciences*, Waikoloa Beach, HI, pp. 5460–5469.

Haffke, I., Kalgovas, B. J., and Benlian, A. 2016. "The role of the CIO and the CDO in an organization's digital transformation," *International Conference of Information Systems*, Dublin, Ireland.

Han, K., Kundisch, D., Weinhardt, C., and Zimmermann, S. 2015. "Economics and value of IS," *Business & Information Systems Engineering* (57:5), pp. 295–297.

Hanelt, A., Nastjuk, I., Krüp, H., Eisel, M., Ebermann, C., Brauer, B., Piccinini, E., Hildebrandt, B., and Kolbe, L. M. 2015a. "Disruption on the way? The role of mobile applications for electric vehicle diffusion," *Wirtschaftsinformatik Conference*, Osnabruck, Germany, pp. 1023–1037.

Hanelt, A., Piccinini, E., Gregory, R. W., Hildebrandt, B., and Kolbe, L. M. 2015b. "Digital transformation of primarily physical industries-exploring the impact of digital trends on business models of automobile manufacturers," *Wirtschaftsinformatik Conference*, Osnabruck, Germany, pp. 1313–1327.

Hansen, A. M., Kraemmergaard, P., and Mathiassen, L. 2011. "Rapid adaptation in digital transformation: A participatory process for engaging IS and business leaders," *MIS Quarterly Executive* (10:4), pp. 175–185.

Hansen, R., and Sia, S. K. 2015. "Hummel's digital transformation toward omnichannel retailing: Key lessons learned," *MIS Quarterly Executive* (14:2), pp. 51–66.

Hartl, E., and Hess, T. 2017. "The role of cultural values for digital transformation: Insights from a Delphi study," *Americas Conference of Information Systems*, Boston, MA.

Hayes, A. 2016. "Decentralized banking: Monetary technocracy in the digital age," *Mediterranean Conference of Information Systems*, Cyprus: Springer, pp. 121–131.

Heilig, L., Schwarze, S., and Voss, S. 2017. "An analysis of digital transformation in the history and future of modern ports," *Hawaii International Conference on System Sciences*, Waikoloa Beach, HI, pp. 1341–1350.

Helfat, C. E., and Martin, J. A. 2015. "Dynamic managerial capabilities: Review and assessment of managerial impact on strategic change," *Journal of Management* (41:5), pp. 1281–1312.

Helfat, C. E., and Raubitschek, R. S. 2018. "Dynamic and integrative capabilities for profiting from innovation in digital platform-based ecosystems," *Research Policy* (47:8), pp. 1391–1399.

Henfridsson, O., and Lind, M. 2014. "Information systems strategizing, organizational sub-communities, and the emergence of a sustainability strategy," *The Journal of Strategic Information Systems* (23:1), pp. 11–28.

Henfridsson, O., Mathiassen, L., and Svahn, F. 2014. "Managing technological change in the digital age: The role of architectural frames," *Journal of Information Technology* (29:1), pp. 27–43.

Henningsson, S., and Hedman, J. 2014. "Technology-based transformation of digital ecosystems: The DETT framework," *International Conference on Information Resources Management*, Ho Chi Minh City, Vietnam.

Hess, T., Matt, C., Benlian, A., and Wiesboeck, F. 2016. "Options for formulating a digital transformation strategy," *MIS Quarterly Executive* (15:2), pp. 123–139.

Hesse, A. 2018. "Digitalization and leadership – How experienced leaders interpret daily realities in a digital world," *Hawaii International Conference on System Sciences*, Waikoloa Beach, HI, pp. 1854–1863.

Hildebrandt, B., Hanelt, A., Firk, S., and Kolbe, L. 2015. "Entering the digital era – the impact of digital technology-related M&As on business model innovations of automobile OEMs," *International Conference of Information Systems*, Fort Worth, TX.

Hjalmarsson, A., Johannesson, P., Jüll-Skielse, G., and Rudmark, D. 2014. "Beyond innovation contests: A framework of barriers to open innovation of digital services," *European Conference of Information Systems*, Tel Aviv, Israel.

Hjalmarsson, A., Juell-Skielse, G., Ayele, W.Y., Rudmark, D., and Johannesson, P. 2015. "From contest to market entry: A longitudinal survey of innovation barriers constraining open data service development," *European Conference of Information Systems*, Munster, Germany.

Holotiuk, F., and Beimborn, D. 2017. "Critical success factors of digital business strategy," *Wirtschaftsinformatik Conference*, St. Gallen, Switzerland AIS Electronic Library, pp. 991–1005.

Hong, J., and Lee, J. 2017. "The role of consumption-based analytics in digital publishing markets: Implications for the creative digital economy," *International Conference of Information Systems*, Seoul, South Korea.

Horlach, B., Drews, P., Schirmer, I., and Böhmann, T. 2017. "Increasing the agility of IT delivery: Five types of bimodal IT organization," *Hawaii International Conference on System Sciences*, Waikoloa Beach, HI, pp. 5420–5429.

Horlacher, A., Klarner, P., and Hess, T. 2016. "Crossing boundaries: Organization design parameters surrounding CDOs and their digital transformation activities," *Americas Conference of Information Systems*, San Diego, CA.

Huang, J., Henfridsson, O., Liu, M. J., and Newell, S. 2017. "Growing on steroids: Rapidly scaling the user base of digital ventures through digital innovation," *MIS Quarterly* (41:1), 301–314.

Huhtamäki, J., Basole, R., Still, K., Russell, M., and Seppänen, M. 2017. "Visualizing the geography of platform boundary resources: The case of the global API ecosystem," *Hawaii International Conference on System Sciences*, Waikoloa Beach, HI, pp. 5305–5314.

Islam, N., Buxman, P., and Eling, N. 2017. "Why should incumbent firms jump on the start-up bandwagon in the digital era? – A qualitative study," *Wirtschaftsinformatik Conference*, St. Gallen, Switzerland: AIS Electronic Library, pp. 1378–1392.

Jarzabkowski, P., Balogun, J., and Seidl, D. 2007. "Strategizing: The challenges of a practice perspective," *Human Relations* (60:1), pp. 5–27.

Jha, S. K., Pinsonneault, A., and Dubé, L. 2016. "The evolution of an ICT platform-enabled ecosystem for poverty alleviation: The case of eKutir," *MIS Quarterly* (40:2), pp. 431–445.

Jiang, W., Mavondo, F.T., and Matanda, M. J. 2015. "Integrative capability for successful partnering: A critical dynamic capability," *Management Decision* (53:6), pp. 1184–1202.

Jöhnk, J., Röglinger, M., Thimmel, M., and Urbach, N. 2017. "How to implement agile IT setups: A taxonomy of design options," *European Conference of Information Systems*, Guimaraes, Portugal, pp. 1521–1535.

Johnson, L. K. 2002. "New views on digital CRM," *MIT Sloan Management Review* (44:1), pp. 10–10.

Joshi, A., Huygh, T., and De Haes, S. 2017. "Examining the association between industry IT strategic role and IT governance implementation," *International Conference of Information Systems*, Seoul, South Korea.

Kahre, C., Hoffmann, D., and Ahlemann, F. 2017. "Beyond business-IT alignment-digital business strategies as a paradigmatic shift: A review and research agenda," *Hawaii International Conference on System Sciences*, Waikoloa Beach, HI, pp. 4706–4715.

Kamel, S. 2015. "Electronic commerce challenges and opportunities for emerging economies: Case of Egypt," *International Conference on Information Resources Management*, Ottawa, Canada.

Kane, G. C. 2014. "The American Red Cross: Adding digital volunteers to its ranks," *MIT Sloan Management Review* (55:4), pp. 1–6.

Kane, G. C. 2015a. "Are you ready for the certainty of unknown?," *MIT Sloan Management Review* (56:3), pp. 1–11.

Kane, G. C. 2015b. "How digital transformation is making health care safer, faster and cheaper," *MIT Sloan Management Review* (57:1), pp. 1–11.

Kane, G. C. 2016a. "The dark side of the digital revolution," *MIT Sloan Management Review* (57:3), pp. 1–9.

Kane, G. C. 2016b. "Digital health care: The patient will see you now," *MIT Sloan Management Review* (57:4), pp. 1–8.

Kane, G. C. 2016c. "How Facebook and Twitter are reimagining the future of customer service," *MIT Sloan Management Review* (55:4), pp. 1–6.

Kane, G. C. 2017a. "Big data and IT talent drive improved patient outcomes at Schumacher Clinical Partners," *MIT Sloan Management Review* (59:1), pp. 1–8.

Kane, G. C. 2017b. "Digital innovation lights the fuse for better health care outcomes," *MIT Sloan Management Review* (59:1), pp. 1–8.

Kane, G. C. 2017c. "Digital maturity, not digital transformation." Retrieved September 1, 2017, from http://sloanreview.mit.edu/article/digital-maturity-not-digital-transformation/

Kane, G. C. 2017d. "In the hotel industry, digital has made itself right at home," *MIT Sloan Management Review* (58:4), pp. 1–9.

Kane, G. C., Palmer, D., Nguyen-Phillips, A., Kiron, D., and Buckley, N. 2017. "Achieving digital maturity," 15329194, Massachusetts Institute of Technology, Cambridge, MA, Cambridge, pp. 1–32.

Kane, G. C., Palmer, D., Phillips, A. N., Kiron, D., and Buckley, N. 2016. "Aligning the organization for its digital future," 15329194, Massachusetts Institute of Technology, Cambridge, MA, Cambridge, pp. 1–30.

Kaplan, S., and Orlikowski, W. J. 2013. "Temporal work in strategy making," *Organization science* (24:4), pp. 965–995.

Karagiannaki, A., Vergados, G., and Fouskas, K. 2017. "The impact of digital transformation in the financial services industry: Insights from an open innovation initiative in Fintech in Greece," *Mediterranean Conference of Information Systems*, Genoa, Italy.

Karimi, J., Somers, T. M., and Bhattacherjee, A. 2009. "The role of ERP implementation in enabling digital options: A theoretical and empirical analysis," *International Journal of Electronic Commerce* (13:3), pp. 7–42.

Karimi, J., and Walter, Z. 2015. "The role of dynamic capabilities in responding to digital disruption: A factor-based study of the newspaper industry," *Journal of Management Information Systems* (32:1), pp. 39–81.

Karpovsky, A., and Galliers, R. D. 2015. "Aligning in practice: From current cases to a new agenda," *Journal of Information Technology* (30:2), pp. 136–160.

Kauffman, R. J., Ting, L., and van Heck, E. 2010. "Business network-based value creation in electronic commerce," *International Journal of Electronic Commerce* (15:1), pp. 113–144.

Kazan, E., and Damsgaard, J. 2014. "An investigation of digital payment platform designs: A comparative study of four European solutions," *European Conference of Information Systems*, Tel Aviv, Israel.

Kiefer, T. 2000. "The future role of banks in electronic commerce-trust as the crucial factor of success in" Business enabling"," *European Conference of Information Systems*, Vienna, Austria.

Klein, K. J., and Kozlowski, S. W. J. 2000. "A multilevel approach to theory and research in organizations: Contextual, temporal, and emergent processes," in *Multilevel Theory, Research, and Methods in Organizations: Foundations, Extensions, and New Directions*, K. J. Klein and S. W. J. Kozlowski (eds.). San Francisco, CA: Jossey-Bass, pp. 3–90.

Kleinschmidt, S., and Peters, C. 2017. "Fostering business model extensions for ICT-enabled human-centered service systems," *Wirtschaftsinformatik Conference*, St. Gallen, Switzerland: AIS Electronic Library, pp. 897–911.

Klötzer, C., and Pflaum, A. 2017. "Toward the development of a maturity model for digitalization within the manufacturing industry's supply chain," *Hawaii International Conference on System Sciences*, Waikoloa Beach, HI, pp. 4210–4219.

Kohli, R., and Johnson, S. 2011. "Digital transformation in latecomer industries: CIO and CEO leadership lessons from Encana Oil & Gas (USA) Inc.," *MIS Quarterly Executive* (10:4), pp. 141–156.

Korpela, K., Hallikas, J., and Dahlberg, T. 2017. "Digital supply chain transformation toward blockchain integration," *Hawaii International Conference on System Sciences*, Waikoloa Beach, HI, pp. 4182–4191.

Krumeich, J., Burkhart, T., Werth, D., and Loos, P. 2012. "Towards a component-based description of business models: A state-of-the-art analysis," *Americas Conference of Information Systems*, Seattle, WA.

Krumeich, J., Werth, D., and Loos, P. 2013. "Interdependencies between business model components – A literature analysis," *Americas Conference of Information Systems*, Chicago, IL.

Le Dinh, T., Phan, T.-C., and Bui, T. 2016. "Towards an architecture for big data-driven knowledge management systems," *Americas Conference of Information Systems*, San Diego, CA.

Lee, E., and Lee, B. 2013. "Changing price elasticity of digital goods: Empirical study from the e-book industry," *International Conference of Information Systems*, Milan, Italy.

Legner, C., Eymann, T., Hess, T., Matt, C., Böhmann, T., Drews, P., Mädche, A., Urbach, N., and Ahlemann, F. 2017. "Digitalization: opportunity and challenge for the business and information systems engineering community," *Business & information systems engineering* (59:4), pp. 301–308.

Leischnig, A., Wölfl, S., Ivens, B., and Hein, D. 2017. "From digital business strategy to market performance: Insights into key concepts and processes," *International Conference of Information Systems*, Seoul, South Korea.

Leonardi, P. M., and Bailey, D. E. 2008. "Transformational technologies and the creation of new work practices: Making implicit knowledge explicit in task-based offshoring," *MIS Quarterly*, (32:2), pp. 411–436.

Leonardi, P. M., Bailey, D. E., Diniz, E. H., Sholler, D., and Nardi, B. A. 2016. "Multiplex appropriation in complex systems implementation: The case of Brazil's correspondent banking system," *MIS Quarterly* (40:2), pp. 461–473.

Leong, C. M. L., Pan, S.-L., Newell, S., and Cui, L. 2016. "The emergence of self-organizing e-commerce ecosystems in remote villages of China: A tale of digital empowerment for rural development," *MIS Quarterly* (40:2), pp. 475–484.

Leonhardt, D., Haffke, I., Kranz, J., and Benlian, A. 2017. "Reinventing the IT function: The role of IT agility and IT ambidexterity in supporting digital business transformation," *European Conference of Information Systems*, Guimaraes, Portugal, pp. 968–984.

Li, L., Su, F., Zhang, W., and Mao, J. Y. 2017. "Digital transformation by SME entrepreneurs: A capability perspective," *Information Systems Journal*, (28:6), pp. 1–29.

Li, W., Liu, K., Belitski, M., Ghobadian, A., and O'Regan, N. 2016. "e-Leadership through strategic alignment: An empirical study of small-and medium-sized enterprises in the digital age," *Journal of Information Technology* (31:2), pp. 185–206.

Lienhard, K., Job, O., Bachmann, L., Bodmer, N., and Legner, C. 2017. "A framework to advance electronic health record system use in routine patient care," *European Conference of Information Systems*, Guimaraes, Portugal, pp. 1114–1128.

Liere-Netheler, K., Packmohr, S., and Vogelsang, K. 2018. "Drivers of digital transformation in manufacturing," *Hawaii International Conference on System Sciences*, Waikoloa Beach, HI, pp. 3926–3935.

Loebbecke, C., and Picot, A. 2015. "Reflections on societal and business model transformation arising from digitization and big data analytics: A research agenda," *The Journal of Strategic Information Systems* (24:3), pp. 149–157.

Lu, N., and Swatman, P. 2008. "The Mobicert mobile information community for organic primary producers: A South Australian prototype," *Bled eConference*, Bled, Slovenia, pp. 91–102.

Lucas Jr, H. C., Agarwal, R., Clemons, E. K., El Sawy, O. A., and Weber, B. 2013. "Impactful research on transformational information technology: An opportunity to inform new audiences," *MIS Quarterly* (37:2), pp. 371–382.

Lucas Jr, H. C., and Goh, J. M. 2009. "Disruptive technology: How Kodak missed the digital photography revolution," *The Journal of Strategic Information Systems* (18:1), pp. 46–55.

Lyytinen, K., and Rose, G. M. 2003a. "Disruptive information system innovation: The case of internet computing," *Information Systems Journal* (13:4), pp. 301–330.

Lyytinen, K., and Rose, G. M. 2003b. "The disruptive nature of information technology innovations: The case of internet computing in systems development organizations," *MIS Quarterly*, (27:4), pp. 557–596.

Maedche, A. 2016. "Interview with Michael Nilles on "What Makes Leaders Successful in the Age of the Digital Transformation?"," *Business & Information Systems Engineering* (58:4), pp. 287–289.

Majchrzak, A., Markus, M. L., and Wareham, J. 2016. "Designing for digital transformation: Lessons for information systems research from the study of ICT and societal challenges," *MIS Quarterly* (40:2), pp. 267–277.

Markus, M. L., and Robey, D. 1988. "Information technology and organizational change: Causal structure in theory and research," *Management Science* (34:5), pp. 583–598.

Matt, C., Hess, T., and Benlian, A. 2015. "Digital transformation strategies," *Business & Information Systems Engineering* (57:5), pp. 339–343.

McGrath, K. 2016. "Identity verification and societal challenges: Explaining the gap between service provision and development outcomes," *MIS Quarterly* (40:2), pp. 485–500.

McGrath, K., Hendy, J., Klecun, E., and Young, T. 2008. "The vision and reality of 'connecting for health': Tensions, opportunities, and policy implications of the UK national programme," *Communications of the Association for Information Systems* (23:1), pp. 603–618.

Medaglia, R., Hedman, J., and Eaton, B. 2017. "Public–private collaboration in the emergence of a national electronic identification policy: The case of NemID in Denmark," *Hawaii International Conference on System Sciences*, Waikoloa Beach, HI, pp. 2782–2791.

Miranda, S. M., Young, A., and Yetgin, E. 2016. "Are social media emancipatory or hegemonic? Societal effects of mass media digitization," *MIS Quarterly* (40:2), pp. 303–329.

Mithas, S., Tafti, A., and Mitchell, W. 2013. "How a firm's competitive environment and digital strategic posture influence digital business strategy," *MIS Quarterly* (37:2), pp. 511–536.

Mohagheghzadeh, A., and Svahn, F. 2016. "Transforming organizational resource into platform boundary resources," *European Conference of Information Systems*, Istanbul, Turkey.

Morakanyane, R., Grace, A. A., and O'Reilly, P. 2017. "Conceptualizing digital transformation in business organizations: A systematic review of literature," *Bled eConference*, Bled, Slovenia, pp. 427–444.

Mueller, B., and Renken, U. 2017. "Helping employees to be digital transformers – The Olympus. Connect case," *International Conference of Information Systems*, Seoul, South Korea.

Myunsoo, K., and Byungtae, L. 2013. "Analysis of advertisement based business model under technological advancements in fair use personal recording services: A law and economics approach," *Pacific Asia Conference on Information Systems*.

Nambisan, S., Lyytinen, K., Majchrzak, A., and Song, M. 2017. "Digital innovation management: Reinventing innovation management research in a digital world," *MIS Quarterly* (41:1), pp. 223–238.

Nastjuk, I., Hanelt, A., and Kolbe, L. M. 2016. "Too much of a good thing? An experimental investigation of the impact of digital technology-enabled business models on individual stress and future adoption of sustainable services," *International Conference of Information Systems*, Dublin, Ireland.

Neate, R. 2018. "Over \$119bn wiped off Facebook's market cap after growth shock," in: *The Guardian*.

Nehme, J. J., Srivastava, S. C., Bouzas, H., and Carcasset, L. 2015. "How Schlumberger achieved networked information leadership by transitioning to a product-platform software architecture," *MIS Quarterly Executive* (14:3), pp. 105–124.

NetworkWorld Asia. 2015. "Virgin Atlantic Airways embarks on digital transformation program," *NetworkWorld Asia* (12:1), pp. 49-49.

Neumeier, A., Wolf, T., and Oesterle, S. 2017. "The manifold fruits of digitalization – determining the literal value behind," *Wirtschaftsinformatik Conference*, St. Gallen, Switzerland: AIS Electronic Library, pp. 484–498.

Newell, S., and Marabelli, M. 2014. "The crowd and sensors era: Opportunities and challenges for individuals, organizations, society, and researchers," *International Conference of Information Systems*, Auckland, New Zealand.

Newell, S., and Marabelli, M. 2015. "Strategic opportunities (and challenges) of algorithmic decision-making: A call for action on the long-term societal effects of 'datification'," *The Journal of Strategic Information Systems* (24:1), pp. 3–14.

Newton, C. 2018. "How Google's China project undermines its claims to political neutrality." Retrieved October 16, 2018, from www.theverge.com/2018/10/11/17962274/google-censorship-presentation-leak-free-speech-china

Ng, V. 2016. "Veritas: Data trends and issues in Asia Pacific," *NetworkWorld Asia* (13:2), pp. 5–6.

Nischak, F., Hanelt, A., and Kolbe, L. M. 2017. "Unraveling the interaction of information systems and ecosystems – A comprehensive classification of literature," *International Conference of Information Systems*, Seoul, South Korea.

Nwankpa, J. K., and Datta, P. 2017. "Balancing exploration and exploitation of IT resources: The influence of Digital Business Intensity on perceived organizational performance," *European Journal of Information Systems* (26:5), pp. 469–488.

Nwankpa, J. K., and Roumani, Y. 2016. "IT capability and digital transformation: A firm performance perspective," *International Conference of Information Systems*, Dublin, Ireland.

Oestreicher-Singer, G., and Zalmanson, L. 2012. "Content or community? A digital business strategy for content providers in the social age," *MIS Quarterly*), pp. 591–616.

Oh, L.-B. 2009. "Managing external information sources in digital extended enterprises: The roles of IT enabled business intelligence competence and network structure strength," *International Conference of Information Systems*, Phoenix, AZ.

Omar, A., and Elhaddadeh, R. 2016. "Structuring institutionalization of digitally-enabled service transformation in public sector: Does actor or structure matters?," *Americas Conference of Information Systems*, San Diego, CA.

Oreglia, E., and Srinivasan, J. 2016. "ICT, intermediaries, and the transformation of gendered power structures," *MIS Quarterly* (40:2), pp. 501–510.

Osmani, M., Weerakkody, V., and El-Haddadeh, R. 2012. "Developing a conceptual framework for evaluating public sector transformation in the digital era," *Americas Conference of Information Systems*, Seattle, WA.

Osterwalder, A., and Pigneur, Y. 2010. *Business model generation: A handbook for visionaries, game changers, and challengers.* Hoboken, NJ: John Wiley & Sons.

Paavola, R., Hallikainen, P., and Elbanna, A. 2017. "Role of middle managers in modular digital transformation: The case of SERVU," *European Conference of Information Systems*, Guimaraes, Portugal.

Pagani, M. 2013. "Digital business strategy and value creation: Framing the dynamic cycle of control points," *MIS Quarterly* (37:2), pp. 617–632.

Peppard, J., Galliers, R. D., and Thorogood, A. 2014. "Information systems strategy as practice: Micro strategy and strategizing for IS," *The Journal of Strategic Information Systems* (23:1), pp. 1–10.

Petrikina, J., Krieger, M., Schirmer, I., Stoeckler, N., Saxe, S., and Baldauf, U. 2017. "Improving the readiness for change-addressing information concerns of internal stakeholders in the Smartport Hamburg," *Americas Conference of Information Systems*, Boston, MA.

Piccinini, E., Gregory, R. W., and Kolbe, L. M. 2015a. "Changes in the producer-consumer relationship-towards digital transformation," *Wirtschaftsinformatik Conference*, Osnabrück, Germany: AIS Electronic Library, pp. 1634–1648.

Piccinini, E., Hanelt, A., Gregory, R., and Kolbe, L. 2015b. "Transforming industrial business: The impact of digital transformation on automotive organizations," *International Conference of Information Systems*, Fort Worth, TX.

Pillet, J.-C., Pigni, F., and Vitari, C. 2017. "Learning about ambiguous technologies: Conceptualization and research agenda," *European Conference of Information Systems*, Guimaraes, Portugal.

Porter, M. E., and Heppelmann, J. E. 2014. "How smart, connected products are transforming competition," *Harvard Business Review* (92:11), pp. 64–88.

Pousttchi, K., Tilson, D., Lyytinen, K., and Hufenbach, Y. 2015. "Introduction to the special issue on mobile commerce: Mobile commerce research yesterday, Today, tomorrow – What remains to be done?," *International Journal of Electronic Commerce* (19:4), pp. 1–20.

Pramanik, M. I., Lau, R. Y., and Chowdhury, M. K. H. 2016. "Automatic crime detector: A framework for criminal pattern detection in big data era," *Pacific Asia Conference on Information Systems*, Chiayi, Taiwan.

Preuveneers, D., Joosen, W., and Ilie-Zudor, E. 2016. "Data protection compliance regulations and implications for smart factories of the future," *12th International Conference on Intelligent Environments (IE' 16)*, London, UK: IEEE Xplore, pp. 40–47.

Prifti, L., Knigge, M., Kienegger, H., and Krcmar, H. 2017. "A competency model for "Industrie 4.0" employees," *Wirtschaftsinformatik Conference*, St. Gallen, Switzerland: AIS Electronic Library, pp. 46–60.

Püschel, L., Röglinger, M., and Schlott, H. 2016. "What's in a smart thing? Development of a multi-layer taxonomy," *International Conference of Information Systems*, Dublin, Ireland.

Rai, A., and Sambamurthy, V. 2006. "Editorial notes – The growth of interest in services management: Opportunities for information systems scholars," *Information Systems Research* (17:4), pp. 327–331.

Rajan, M. T. S. 2002. "Moral rights in the digital age: New possibilities for the democratization of culture," *International Review of Law, Computers & Technology* (16:2), pp. 187–197.

Ramasubbu, N., Woodard, C. J., and Mithas, S. 2014. "Orchestrating service innovation using design moves: The dynamics of fit between service and enterprise IT architectures," *International Conference of Information Systems*, Auckland, New Zealand: Association for Information Systems.

Rauch, M., Wenzel, M., and Wagner, H.-T. 2016. "The digital disruption of strategic paths: An experimental study," *International Conference of Information Systems*, Dublin, Ireland.

Reinartz, W., and Imschloß, M. 2017. "From point of sale to point of need: How digital technology is transforming retailing," *GfK-Marketing Intelligence Review* (9:1), pp. 43–47.

Reinhold, O., and Alt, R. 2009. "Enhancing collaborative CRM with mobile technologies," *Bled eConference*, Bled, Slovenia, pp. 97–116.

Reitz, A., Jentsch, C., and Beimborn, D. 2018. "How to decompress the pressure – The moderating effect of IT flexibility on the negative impact of governmental pressure on business agility," *Hawaii International Conference on System Sciences*, Waikoloa Beach, HI, pp. 4613–4620.

Remane, G., Hanelt, A., Hildebrandt, B., and Kolbe, L. 2016a. "Changes in digital business model types – A longitudinal study of technology startups from the mobility sector," *Americas Conference of Information Systems*, San Diego, CA.

Remane, G., Hanelt, A., Wiesboeck, F., and Kolbe, L. 2017. "Digital maturity in traditional industries – an exploratory analysis," *European Conference of Information Systems*, Guimaraes, Portugal, pp. 143–157.

Remane, G., Hildebrandt, B., Hanelt, A., and Kolbe, L. M. 2016b. "Discovering new digital business model types – A study of technology startups from the mobility sector," *Pacific Asia Conference on Information Systems*, Chiayi, Taiwan.

Resca, A., Za, S., and Spagnoletti, P. 2013. "Digital platforms as sources for organizational and strategic transformation: A case study of the Midblue project," *Journal of Theoretical & Applied Electronic Commerce Research* (8:2), pp. 71–84.

Riasanow, T., Galic, G., and Böhm, M. 2017. "Digital transformation in the automotive industry: Towards a generic value network," *European Conference of Information Systems*, Guimaraes, Portugal, pp. 3191–3201.

Richter, A., Vodanovich, S., Steinhüser, M., and Hannola, L. 2017. "IT on the shop floor-challenges of the digitalization of manufacturing companies," *Bled eConference*, Bled, Slovenia, pp. 483–500.

Riedl, R., Benlian, A., Hess, T., Stelzer, D., and Sikora, H. 2017. "On the relationship between information management and digitalization," *Business & Information Systems Engineering* (59:6), pp. 475–482.

Rizk, A., Bergvall-Kåreborn, B., and Elragal, A. 2018. "Towards a taxonomy of data-driven digital services," *Hawaii International Conference on System Sciences*, Waikoloa Beach, HI, pp. 1076–1085.

Roecker, J., Mocker, M., and Novales, A. 2017. "Digitized products: Challenges and practices from the creative industries," *Americas Conference of Information Systems*, Boston, MA.

Ross, J. W., Sebastian, I., Beath, C., Mocker, M., Moloney, K., and Fonstad, N. 2016. "Designing and executing digital strategies," *International Conference of Information Systems*, Dublin, Ireland.

Rowe, F. 2014. "What literature review is not: Diversity, boundaries and recommendations," *European Journal of Information Systems* (23:3), pp. 241–255.

Sachse, S., Alt, R., and Puschmann, T. 2012. "Towards customer-oriented electronic markets: A survey among digital natives in the financial industry," *Bled eConference*, Bled, Slovenia, pp. 333–354.

Saldanha, T. J., Mithas, S., and Krishnan, M. S. 2017. "Leveraging customer involvement for fueling innovation: The role of relational and analytical information processing capabilities," *MIS Quarterly* (41:1), pp. 267–286.

Sambamurthy, V., Bharadwaj, A., and Grover, V. 2003. "Shaping agility through digital options: Reconceptualizing the role of information technology in contemporary firms," *MIS Quarterly* (27:2), pp. 237–263.

Sambamurthy, V., and Zmud, R. W. 2000. "Research commentary: The organizing logic for an enterprise's IT activities in the digital era – A prognosis of practice and a call for research," *Information Systems Research* (11:2), pp. 105–114.

Schellhorn, L. K. 2016. "Developing a benefits assessment framework for sea traffic management systems," *European Conference of Information Systems*, Istanbul, Turkey.

Schilke, O., Hu, S., and Helfat, C. E. 2018. "Quo vadis, dynamic capabilities? A content-analytic review of the current state of knowledge and recommendations for future research," *Academy of Management Annals* (12:1), pp. 390–439.

Schmid, A. M., Recker, J., and vom Brocke, J. 2017. "The socio-technical dimension of inertia in digital transformations," *Hawaii International Conference on System Sciences*, Waikoloa Beach, HI, pp. 4796–4805.

Schmidt, J., Drews, P., and Schirmer, I. 2017. "Digitalization of the banking industry: A multiple stakeholder analysis on strategic alignment," *Americas Conference of Information Systems*, Boston, MA.

Scott, J. E. 2007. "An e-transformation study using the technology-organization-environment framework," *Bled eConference*, Bled, Slovenia, pp. 50–61.

Sebastian, I. M., Ross, J. W., Beath, C., Mocker, M., Moloney, K. G., and Fonstad, N. O. 2017. "How big old companies navigate digital transformation," *MIS Quarterly Executive* (16:3), pp. 197–213.

Selander, L., Henfridsson, O., and Svahn, F. 2010. "Transforming ecosystem relationships in digital innovation," *International Conference of Information Systems*, St. Louis, MN.

Selander, L., and Jarvenpaa, S. L. 2016. "Digital action repertoires and transforming a social movement organization," *MIS Quarterly* (40:2), pp. 331–352.

Seo, D. 2017. "Digital business convergence and emerging contested fields: A conceptual framework," *Journal of the Association for Information Systems* (18:10), pp. 687–702.

Serrano, C., and Boudreau, M.-C. 2014. "When technology changes the physical workplace: The creation of a new workplace identity," *International Conference of Information Systems*, Auckland, New Zealand.

Setia, P., Venkatesh, V., and Joglekar, S. 2013. "Leveraging digital technologies: How information quality leads to localized capabilities and customer service performance," *MIS Quarterly* (37:2), pp. 565–590.

Shahlaei, C., Rangraz, M., and Stenmark, D. 2017. "Transformation of competence – The effects of digitalization on communicators' work," *European Conference of Information Systems*, Guimaraes, Portugal, pp. 195–209.

Shivendu, S., and Zhang, R. 2016. "The impact of digitization on information goods pricing strategy," *Americas Conference of Information Systems*, San Diego, CA.

Sia, S. K., Soh, C., and Weill, P. 2016. "How DBS Bank pursued a digital business strategy," *MIS Quarterly Executive* (15:2), pp. 105–121.

Singh, A., and Hess, T. 2017. "How chief digital officers promote the digital transformation of their companies," *MIS Quarterly Executive* (16:1), pp. 1–17.

Smith, C., and Webster, C. W. R. 2006. "Interactive digital television and electronic public services: Emergent issues," *European Conference of Information Systems*, Gothenburg, Sweden: European Conference on Information Systems, pp. 1932–1948.

Smith, H. J. 2002. "Ethics and information systems: Resolving the quandaries," *the DATABASE for Advances in Information Systems* (33:3), pp. 8–22.

Smith, H. J., and Hasnas, J. 1999. "Ethics and information systems: The corporate domain," *MIS Quarterly* (23:1), pp. 109–127.

Soava, G. 2015. "Development prospects of the tourism industry in the digital age," *Young Economists Journal / Revista Tinerilor Economisti* (12:25), pp. 101–116.

Sørensen, C. 2016. "The curse of the smart machine? Digitalisation and the children of the mainframe," *Scandinavian Journal of Information Systems* (28:2), pp. 57–68.

Sørensen, C., De Reuver, M., and Basole, R. C. 2015. "Mobile platforms and ecosystems," *Journal of Information Technology* (30), pp. 195–197.

Srivastava, S. C., and Shainesh, G. 2015. "Bridging the service divide through digitally enabled service innovations: Evidence from Indian health care service providers," *MIS Quarterly* (39:1), pp. 245–267.

Srivastava, S. C., Teo, T. S., and Devaraj, S. 2016. "You can't bribe a computer: Dealing with the societal challenge of corruption through ICT," *MIS Quarterly* (40:2), pp. 511–526.

Stahl, B. C. 2012. "Morality, ethics and reflection: A categorisation of normative IS research," *Journal of the Association for Information Systems* (13:8), pp. 636–656.

Stahl, B. C., Eden, G., Jirotka, M., and Coeckelbergh, M. 2014. "From computer ethics to responsible research and innovation in ICT: The transition of reference discourses informing ethics-related research in information systems," *Information & Management* (51:6), pp. 810–818.

Standaert, W., and Jarvenpaa, S. 2017. "Emergent ecosystem for radical innovation: Entrepreneurial probing at Formula E," *Hawaii International Conference on System Sciences*, Waikoloa Beach, HI, pp. 4736–4745.

Staykova, K. S., and Damsgaard, J. 2015a. "Introducing reach and range for digital payment platforms," *International Conference on Mobile Business*, Fort Worth, TX.

Staykova, K. S., and Damsgaard, J. 2015b. "A typology of multi-sided platforms: The core and the periphery," *European Conference of Information Systems*, Munster, Germany.

Suddaby, R. 2010. "Editor's comments: Construct clarity in theories of management and organization," *The Academy of Management Review*), pp. 346–357.

Svahn, F., Mathiassen, L., and Lindgren, R. 2017a. "Embracing digital innovation in incumbent firms: How Volvo Cars managed competing concerns," *MIS Quarterly* (41:1), pp. 239–253.

Svahn, F., Mathiassen, L., Lindgren, R., and Kane, G. C. 2017b. "Mastering the digital innovation challenge," *MIT Sloan Management Review* (58:3), pp. 14–16.

Tan, B., Pan, S. L., Lu, X., and Huang, L. 2015a. "The role of IS capabilities in the development of multi-sided platforms: The digital ecosystem strategy of Alibaba.com," *Journal of the Association for Information Systems* (16:4), p. 248.

Tan, F. T. C., Tan, B., Lu, A., and Land, L. 2017. "Delivering disruption in an emergent access economy: A case study of an e-hailing platform," *Communications of the Association for Information Systems* (41), pp. 497–516.

Tan, T. C. F., Tan, B., Choi, B. C., Lu, A., and Land, L. P. W. 2015b. "Collaborative consumption on mobile applications: A study of multi-sided digital platform GoCatch," *International Conference on Mobile Business*, Fort Worth, TX.

Tanniru, M., Khuntia, J., and Weiner, J. 2016. "Aligning digital health to services: A case of leadership for transformation," *Americas Conference of Information Systems*, San Diego, CA.

Tanriverdi, H., and Lim, S.-Y. 2017. "How to survive and thrive in complex, hypercompetitive, and disruptive ecosystems? The roles of IS-enabled capabilities," *International Conference of Information Systems*, Seoul, South Korea.

Teece, D. J. 2007. "Explicating dynamic capabilities: The nature and microfoundations of (sustainable) enterprise performance," *Strategic Management Journal* (28:13), pp. 1319–1350.

Teece, D. J. 2014. "The foundations of enterprise performance: Dynamic and ordinary capabilities in an (economic) theory of firms," *The Academy of Management Perspectives* (28:4), pp. 328–352.

Terrenghi, N., Schwarz, J., Legner, C., and Eisert, U. 2017. "Business model management: Current practices, required activities and IT support," *Wirtschaftsinformatik Conference*, St. Gallen, Switzerland: AIS Electronic Library, pp. 972–986.

The New York Times. 2004. "Letter from the founders." Retrieved September 15, 2015, from www.nytimes.com/2004/04/29/business/letter-from-the-founders.html

Tiefenbacher, K., and Olbrich, S. 2016. "Developing a deeper understanding of digitally empowered customers – A capability transformation framework in the domain of customer relationship management," *Pacific Asia Conference on Information Systems*, Chiayi, Taiwan.

Tilson, D., Lyytinen, K., and Sørensen, C. 2010. "Research commentary – Digital infrastructures: The missing IS research agenda," *Information Systems Research* (21:4), pp. 748–759.

Tiwana, A. 2015. "Platform desertion by app developers," *Journal of Management Information Systems* (32:4), pp. 40–77.

Tiwana, A., Konsynski, B., and Bush, A. A. 2010. "Research commentary – platform evolution: Coevolution of platform architecture, governance, and environmental dynamics," *Information Systems Research* (21:4), pp. 675–687.

Töytäri, P., Turunen, T., Klein, M., Eloranta, V., Biehl, S., Rajala, R., and Hakanen, E. 2017. "Overcoming institutional and capability barriers to smart services," *Hawaii International Conference on System Sciences*, Waikoloa Beach, HI, pp. 1642–1651.

Trantopoulos, K., von Krogh, G., Wallin, M. W., and Woerter, M. 2017. "External knowledge and information technology: Implications for process innovation performance," *MIS Quarterly* (41:1), pp. 287–300.

Tumbas, S., Berente, N., Seidel, S., and vom Brocke, J. 2015. "The 'digital façade' of rapidly growing entrepreneurial organizations," *International Conference of Information Systems*, Fort Worth, TX.

Urquhart, C., and Vaast, E. 2012. "Building social media theory from case studies: A new frontier for IS research," *International Conference of Information Systems*, Orlando, FL.

Utesheva, A., Cecez-Kecmanovic, D., and Schlagwein, D. 2012. "Understanding the digital newspaper genre: Medium vs. Message," *European Conference of Information Systems*, Barcelona, Spain.

Venkatesh, V., Rai, A., Sykes, T. A., and Aljafari, R. 2016. "Combating infant mortality in rural India: Evidence from a field study of eHealth kiosk implementations," *MIS Quarterly* (40:2), pp. 353–380.

Wacker, J. G. 2004. "A theory of formal conceptual definitions: developing theory-building measurement instruments," *Journal of Operations Management* (22:6), pp. 629–650.

Watson, H. J. 2017. "Preparing for the cognitive generation of decision support," *MIS Quarterly Executive* (16:3), pp. 153–169.

Weill, P., and Woerner, S. 2018. "Is your business ready for a digital future?" *MIT Sloan Management Review* (59:2), pp. 21–24.

Weinrich, T., Muntermann, J., and Gregory, R. W. 2016. "Exploring principles for corporate digital infrastructure design in the financial services industry," *Pacific Asia Conference on Information Systems*, Chiayi, Taiwan, p. 285.

Weiß, P., Zolnowski, A., Warg, M., and Schuster, T. 2018. "Service dominant architecture: Conceptualizing the foundation for execution of digital strategies based on S-D logic," *Hawaii International Conference on System Sciences*, Waikoloa Beach, HI, pp. 1630–1639.

Weissenfeld, K., Abramova, O., and Krasnova, H. 2017. "Understanding storytelling in the context of information systems," *Americas Conference of Information Systems*, Boston, MA.

Wenzel, M., Wagner, D., Wagner, H.-T., and Koch, J. 2015. "Digitization and path disruption: An examination in the funeral industry," *European Conference of Information Systems*, Munster, Germany.

Wessel, M., Thies, F., and Benlian, A. 2017. "Opening the floodgates: The implications of increasing platform openness in crowdfunding," *Journal of Information Technology* (32:4), pp. 344–360.

Westerman, G. 2016. "Why digital transformation needs a heart," *MIT Sloan Management Review* (58:1), pp. 19–21.

Westerman, G., and Bonnet, D. 2015. "Revamping your business through digital transformation," *MIT Sloan Management Review* (56:3), pp. 10–13.

Westerman, G., Bonnet, D., and McAfee, A. 2014. "The nine elements of digital transformation." 7, from https://sloanreview.mit.edu/article/the-nine-elements-of-digital-transformation/

Westerman, G., Calméjane, C., Bonnet, D., Ferraris, P., and McAfee, A. 2011. "Digital transformation: A roadmap for billion-dollar organizations," MIT Center for Digital Business and Capgemini Consulting, pp. 1–68.

Winkler, M., Huber, T., and Dibbern, J. 2014. "The software prototype as digital boundary object – A revelatory longitudinal innovation case," *International Conference of Information Systems*, Auckland, New Zealand.

Wolfswinkel, J. F., Furtmueller, E., and Wilderom, C. P. 2013. "Using grounded theory as a method for rigorously reviewing literature," *European Journal of Information Systems* (22:1), pp. 45–55.

Woodard, C., Ramasubbu, N., Tschang, F. T., and Sambamurthy, V. 2012. "Design capital and design moves: The logic of digital business strategy," *MIS Quarterly* (37:2), pp. 537–564.

Woon, J. 2016. "Challenger's digital transformation enabled by the cloud," *Network World Asia* (13:2), pp. 47–48.

Wörner, D., Von Bomhard, T., Schreier, Y.-P., and Bilgeri, D. 2016. "The Bitcoin ecosystem: Disruption beyond financial services?," *European Conference of Information Systems*, Istanbul, Turkey.

Wulf, J., Mettler, T., and Brenner, W. 2017. "Using a digital services capability model to assess readiness for the digital consumer," *MIS Quarterly Executive* (16:3), pp. 171–195.

Xie, K., Xiao, J., Wu, Y., and Hu, Q. 2014. "Ecommerce: Channel or strategy? Insights from a comparative case study," *International Conference of Information Systems*, Auckland, New Zealand.

Yang, X., Liu, L., and Davison, R. 2012. "Reputation management in social commerce communities," *Americas Conference of Information Systems*, Seattle, WA.

Yee, T. M., and Ng, V. 2015. "From digital transformation to disruption, from customer experience to obsession," *Network World Asia* (12:2), pp. 2–3.

Yeow, A., Soh, C., and Hansen, R. 2017. "Aligning with new digital strategy: A dynamic capabilities approach," *The Journal of Strategic Information Systems* (27:1), pp. 43–58.

Yoo, Y. 2013. "The tables have turned: How can the information systems field contribute to technology and innovation management research?," *Journal of the Association for Information Systems* (14:5), pp. 227–236.

Yoo, Y., Bryant, A., and Wigand, R. T. 2010a. "Designing digital communities that transform urban life: Introduction to the special section on digital cities," *Communications of the Association for Information Systems* (27), pp. 637–640.

Yoo, Y., Henfridsson, O., and Lyytinen, K. 2010b. "Research commentary – the new organizing logic of digital innovation: An agenda for information systems research," *Information Systems Research* (21:4), pp. 724–735.

Zhu, K., Dong, S. T., Xu, S. X., and Kraemer, K. L. 2006. "Innovation diffusion in global contexts: Determinants of post-adoption digital transformation of European companies," *European Journal of Information Systems* (15:6), pp. 601–616.

Zolnowski, A., and Warg, M. 2018. "Conceptualizing resource orchestration – The role of service platforms in facilitating service systems," *Hawaii International Conference on System Sciences*, Waikoloa Beach, HI, pp. 1036–1045.

Zuboff, S. 1988. *In the age of the smart machine: The future of work and power.* New York, NY: Basic books.

Chapter 3

The three pillars of digital transformation

Improving the core, building new business models, and developing digital capabilities

Andreas Hinterhuber and Stefan Stroh

This interview discusses key elements of the digital transformation of Deutsche Bahn: Stefan Stroh, Chief Digital Officer, highlights how the digital transformation allows and requires improving core products and processes, building new business models and developing new digital capabilities: digital capabilities include data, IT infrastructure, innovation ecosystem and new work styles.

Andreas Hinterhuber: How does the digital transformation fit within the overall strategy of Deutsche Bahn?

Stefan Stroh: In the context of our digital transformation, we focus on three topics (see Figure 3.1): *improving the core, building new business models,* and *developing digital capabilities.*

The first issue regards the use of digital technologies to enhance our core product and improve our current processes. The second one deals with the way we invest and develop new business models to identify new business opportunities. The third point is building the capabilities that drive us to successful digital transformation, capabilities that are related to how we work inside the company, how we work with partners, and how we use data to improve products and processes.

We see these three elements as stacks that have to be present to make the digital transformation successful.

Andreas Hinterhuber: How, specifically, do these three elements allow you to develop a digital strategy in Deutsche Bahn?

Stefan Stroh: I have to start by saying that as Europe's largest railway operator our main physical asset is a huge infrastructure consisting of 30,000 km of rail tracks and 6000 railway stations. Developing a digital strategy thus also means digitalizing our infrastructure, not only our passenger and cargo train operators like DB Fernverkehr or DB Cargo.

Andreas Hinterhuber: How do you prioritize between digitalizing your B2C business, Deutsche Bahn, versus digitalizing your B2B business, DB Cargo? Which comes first?

Stefan Stroh: We cannot prioritize between our B2C and our B2B business, they are all critically important to our business and we have to deal with digitalizing our entire operations at the same time. All of our businesses face new competition. Here, at Group level, our target is to provide corporate level guidance and capabilities to help our operating units, involved in the digitalization of products and processes. We, of course, also have digital teams in our operating businesses.

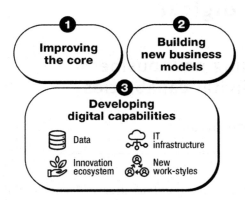

Figure 3.1 Deutsche Bahn: three building blocks of the digital transformation.

Andreas Hinterhuber: Three pillars of digitalization: Can you explain how you enhance the current core product and improve core processes?

Stefan Stroh: Mobility, in essence, in an information product. In our mobility business, Deutsche Bahn, we enhance the customer experience through digital technologies. We are already very digital: 50% of sales are on digital channels, we offer free on-board internet on our high-speed trains and invest to expand this to our entire fleet; digital technologies allow us to provide real-time information to customers on travel time, delays, and alternatives in case of disruptions. We offer self-check-in so a train conductor is not needed anymore and this improves our processes. We are really covering the whole customer journey, from the booking to the overall experience in the station, on the train, up to the last mile.

This focus on improving the customer experience via digital technologies applies also to our B2B business where digital technologies allow real-time track and trace. We use data to make our core product much better.

Andreas Hinterhuber: How do digital technologies influence your pricing strategy?

Stefan Stroh: Digital technologies enable the collection of large datasets that help to improve pricing. Data science allows us to better understand customer preferences and to implement even more sophisticated approaches to revenue management. Depending on the expected capacity utilization, we are able to offer heavily discounted tickets to customers. In the end, this allows to improve our own performance, capacity utilization and profits, as well as customer satisfaction, through lower prices, conjointly. Going forward, we expect to see even more innovations in pricing.

Andreas Hinterhuber: At the foundation of any digital strategy is a strong IT infrastructure. Is this proprietary and thus considered part of your own competitive advantage or, by contrast, open?

Stefan Stroh: We tend to take an open approach. We have sold our own data centers and our computing infrastructure will be in the cloud, as opposed to having it on premises.

Moreover, we use a lot of external players that allow us to learn and implement. In terms of programs, we use, like any other company, both commercial software and software that we have developed, for example for ticketing. But we do this like many other companies, and we are not in any way special in this. There is not one software that can do it all and we are not different from other companies of similar size. We have many providers, and we have to manage very complex technology operations.

Andreas Hinterhuber: Let us move now to the second element of your digital strategy, developing new business models.

Stefan Stroh: The mobility and logistics markets are changing at an unprecedented pace: there are new technology competitors – Uber, Lyft – entering this market and we have incumbents – BMW, Mercedes, Tesla – that are developing radically new services, such as ride-sharing or autonomous driving. There is a huge disruption that is happening.

We build new business models in two different ways: on the one hand, we build new companies around these platforms that offer new forms of mobility, such as ride-sharing; on the other hand, we invest in start-ups like a venture capital fund. In sum: we fully recognize that the mobility market of the future will be radically different than the market you see today, and we have a very clear intent of competing in this future market as well. We recognize that we cannot do this on our own and so we build partnerships with external players. But we also take money and build companies.

Andreas Hinterhuber: So, your ambition is clearly to be able to compete with non-traditional players?

Stefan Stroh: Yes, but this is not a plan. We are already doing this. For example, we are a sizeable provider of ride-sharing services in Germany with CleverShuttle, 80% owned by Deutsche Bahn.

We are increasingly interested in offering a variety of services and connecting these new services to our core business in a customer-based approach. So, for example, we can offer customers the possibility to cover long distances by high-speed train and make the first or the last mile by ride pooling.

Andreas Hinterhuber: So, this leads to the question: Where do you draw the line? How far will you spread from your current core?

Stefan Stroh: We are definitely remaining in the mobility, logistics, and infrastructure area: we transport people, we transport goods, and we build the infrastructure. We are not going to invest in other fields, such as healthcare or fintech, because we have a clear investment map and are focusing on it, without wasting energy on markets that are not relevant to us. We are not taking the approach of a venture capitalist who is looking at the whole market: specific elements of new technologies are simply not relevant for us.

Andreas Hinterhuber: To what extent are you willing to cannibalize your core business?

Stefan Stroh: You have to be willing to cannibalize your own business. This is almost a trivial statement. We are looking at these new markets and if there is cannibalization, so be it.

But we really believe that there are a lot of synergies between what we are currently doing and what the new business models are promising of doing. In 50 years, we might have a fleet of autonomous cars or flying cars and we may not need railways anymore, but I don't see this future rising in the short or medium term, this is clear.

Andreas Hinterhuber: What is the role of autonomous cars?

Stefan Stroh: Yes, we operate autonomous cars. We have ioki, a bus shuttle service, and we have two roads where these shuttles operate autonomously: these are dedicated lanes, so that this shuttle does not interfere with any other traffic. My take is that the future of fully autonomous, self-driving cars is a couple of years ahead. The car industry itself is pulling back and taking a more conservative approach than before.

Andreas Hinterhuber: The final element in your digital strategy is the development of new capabilities. How does the digital transformation change the way you work?

Stefan Stroh: I would say that there are four key elements: data, IT infrastructure, innovation ecosystem and new work styles. One: Data is important to improve product and processes; so, we are building capabilities concerning a data platform that all our business units can access and use seamlessly. We make the access to data much easier across the entire organization and we are leveraging new technologies, such as artificial intelligence and machine learning to actually make use of this data.

Two: IT infrastructure implies moving all data into the cloud. This is essential for any kind of business involved in the digital transformation, in order to be able to rapidly scale and adapt data via standardization.

Three: Building your innovation ecosystem is about the development of programs to bring innovation inside. We are working with the full spectrum of available approaches: we build prototypes and scale them, we work with start-ups, we develop new products and improve processes with large tech companies, we have our entrepreneurship program that creates entrepreneurs within our company, and we have digital ventures. We are also building product factories that create products in a dev-ops environment as many tech firms are doing.

Four: New work styles. The digital transformation implies a different way of working together. We will work much more in teams in the future: cross-functional, agile teams that are not just IT or business teams, but integrated teams that will have much more ownership in the final product. We are also experimenting with self-organizing teams that really do not have a boss any more. We are experimenting with agile teams, with sprint-teams to make the organization faster and work in a much more inclusive way towards and common goal, as opposed of working in silos. We will be leveraging many external forces in this transformation, e.g. start-ups, large companies. We cannot do this transformation on our own, so we are really opening up and want to become a much more collaborative company than maybe what we have been so far.

Andreas Hinterhuber: Is the future of work project-based? You work on a project, get it done, the team dissolves, and you do something new?

Stefan Stroh: That can be the case, but I do not think that this model will apply to the entire organization. There will always be parts of the organization that will run in a more traditional, Tayloristic, way. The big challenge is to make two systems, a hierarchical and a team-based and more integrated approach, work together and to organize the interface among them, so that both work styles will be equally accepted in the organization. At the moment the vast majority of our organization is still organized in the traditional way and you have to qualify yourself for the team-based approach to work: working in self-organizing teams is not always fun; teams elect their own boss, and this creates new, unpredictable dynamics.

Andreas Hinterhuber: The need to develop the innovation ecosystem is a common element in any digital transformation. How do you develop relationships with other partners?

Stefan Stroh: I think we have a very successful start-up program: we see several hundred start-ups every year and select around 30 of them to work on projects. Our business units look at this as a funnel, as a way to learn more about a specific technology, let us say artificial intelligence, or about a new business model. For three months, the selected start-ups have access to our data and assets, and they can showcase their product. If their product is successful and we see further potential, we ask the business units to take over and continue with them on their own. We are thus a matchmaker between the corporate world and the start-up world.

Andreas Hinterhuber: So, you acquire the start-ups?

Stefan Stroh: Typically, not. Start-ups do not want to be acquired by corporates. If we acquire start-ups, we acquire them through Deutsche Bahn Digital Ventures, but not through Deutsche Bahn. Our model is: start-ups can always work with us, and we do not insist on equity. Equity is an afterthought. If we really see a great firm that can drive us into a completely new future, then we consider an acquisition. But again: start-ups have enough money, being acquired by DB Digital Ventures is not really what their aspiration is.

Andreas Hinterhuber: But at least you sign an exclusivity agreement?

Stefan Stroh: No, nothing. We do not need exclusivity. We simply use their product and if the product is good, we continue with them. If not, they can work with someone else. And only when we see that we really can help these guys and when we see a big, growing new market coming, we take a share, but as Deutsche Bahn Digital Ventures, our venture capital arm. It is easier for a start-up to accept money from a venture capital arm than from a corporation.

Andreas Hinterhuber: I understand that many of your initiatives related to start-ups may be confidential, but nevertheless there is probably an example that you want to share.

Stefan Stroh: Conox is a company that does sensors for predictive maintenance for switches. We incubated them and this company is now a €30 million start-up. Now we use their product, but we do not use it exclusively. If we used exclusivity, they would be depending on us and if we didn't buy from them they would go bankrupt: so what's the deal?

Andreas Hinterhuber: If you had exclusivity, then this would prevent the start-up from selling their product to Ferrovie dello Stato Italiane, for example.

Stefan Stroh: Why is that a problem? They can improve their product by selling it to someone else. We need exclusivity on specific things, but not on technology, which gets better if it is shared and worse if it is not shared.

Andreas Hinterhuber: A comment on artificial intelligence (AI). AI is an integral part of the digital transformation.

Stefan Stroh: Yes, absolutely. I truly believe that there is no digital transformation without mastering data. The majority of product and process innovation is really very much data driven. So, if do not master your data infrastructure and if you do not have the skills to leverage your data, you have a serious competitive disadvantage.

Andreas Hinterhuber: Data is abundant, but not many companies have developed the skills to make intelligent use of the available data.

Stefan Stroh: I would not say that we have cracked the nut entirely, but we have understood that we cannot continue operating as we have been doing, so we started a data program at Deutsche Bahn, where we make data available, we categorize data, we add external data to build better models, and we build a data community. We have to bring people together to create a critical mass. I do not believe that you can hire 200 people here at corporate and expect that they can tell people in the business units how to use data better. Our business units will push them away and do their own thing. It has to be an integrated approach. It is a very fine line between a corporate directivity and business–unit specific enablement.

Andreas Hinterhuber: How to you measure the success of the digital transformation? Do you use different metrics for traditional products than for digital products/initiatives?

Stefan Stroh: No. In the end we really use digital to improve our products and operations. So, the outcome of digital initiatives must be measured very much like any other business opportunity: increase in revenues or customer reach, customer satisfaction and efficiency improvement.

Andreas Hinterhuber: What are the main obstacles to a successful implementation of the digital transformation? How did you overcome internal (= other departments, sales manager) or external (= customers, distribution channels) obstacles?

Stefan Stroh: A digital transformation is as much as technological as a cultural task. And I would argue that the cultural aspects of digital transformation are much harder to tackle than the technological. So, this needs a lot of communication, capability building, and setting the right incentives to promote customer-centric thinking, an open innovation approach and working in integrating teams breaking down traditional siloes.

Andreas Hinterhuber: Mr. Stroh, thank you for this interesting exchange of thoughts. I appreciate the clarity of your ideas on essential elements to make the digital transformation work for traditional companies.

Stefan Stroh: Welcome, I also enjoyed our exchange of thoughts.

Chapter 4

Big data and analytics

Opportunities and challenges for firm performance

Marisa Agostini

Introduction

Companies generate, communicate, share, acquire, protect, and analyze large amounts of data. At the same time, they should adapt to the changes produced by the data themselves. This phenomenon has taken the name of *big data* (BD), considering both structured and unstructured data, coming from heterogeneous sources both internal and external to the companies (Clemons et al., 2017). There are around 2.5 quintillion bytes of data generated every day and over 90% of all existing data created only in recent years (IBM, 2017). Companies must coexist with such amounts of data, affording challenges and taking opportunities to improve their performance. PwC recently conducted a survey of around 2000 industrial companies in 26 countries and found that 33% of companies had already reached "advanced digitization" in 2017, but 72% plan to do so by the end of 2020 (PwC, 2017). However, many companies experience difficulties in processing, analyzing, and using the large amounts of data they are acquiring. PwC (2017) estimates that companies are willing to spend 5% of their annual revenues on processes, data analysis technologies, and expert personnel in this regard. There is a growing demand for data processing and analysis: BD are complex datasets, formed by heterogeneous sources, that may be analyzed only using *big data analytics* (BDA) systems (Rialti et al., 2018). Although there is considerable and growing attention on BD-BDA, the analysis of their implications on firm performance still appears fragmented and not very detailed (Grover et al., 2018; Côrte-Real et al., 2017; Rialti et al., 2019). The objective of this chapter is precisely to systematize the recent literature about BD–BDA and firm performance, providing an overview of their reciprocal implications, highlighting both opportunities and challenges. Since the beginning of the analysis, the main opportunity seems related to the transformation of data (uncorrelated or collected for different purposes) into new information useful for supporting decision-making processes at various corporate levels. The main features of BD-BDA (i.e., large amounts of data and specific analysis methods) may overcome some related problems, such as the (poor) quality of the same (single) data. In this sense, the main challenge is due to the relevant difference between the collection of data and the provision of information: only the second is impactful for the corporate decision-making process. The passage from raw data to proper information requires the identification of suitable processes (for data storage, management, and synthesis), the consolidation of new routines, and the overcoming of specific cultural, technological, ethical, organizational barriers. All these aspects will be discussed and further explored in the following paragraphs.

Big data and analytics

Some confusion still persists on the BD-BDA theme: despite the widespread use in practice, the expression *big data* does not have a commonly accepted definition (Gandomi & Haider, 2015), raising doubts about their own ontology. The initial definition (Laney, 2001) outlines three main characteristics of BD: high volume, high speed, and high variety. Since 2001, further features have been highlighted to define BD. According to the recent "Seven Vs of Big Data" (Mishra et al., 2017), BD differ from traditional large datasets in terms of:

1. *volume*, the huge dimensions of datasets that frequently exceed the terabyte;
2. *velocity*, the "rate at which data are generated and the speed at which it should be analysed and acted upon" (Gandomi & Haider, 2015, p. 138);
3. *variety*, the "heterogeneous sources of big data (i.e., sensors embedded in machines, consumers' activities on social media, B2C or B2B digital interactions, etc.) and the consequent heterogeneous formats that the files composing big data may assume" (Rialti et al., 2018, p. 7);
4. *veracity*, the truthfulness and necessary trustworthiness of BD sources;
5. *value*, the potential intrinsic value of information derived from BD-BDA;
6. *variability*, the possible variations in data flow rate, processing, and data sources (Wamba & Mishra, 2017); and
7. *visualization*, the visual insights derived from BD-BDA.

As already emphasized in the introduction, the extraction of insights from the so-characterized BD sets poses unprecedented challenges to organizations that cannot use traditional database software and systems. Indeed, BD are today conceived only together with BDA because the implementation of data lakes, machine learning, and other tools based on artificial intelligence paradigms are necessary to collect, store, and analyze such data. This combination (BD and BDA) represents an essential condition to extract information from both structured and unstructured data. These two types of data have been called in different ways. On the one hand, structured data are generally quantitative, and generate "concrete" (or hard) information. They are often numeric (i.e., communicated with numbers), more objective and comparable, easier to store and process. On the other hand, unstructured data are generally qualitative and generate soft information. They are more expensive to memorize, subjective, and difficult to transmit without loss of content. Despite these difficulties, numerous studies have stated that both quantitative and qualitative data are relevant for evaluating firm performance (Liberti & Petersen, 2019). Examples of such data regarding firm performance are accounting items reported in annual statements and financial indices (as structured data) and also soft accounting data derived from social media and textual information contained in the management report (as unstructured data). They are all crucial for informed investment decisions and for forecasting the company's future performance. Despite containing valuable and relevant information, unstructured (qualitative-soft) data are often ignored by analysts because of the dreaded inconsistency between corporate narrative disclosure and its economic-financial performance (Chou et al., 2018). These difficulties are related to the persisting confusion about the BD-BDA definition mentioned at the beginning of this paragraph. Moreover, there is a lack of univocal consent regarding BD valorization because of the gap between accounting evaluation rules and company needs. Stakeholders ask for the right assessment of company resources, but BD characteristics (i.e., intangible, limited useful life, ambiguous value) do not lend BD-BDA to the traditional accounting rules for financial reporting (Rivera et al., 2018). There are no official guidelines to determine exactly how much information is worth. This difficulty to properly estimate and communicate the value of BD-BDA makes companies question the real economic benefits of information acquired or produced through BD-BDA. Despite this difficulty that deserves careful consideration and the development of adequate assessment criteria, the recent literature, which is going to be deeply examined and recalled in the following paragraphs, shows that the exploitation of BD-BDA impacts on firm performance also in terms of value.

Methodology

This study concerns the impact of BD-BDA on firm performance. In order to examine how the scientific literature dealt with this issue and to understand the effects of such digital transformation, a systematic literature review (SLR) has been carried out by searching the Scopus database and following a consecutive-steps approach (Littell et al., 2008; Massaro et al., 2016; Sivarajah et al., 2017). First, to identify the relevant articles through the Scopus database, the following keywords search criteria were used: *"big data" OR analytics AND accounting AND performance OR result OR profit OR loss OR earnings*. The use of the search keyword *accounting* within the query serves to concentrate the research in this chapter's specific field of relevance, meaning performance only as a firm parameter. In this way, 288 documents of different types were identified. The works selected on the basis of such bibliographic research were further examined according to several criteria (Table 4.1).

Indeed, only the articles (through the *document type: articles* option) printed or already accepted (in print) by scientific journals (through the *source type: journals* option) and written in English were selected: six articles were not printed or already accepted (for print) in scientific journals, six articles were published in Chinese and one in German.

Table 4.1 Steps of the implemented systematic literature review (SLR)

Step number	Option type	Step description
1	Keywords input	Specific keywords search criteria: *"big data" OR analytics AND accounting AND performance OR result OR profit OR loss OR earnings*
2	Document type	Only articles
3	Source type	Articles printed or already accepted (for print) by scientific journals
4	Language	Articles published in English
5	Year	Not articles published in 2020 (not yet final in Scopus at the time of analysis)
6	Subject area	Articles related to the Business, Management and Accounting area
7	Abstract	Relevant articles by reading the abstracts
8		Final sample of articles to be analyzed

Table 4.2 Scientific journals publishing the largest number of selected articles

Journal title	No. of published articles
Journal of Information Systems	7
Journal of Accounting Education	5
Accounting Auditing and Accountability Journal	4
Journal of Cleaner Production	4
Accounting Review	2
Business Horizons	2
International Journal of Accounting and Information Management	2
International Journal of Accounting Information Systems	2
International Journal of Economics and Business Research	2
Journal of Emerging Technologies in Accounting	2
Managerial Auditing Journal	2

In this way, 154 articles remained to be analyzed. The last selection criterion concerned the thematic area (through the *subject area* option): the analysis focused on "Business, Management and Accounting" that was relevant to the topic covered in this volume and was also the most populated, selecting 69 items. These were published in 46 scientific journals: 11 of these journals published 34 articles on the analyzed topic (Table 4.2), while all the other journals presented only one of the selected contributions.

In order to better understand which main topics these 69 articles focused on, the keywords indicated in the articles (in addition to those used for bibliographic research in Scopus) were also taken into consideration. Most of such keywords could be grouped into five main groupings: *accounting* (e.g., financial reporting, disclosure, cost accounting, forecasting), *audit* (e.g., auditors, audit quality, big 4, external audit), *sustainability* (e.g., sustainable development, carbon, carbon dioxide, ecosystems, energy utilization, environmental impact, natural disasters, value relevance), *corporate governance*, and *social media*.

The analysis carried out so far considered all the articles selected on the subject in Scopus and published before 2020. It is noted, however, that some articles are dated, while an important objective of this chapter is to outline the topics currently most debated in the literature regarding the impact of BD–BDA on firm performance. It is therefore important to underline the strong recent interest of the scientific literature on the subject examined here. The manuscripts' distribution over the years shows that the majority of the selected papers are two years old: more than half of the selected contributions (i.e., 38 out of 69 articles) were published from 2018 to 2019 (Table 4.3).

Therefore, the analysis focused on the 38 articles thus selected, going on then to verify the relevance of each work to the specific topic of this chapter through the analysis of the abstracts. For this reason, 17 articles were finally discarded because they did not fit properly to the considered issue, while the others (21 articles) are examined in the contents thoroughly.

Analysis and results

The full reading and textual analysis of the sampled articles led to grouping them according to different perspectives of analysis (Table 4.4). In this way, four subtopics are identified to

Table 4.3 Number of papers among the years

Year of publication	Number of published articles
2019	21
2018	17
2017	10
2016	7
2015	3
2014	3
2013	2
2012	1
2011	1
2010	1
2007	3

better examine and understand the impact of BD–BDA on firm performance through the analysis of the recent literature. They regard:

1. *decision-making process*, analyzing the changes made to such process thanks to BD availability and BDA implementation;
2. *education and training*, highlighting their importance for both workers and professionals (especially, but not limited to, accountants) to understand and exploit the impact of BD–BDA on firm performance;
3. *risks and barriers*, discussing what companies should face and overcome in order to benefit and derive value from BD–BDA; and
4. *tools of implementation*, making reference to methods and systems improved through recent developments in information technologies in order to enhance firm performance.

The sampled articles have thus been grouped, and are then examined and recalled below (through the ID code that distinguishes them in Table 4.4), also analyzing the literary strand in which each of such works has been included. The categorization of the articles highlights the main perspective of analysis adopted in each article (X), but it also emphasizes that many contributions address multiple issues together, proving that the analysis of BD–BDA has influenced research in an interdisciplinary and pervasive way, as well as the different business areas. Each category, corresponding to a perspective of analysis, is examined below.

Decision-making process

As their name suggests, BD are first of all characterized by the "big" dimension (or huge amount). The quantity of data can be overwhelming to organize and interpret: Managers should be able to sort the data and determine what is really important. There emerges the fundamental challenge to decodify datasets into significant insights, to transform raw data (also unstructured BD) into valuable information, useful decisions, and improved performance. For this reason, companies search for atomized raw data, deprived of any external

Table 4.4 Categorization of the sampled articles according to four different perspectives

*ID	**Authors	Year	***CAT 1. Decision-making process	CAT 2. Education	CAT 3. Risks and barriers	CAT 4. Tools of implementation
ID1	Al Chahadah,A., El Refae, & G.A., Qasim,A.	2018	X̲			X
ID2	Ballou, B., Heitger, D. L., & Stoel, D.	2018		X̲		
ID3	Chang,Y. T., & Stone, D. N.	2019		X		X̲
ID4	Chou, C. C., Chang, C. J., Chin, C. L., & Chiang,W. T.	2018	X			X̲
ID5	Chung,W., Rao, B., & Wang, L.	2019			X	X̲
ID6	Cong,Y., Du, H., & Vasarhelyi, M. A.	2018	X			X̲
ID7	Dzuranin,A. C., Jones,J. R., & Olvera, R. M.	2018		X̲		
ID8	Green, S., McKinney, E., Jr., & Heppard, K., Garcia, L.	2018	X̲			
ID9	Hamilton, R. H., & Sodeman, W. A.	2019	X̲		X	
ID10	Hoelscher, J., & Mortimer,A.	2018		X̲	X	
ID11	La Torre, M., Dumay, J., & Rea, M. A.	2018			X̲	
ID12	Neu, D., Saxton, G., Rahaman,A., & Everett,J.	2019				X̲
ID13	Obschonka, M., Lee, N., Rodríguez-Pose,A., Eichstaedt, J.C., & Ebert,T.	2019				X̲
ID14	Oesterreich,T. D., & Teuteberg, F.	2019		X̲		
ID15	Paulevich, S. V., Nikolaevich, K. A., & Mikhailovna, P. Y.	2019	X̲			
ID16	Perdana,A., Robb,A., & Rohde, F.	2018	X		X̲	X
ID17	Perdana,A., Robb,A., & Rohde, F.	2019				X̲
ID18	Perkhofer, L. M., Hofer, P., Walchshofer, C., Plank,T., & Jetter, H. C.	2019	X	X	X	X̲
ID19	Rezaee, Z., & Wang,J.	2019		X̲	X	
ID20	Rivera, S. I., Román,J., & Schaefer,T.	2018			X̲	X
ID21	Widaningsih, M., Rusli B., Punomo, M., & Candradewini	2018			X̲	X

The underlining (X̲) indicates the main perspective of analysis adopted in each article.

  * The ID code of each of the sampled articles is recalled in this paragraph during the analysis of the literature.
 ** The complete citation of the article is reported in the final bibliography.
*** The abbreviation "CAT" indicates the category or perspective of analysis.

transformation or aggregation: Such data are reduced to discrete units, reconfigurable and transparent, and can be combined by the company to answer its precise decision-making needs and better evaluate company performance (Green et al., 2018, ID8). The use of BD-BDA makes it possible to dispose also of information about the external environment (e.g., competitors and market conditions) of the company, supporting both strategic decisions and communication targeted to external users interested in both current business results and performance prospects in comparison to the corporate external competitive environment. The interests of the stakeholders (also those not involved in the day-to-day business) seem to be taken into more account. The use of BD-BDA could also allow decision-makers to anticipate possible responses and reactions to unexpected situations. In this way, the use of BD-BDA transforms management assessment of corporate performance and (in case of proper use) allows for evaluating the economic-financial situation more precisely and making better-informed decisions for business planning (Paulevich et al., 2019, ID15). Such great potential for the development of business analysis involves different business areas. The impact of BD-BDA appears as pervasive in the entire firm, regarding marketing, accounting and finance, supply chain management, and even strategic choices regarding human capital. Indeed, BD-BDA may allow managers to be increasingly informed about workforce performances and permit the HR (i.e., human resources) function to improve the general performance of the company (Hamilton & Sodeman, 2019, ID9). They may help to overcome the limits of the traditional HR decisional system, which is based on past performance, associated with final individual evaluations, and often related to supervisors' personal perspectives. They highlight the role of predictive analysis to determine appropriate actions and resource allocations. In this regard, HR function needs data also generated and gathered by other parts of the organization: BD-BDA require a corporate strategic approach where both data and their (subsequent) analyses are available centrally in the company, without distinctions between business units. This approach implies several opportunities, for instance, employees may better understand the importance of their work also in relation to the overall performance of the company, and employees' training may concern specific areas of improvement once systematic difficulties and critical positions have been identified. On the other hand, legal and ethical challenges may also take over. The use of BD-BDA could be interpreted as a threat or a tendency to reduce the workforce currently employed. Some employees may be motivated to distort or sabotage such data collection if they perceive it as invasive of their privacy. Corporate transparency and workers' consent become essential. Transparency regards the intended purpose of both the data collected and the strategic nature of BD-BDA. Explicit, written, and voluntary employees' consent must be obtained before the collection of data through adequate communication of how and why data are collected. Such a corporate approach should be based on sharing both data and results to capture the reciprocal link between human capital and firm performance, determining how the HR function can improve employee skills and knowledge to develop a lasting competitive advantage. This again emphasizes the need to interpret and derive meaning from BD flow and BDA implementation that in turn require education and training, as illustrated in the following paragraph.

Education and training

The main peculiarities of BD-BDA have already been described above: these new types of data induce organizations to change routines and require additional skills, focused on innovative data acquisition and analysis practices, to provide a competitive advantage and thus improve corporate performance. Data users (people) remain central in this process of change if they are and will increasingly be able to adequately manage data. In this sense, the "human element" represents one of the greatest challenges to the growth and strengthening of BD-BDA use inside companies (Hoelscher & Mortimer, 2018, ID10). In order to obtain the maximum benefit from the analysis, all the workers (not just some experts) should have (at least) basic training on data analysis. Without an embedded understanding of the usefulness of BD-BDA, advanced analysis tools risk being underutilized, and the advantages will be small or non-existent. In particular, in line with these objectives, accounting functions should play a key role in these data-based decision-making processes and be able to identify problems, consider alternative routes and effectively present the results of the undertaken actions. Consequently, accounting education programs should provide the skills necessary to face an increasingly data-driven decision-making environment: data analysis, holistic understanding of business, problem-solving, the use of technical tools, effective communication seem to be examples of skills to be possessed in addition to technical accounting knowledge (Ballou et al., 2018, ID2). All these skills are not new, but they find renewed emphasis in a decision-making environment based on BD-BDA. Indeed, the professional literature (PwC, 2016) highlights three sets of skills necessary for accountants, especially in corporate key roles. First, accountants must be able to ask the right questions (critical thinking skills). Second, they must understand the data and perform appropriate analyses. Third, they must be able to interpret and effectively communicate the results of the implemented analyses. Accounting functions are increasingly asked to go beyond traditional roles and skills to analyze large datasets in order to provide management support. In this regard, the empirical literature highlights how often accountants' profiles (especially those of controllers and management accountants) do not comply with the requirements that have recently become necessary in terms of business analytics competencies. This skills gap must be assessed, mainly taking into account three items: the type of education, the degree of training, and the specific organizational context. This is affected by a multiplicity of factors, including the level of IT adoption and the degree of specialization of the work (Oesterreich & Teuteberg, 2019, ID14). Second, the value of human capital may really increase through specialized training capable of promoting greater efforts, deeper knowledge, or reduction of human capital costs. To achieve these objectives, such training should evolve and adapt to changes due to the transition to a data-driven decision paradigm. Finally, the tendency in accounting education programs to assign the greatest relevance to technical knowledge represents another challenge for BD-BDA exploitation (Ballou et al., 2018, ID2). There emerges a need for an adequate university response based on precise choices on how to insert BD-BDA into the academic curriculum. The options available are manifold (e.g., three approaches described in Dzuranin et al., 2018, ID7) to decide which courses can or should include data analysis, and which skills and tools can be effectively taught. The adoption of a precise approach for the transposition of the required skills through the degree course programs responds primarily to the requests of the professional and business community. For instance, in the USA, the Association to Advance Collegiate Schools of Business (AACSB) has required the updating of curricular accounting programs because a specific accreditation standard ("Accreditation Standard A7: Information Technology Skills and Knowledge for Accounting Graduates") states the inclusion of content and learning objectives associated with data analysis and IT skills.

Therefore, on the one hand, universities must identify the data analysis skills and tools relevant to the accounting function, deciding how and when to incorporate these issues into a curriculum. On the other hand, firms must provide training capable of enforcing the skills necessary to extract value from data through advanced analyzes, being a unique opportunity for the accounting function to become a forward-looking strategic partner of the corporate organization (Dzuranin et al., 2018, ID7). This call for BD-BDA use with predictive purposes is particularly relevant for forensic accounting, which emerges as one of the main accounting areas, including the examination of fraud, anti-corruption, business, and performance evaluation. There is a shortage of forensic accountants who possess adequate BD-BDA competences and are able to use sophisticated tools to effectively and accurately identify potential risks, proactively search for irregularities, and evaluate risk profiles (Rezaee & Wang, 2019, ID19).

Risks and barriers

The literature insists a lot on the need for full awareness about the risks and the barriers that companies face to take advantage of a data-driven decision-making process and to exploit BD-BDA opportunities as intrinsic sources of value. Such risks and barriers are related to three main items: people, culture, and technology (Perdana et al., 2018, ID16). The human capital should provide the precious insights and knowledge extracted from BD-BDA (La Torre et al., 2018, ID11): it is important to emphasize again the importance (and the need) to change the competences of BD-BDA (potential) users, introducing new approaches to innovation, organizational culture, internal procedures, information systems, and decision-making processes. These are the main challenges that organizations face to consciously exploit the value that can be extracted from BD-BDA. This extraction may improve corporate performance thanks to a more accurate knowledge of the workforce, an increase in productivity and customer services, the introduction of targeted business processes, and communication. The impact on the performance of these advantages is also highlighted (by contrast) in the studies examining companies that are not exploiting BD-BDA (Widaningsih et al., 2018, ID21). In addition to the human element and cultural factors, another item is central to the analysis here proposed: technology. Technological barriers are mainly due to the infrastructures that are needed for data acquisition, management, and archiving. The cost of these infrastructures is still significant, and it prevents the spread of BDA, especially in small and medium-sized enterprises (SMEs). Moreover, evolution and development should regard not only machines (as processor power or storage capacity) but also the methods applied for data analysis and knowledge extraction (as described in the next paragraph). These methods should also favor the evaluation of data sources and of the ways in which such data were collected. Indeed, data is continuously generated also through the use of mobile devices and digital services that release continuous digital traces. This continuous process generates a high risk of violations in both data security and privacy. Security problems represent one of the main concerns for most organizations: the aforementioned BD features (especially volume, velocity, variety, and variability) explain the harmful effects associated with cyber threats and data security problems. They in turn generate the increase of investments in cybersecurity and the growth of security-related disciplines, such as computational criminology and information technology. Data theft, data loss, cyber espionage, and data sabotage threaten relational, structural, and human capital. The reputational risk, the damage to the image and the brand, the loss of competitiveness, innovation and knowledge for the decision-making process are all examples of risks that can derive from data security problems and must be taken into account (La Torre et al., 2018, ID11), together with privacy risks. Access to and acquisition of data generally take place with the awareness and consent of people, but this is not always true. Since data are acquired with the aim of extracting useful knowledge to create value and improve business performance, there is a strong need to protect and enforce users' privacy. Concluding, companies face two main challenges in this regard: on the one hand, they are called to respect people's privacy, and this is a concern at the individual level (i.e., individuals who generate data, sometimes unconsciously). On the other hand, companies must protect their (acquired) data from cybersecurity threats and security incidents, conceived as (corporate and social) challenges that can undermine the advantages deriving from BD–BDA.

Tools of implementation

As indicated in the previous paragraph, the potential barriers to the adoption of BD–BDA may be mainly due to users and/or technology (Perdana et al., 2018, ID16). With reference to the second, the literature indicates that there is a significant gap between what is possible and recommended by the experts (as presented in the literature) and what is actually implemented by decision-makers and professionals. Although the value of BD–BDA is recognized by most, it still seems a topic of discussion aimed at future implementation, rather than something widely used and easily adoptable. As a result, users continue to rely on simple tools that were valid only for traditional datasets. This erroneously reduces alternative options, hides interesting data relationships, encourages preconceptions, and modifies the extraction of information during the decision-making process. For this reason, the recent technological advances aim to introduce methods and techniques to support intuition and human abilities, facilitating the possibility of extracting complete and useful information from BD–BDA for improving corporate decision-making and performance. In this regard, three main types of innovative tools may be analyzed according to their main purposes, i.e., data provision, readability, and processing.

With reference to data provision, in recent years, social media have quickly emerged as a source of data regarding both organizations and society, capturing information about the agents and between the agents (i.e., their relationships). Since information and relationships change continuously, BD detection through social media can facilitate the projection of future trends, predicting the effects and potential evolution of current phenomena. In this way, changes can be made in the products' life cycle and in the stages of innovation adoption. For this reason, scholars are examining various aspects of dynamic social networks (Chung et al., 2019, ID5). This stream of research is going toward previously unexplored and unthinkable fronts: two of them are here proposed, i.e., social accountability and Big Five model. Social media may imply a democratization of the accountability process, becoming important also for the dissemination of economic and financial information of public interest. They allow individuals to participate and react to "public" events, merging individual reactions into something new and collective. This process has encouraged some international organizations (e.g., World Bank, 2017) to implement "social accountability" initiatives, disclosing economic and financial information previously kept only private. This innovative use of social media leads to the disclosure and public evaluation of accounting data, also favoring a new fervor in this discipline (Neu et al., 2019, ID12). A second research front proposed here and concerning social media as a source of BD-BDA having an impact on corporate performance refers to the effects of entrepreneurial personality traits (Obschonka et al., 2019, ID13). The "big five" is the most established model that has received considerable attention from research on entrepreneurship in recent years. It uses and tests data available on social media for the evaluation of the differences in the personality at the local level, which may then reflect regional differences in entrepreneurial activity. It is the first attempt to explore the potential and validity of BD-BDA derived from social media to evaluate their predictive capacity in relation to successful entrepreneurship and corporate performance.

With reference to data readability, it is connected with the interpretation, especially of unstructured data (Chang & Stone, 2019, ID3), and represents the extent to which a reader understands the author's messages. The format and display of information both play essential roles in favoring this readability and mitigating users' limits (i.e., being inexperienced, having limited knowledge, and relying on simplified information), allowing them to focus more on the information message rather than on single data (Perdana et al., 2018, ID16). In particular, the readability of unstructured accounting data promotes a better evaluation of firm performance and is certainly facilitated by the format for displaying information. It represents a central theme in recent literature, concerning the quality of data and information provided through XBRL (i.e., "eX-tensible Business Reporting Language"). This is the international standard for digital communication of financial information relating to performance, risks, and compliance (Perdana et al., 2019, ID17). The literature states that XBRL (despite the adoption costs) has a positive impact on the performance also of SMEs: Financiers and small business investors use XBRL files and prefer them to non-XBRL files when they are both available for download (Cong et al., 2018, ID6). XBRL files use a markup language that facilitates the download and analysis of financial statement information. Such markup language, specific for the disclosure of accounting information, has also contributed to advances in IDV (i.e., "Interactive Data Visualization") applications (Perkhofer et al., 2019, ID18).

With reference to data processing, the literature illustrates several tools useful to facilitate the effective prediction of future trends in business development, helping managers to improve their decision-making process and performance evaluation. These include the use of "data mining" (Al Chahadah et al., 2018, ID1), which represents a set of strategic management tools deriving from the combination of both "business intelligence" and "knowledge management". Indeed, "data mining" is a powerful "business intelligence" tool for "knowledge management": it represents the process of discovering significant patterns within huge databases, allowing the extraction of knowledge from BD-BDA. The effective and efficient use of "data-mining" offers significant competitive advantages in the decision-making process. Both the "American Institute of Certified Public Accountants" (AICPA) and the "Institute of Internal Auditing" (IIA) have recognized "data mining" as one of the best techniques that will shape the future of the accounting profession. In this regard, there are already attempts to develop the same "data mining" (Chou et al., 2018, ID4). For instance, the combination of text analysis and data mining techniques should help to evaluate the tone of unstructured quality data, and their coherence with quantitative economic and financial data taken from the same annual report. In this way, information extracted from both structured and unstructured BD, from both quantitative and qualitative data inherent to corporate performance, can be compared.

Concluding remarks

The present work has systematized the recent literature about the impact of BD and BDA on firm performance. Companies must coexist with such amounts of data, affording challenges, and taking opportunities to improve their performance. The first challenge is due to the same definition and evaluation of BD–BDA. Indeed, BD definition has been progressively specified, introducing more features ("Seven Vs of Big Data"; Mishra et al., 2017) able to describe BD and distinguish them from traditional large datasets. In this sense, especially the presence of unstructured BD (and the consequent qualitative, soft, textual information) makes the implementation of BDA and essential condition to extract both insights and value from data. The evaluation of BD–BDA generates another challenge for companies because such data represent intangibles with limited useful life and ambiguous value that is difficult to both state and report (communicate) to stakeholders according to the traditional accounting rules (Rivera et al., 2018, ID20). Therefore, BD–BDA appear as intrinsic sources of value, but taking advantage of such value depends on the corporate ability and possibility to know how to benefit from such an opportunity. In this regard, four main perspectives of analysis have been introduced and examined in this work. First, the corporate decision-making process needs to change because of BD–BDA. It may improve thanks to the availability of more information also about the external environment of the company (e.g., competitors and market conditions), the greater consideration of (more) stakeholders' interests, the better awareness of each worker's contribution to the overall performance, the prompter improvement of specific critical business areas and positions, the more precise evaluation of firm performance. On the other hand, the decision-making process based on BD–BDA faces new challenges, such as the described legal and ethical issues. The new types of data and analytics require greater transparency and consent, inducing organizations to change routines and require additional skills: education and training represent the second analyzed perspective. On the one hand, data analysis skills and operative tools must be incorporated in degree course programs. On the other hand, firms must provide training capable of enforcing employees' skills necessary to extract insights from BD through BDA, being a unique opportunity of promoting greater effort and deeper knowledge, increasing the value of human capital, and improving performance (also through the reduction of human capital costs). Data users (people) remain central in the process of change implied by the digital transformation if they are (thanks to their education) and will increasingly be able (thanks to training and specific organizational context) to adequately manage data. In this sense, the "human element" represents one of the greatest challenges to BD–BDA exploitation (Hoelscher & Mortimer, 2018, ID10). Indeed, three main items (i.e., people, culture, and technology) represent the main barriers (and the potential solutions) to the examined data-driven process of change. In particular, technological barriers are related to infrastructural costs and systems development. The methods applied for data analysis and knowledge extraction should also favor the evaluation of data sources and collection in order to reduce the risks of violations regarding data security and privacy. Indeed, continuous technological advances introduce tools to support data provision, readability, and processing, going toward previously unexplored and unthinkable frontiers, as emphasized in the analysis of the last (fourth) perspective. In this regard, the literature recognizes a significant gap between what is possible according to the scholars and what is actually implemented by corporate decision-makers and professionals. Although the intrinsic value of BD–BDA and their impact on firm performance are recognized by most, the data-driven process of change still seems a topic of discussion for future implementation, rather than an actual opportunity and challenge for most organizations.

References

Al Chahadah, A., El Refae, G. A., & Qasim, A. (2018). The use of data mining techniques in accounting and finance as a corporate strategic tool: an empirical investigation on banks operating in emerging economies. *International Journal of Economics and Business Research*, *15*(4), 442–452.

Ballou, B., Heitger, D. L., & Stoel, D. (2018). Data-driven decision-making and its impact on accounting undergraduate curriculum. *Journal of Accounting Education*, *44*, 14–24.

Chang, Y. T., & Stone, D. N. (2019). Why does decomposed audit proposal readability differ by audit firm size? A Coh-Metrix approach. *Managerial Auditing Journal*, *34*(8), 895–923.

Chou, C. C., Chang, C. J., Chin, C. L., & Chiang, W. T. (2018). Measuring the consistency of quantitative and qualitative information in financial reports: A design science approach. *Journal of Emerging Technologies in Accounting*, *15*(2), 93–109.

Chung, W., Rao, B., & Wang, L. (2019). Interaction models for detecting nodal activities in temporal social media networks. *ACM Transactions on Management Information Systems*, *10*(4), 1–30.

Clemons, E. K., Dewan, R. M., Kauffman, R. J., & Weber, T. A. (2017). Understanding the information-based transformation of strategy and society. *Journal of Management Information Systems*, *34*(2), 425–456.

Cong, Y., Du, H., & Vasarhelyi, M. A. (2018). Are XBRL files being accessed? Evidence from the SEC EDGAR log file dataset. *Journal of Information Systems*, *32*(3), 23–29.

Côrte-Real N., Oliveira T., & Ruivo, P. (2017). Assessing business value of big data analytics in European firms. *Journal of Business Research*, *70*, 379–390.

Dzuranin, A. C., Jones, J. R., & Olvera, R. M. (2018). Infusing data analytics into the accounting curriculum: A framework and insights from faculty. *Journal of Accounting Education*, *43*, 24–39.

Gandomi, A., & Haider, M. (2015). Beyond the hype: Big data concepts, methods, and analytics. *International Journal of Information Management*, *35*(2), 137–144.

Green, S., McKinney E., Jr., Heppard, K., & Garcia, L. (2018). Big data, digital demand and decision-making. *International Journal of Accounting & Information Management*, *26*(4), 541–555.

Grover V., Chiang R. H. L., Liang T.-P., & Zhang, D. (2018). Creating strategic business value from big data analytics: A research framework. *Journal of Management Information Systems*, *35*(2), 388–423.

Hamilton, R., & Sodeman, W. A. (2019). The questions we ask: Opportunities and challenges for using big data analytics to strategically manage human capital resources. *Business Horizons*, *63*(1), 85–95.

Hoelscher, J., & Mortimer, A. (2018). Using tableau to visualize data and drive decision-making, *Journal of Accounting Education*, *44*, pp. 49–59.

IBM. (2017). *Infographics and animations: The four v's of big data.* IBM Big Data & Analytics Hub. Retrieved January 27, 2020, from www.ibmbigdatahub.com/infographic/four-vs-big-data

La Torre, M., Dumay, J., & Rea, M. A. (2018). Breaching intellectual capital: Critical reflections on big data security. *Meditari Accountancy Research*, *26*(3), 463–482.

Laney, D. (2001). 3D data management: Controlling data volume, velocity and variety. *META Group Research Note*, *6*(70), 1.

Liberti, J. M., & Petersen, M. A. (2019). Information: Hard and soft. *Review of Corporate Finance Studies*, *8*(1), 1–41.

Littell, J. H., Corcoran, J., & Pillai, V. (2008). *Systematic reviews and meta-analysis*. Oxford: Oxford University Press.

Massaro, M., Dumay, J., & Guthrie, J. (2016). On the shoulders of giants: Undertaking a structured literature review in accounting. *Accounting, Auditing & Accountability Journal*, *29*(5), 767–801.

Mishra, D., Luo, Z., Jiang, S., Papadopoulos, T., & Dubey, R. (2017). A bibliographic study on big data: concepts, trends and challenges. *Business Process Management Journal*, *23*(3), 555–573.

Neu, D., Saxton, G., Rahaman, A., & Everett, J. (2019). Twitter and social accountability: Reactions to the Panama Papers. *Critical Perspectives on Accounting*, *61*, 38–53.

Obschonka, M., Lee, N., Rodríguez-Pose, A., Eichstaedt, J. C., & Ebert, T. (2019). Big data methods, social media, and the psychology of entrepreneurial regions. *Small Business Economics*, *55*, 567–588.

Oesterreich, T. D., & Teuteberg, F. (2019). The role of business analytics in the controllers and management accountants' competence profiles: An exploratory study on individual-level data. *Journal of Accounting and Organizational Change*, *15*(2), 330–356.

Paulevich, S. V., Nikolaevich, K. A., & Mikhailovna, P. Y. (2019). Change from economic analysis to operational analytics and corporate analysis in innovative entrepreneurship. *Academy of Entrepreneurship Journal*, *25*(1S), 1–5.

Perdana, A., Robb, A., & Rohde, F. (2018). Does visualization matter? The role of interactive data visualization to make sense of information. *Australasian Journal of Information Systems*, *22*, 1–34.

Perdana, A., Robb, A., & Rohde, F. (2019). Textual and contextual analysis of professionals' discourses on XBRL data and information quality. *International Journal of Accounting and Information Management*, *27*(3), 492–511.

Perkhofer, L. M., Hofer, P., Walchshofer, C., Plank, T., & Jetter, H. C. (2019). Interactive visualization of big data in the field of accounting: A survey of current practice and potential barriers for adoption. *Journal of Applied Accounting Research*, *20*(4), 497–525.

PwC. (2016). *Industry 4.0: Building the digital enterprise*. Retrieved January 27, 2020, from www.pwc.com/gx/en/industries/industries-4.0/landing-page/industry-4.0-building-your-digital-enterprise-april-2016.pdf

PwC. (2017). *Digital factories 2020: Shaping the future of manufacturing*. Retrieved January 27, 2020, from www.pwc.de/de/digitale-transformation/digital-factories-2020-shaping-the-future-of-manufacturing.pdf

Rezaee, Z., & Wang, J. (2019). Relevance of big data to forensic accounting practice and education. *Managerial Auditing Journal*, *34*(3), 268–288.

Rialti, R., Marzi, G., Ciappei, C., & Busso, D. (2019). Big data and dynamic capabilities: A bibliometric analysis and systematic literature review. *Management Decision*, *57*(8), 2052–2068.

Rialti, R., Marzi, G., Silic, M., & Ciappei, C. (2018). Ambidextrous organization and agility in big data era. *Business Process Management Journal*, *24*(5), 1091–1109.

Rivera, S. I., Román, J., & Schaefer, T. (2018). An application of the Ohlson model to explore the value of big data for AT & T. *Academy of Accounting and Financial Studies Journal*, *22*(1), 1–9.

Sivarajah, U., Kamal, M. M., Irani, Z., & Weerakkody, V. (2017). Critical analysis of big data challenges and analytical methods. *Journal of Business Research*, *70*, 263–286.

Wamba, S. F., & Mishra, D. (2017). Big data integration with business processes: a literature review. *Business Process Management Journal*, *23*(3), 477–492.

Widaningsih, M., Rusli, B., Punomo, M., & Candradewini. (2018). An empirical investigation of the relationship between institutional aspect and supply chain strategy in relation to investment policy in Indonesia. *International Journal of Supply Chain Management*, *7*(5), 396–401.

World Bank. (2017). *Global partnership for social accountability*. Retrieved January 27, 2020, from www.thegpsa.org/

Chapter 5

Technology is just an enabler of digital transformation

Interview with Gianfranco Chimirri, HR communication director of Unilever Italy

Francesca Checchinato

Introduction

With a portfolio of 400 brands distributed in over 190 countries, Unilever is one of the leading companies in the FMCG sector. At Unilever, the digital shift began under the CEO Paul Polman, but it was with the new CEO, Alan Jope, that digital transformation was put at heart of Unilever's strategy.[1] Therefore, all of the subsidiaries in various countries work with the aim of transforming the company to successfully compete in the digital era.

Even if a huge investment in technology has been made, the key driver of digital transformation is the new approach and a new mission and vision to support the new multi-stakeholder business model with the aim of becoming a sustainable company.

To investigate the digital transformation of Unilever Italy, Gianfranco Chimirri, the HR communication director of Unilever Italy, was interviewed.

The interview discusses the digital transformation process in Unilever and highlights opportunities, challenges, and solutions.

Some final remarks derived from the interview summarize the key points and contribute to developing our roadmap to succeed in digital transformation.

Francesca Checchinato: How do you define and interpret "digital transformation"?

Gianfranco Chimirri: We connect digital transformation to the fourth industrial revolution, because technology has been affecting not only the economic and organizational context but also demographic flows, consumers' behaviour, and thus the environment and the planet. We are a global company with more than two billion daily consumers; we cannot ignore this revolution. Our digital transformation has taken into account this shift and the impact of the company on the planet. A new business model, a definition of a new set of capabilities, and a new cultural mindset have been adopted.

Due to the digital transformation, the mission and vision of the company were redefined, and sustainability was put at the heart of the new business model. We define it as "a multi-stakeholder business model" because the aim of Unilever is to satisfy many different stakeholders, not only employees, suppliers, shareholders, consumers, and citizens but also the planet. Our business model is only sustainable if we are able to satisfy all of these stakeholders. This is a long-term strategy because we need to balance all of these effects. Sometimes, we need to suffer in the short term, but it is necessary. This model is an evolution of the past model, the Unilever Sustainable Living Plan (USLP).

It is important to notice that social responsibility is embedded in the business model; we do not just take business actions and then some social responsibility actions.

Unilever is a global company, and the impact of its business is huge compared to the impact of one small enterprise.

Francesca Checchinato: When and why did the company start the digital transformation?

Gianfranco Chimirri: Three years ago, the company realized that technologies are changing our way of life, how people live and work, and how we interact each other. Technologies were and are affecting everything, and thus the business model had to change accordingly. The company changed the business model and the organizational culture to compete in this new context. This is the point – Unilever is a global company that competes in a global market, and its size has been one of its competitive advantages. When the company understood that the digital transformation could remove many entry barriers (for smaller players), it started to deeply change its DNA.

Economy of scale and standardization are no longer competitive advantages. Fast decision-making, agility, adaptability, and fast innovation are the new real competitive advantages. Unilever has been working to become an agile organization by personalizing its propositions and developing an entrepreneurial mindset among its employees, changing from a McDonald's company (big size, high standardization, and supply chain control) to an Uber company, even if we have some physical assets.

Our organization works as a platform driven by technology. Technology helps the company to adapt its propositions and change or personalize an offer quickly.

Once we developed our products for a global consumer, no matter where they lived. Now, we consider every consumer as unique.

Francesca Checchinato: Can you describe the digital transformation process?

Gianfranco Chimirri: The company made some analysis about this new era. When the digital transformation started, we had a clear strategy to follow, and we were aware of the components of the new business, the new organizational flows, and how to cope with the consumers' behaviour and competitors. Unilever changed all together.

Consumers started to want more customized solutions. They continuously switched from online to offline, and vice versa, and we cannot underestimate this new pattern.

We implemented a new strategy based on three drivers. First, "company is part of life" refers to our sustainability idea and our responsibility towards the planet. Second, brand values must be clear and recognizable and should be the reason why consumers choose the brand. For example, Dove is a brand recognized for supporting girls' self-esteem, and this is more important than the functional attributes. The same is for Ben & Jerry's supporting refugees. The third driver (and most related to the digital transformation) is people. In our new model, we have a lot of teams that work with a start-up logic. Motivated employees with broad minds and an entrepreneurial attitude are key assets for us. Thanks to this model, we are flexible and fast, but at the same time, when an innovation works, we can scale to the market immediately because we still are a global company. We have adopted a fail and try again culture, and it works.

We are still in a transition phase, but we overcame the starting point. Why is it working? It is because of the company's commitment and a clear strategy.

Technology without a strategy doesn't perform well. We need technology to support our strategy, and the strategy can be implemented based on the technology.

Francesca Checchinato: How much is technology important in your digital transformation?

Gianfranco Chimirri: It is important, but digital transformation has a double nature. On one side, we have to deal with technology and on the other with people. As far as the technology is concerned, Unilever created the so-called digital hubs where specialists work together to provide solutions and data to make data-driven marketing a reality. At the same time, digital hubs are important to spread digital competences to all of the company functions. Digital hubs are like an open structure where marketers, HR, IT, media, and so on work together. It's like a sun from which technological competences radiate to others. Regarding the second point, people are even more important; traditional organizational structures do not exist anymore.

Francesca Checchinato: So organization plays a big role in the digital transformation, can you explain the main changes to us?

Gianfranco Chimirri: The way of working is different from the past. Before the digital transformation process, we dealt with a traditional organization based on hierarchy and functions (marketing, sales …). Now we are an agile company; cross functional teams are at the organization's centre. The team is a de facto small company; all competences and functions are represented. Thanks to this new organizational model, our time to market has decreased in an impressive way, from 18 months to 3 months. Why? Because there are no inter-function approvals needed. Now, we have eight-person teams with all the needed competences and skills, and they are empowered.

Francesca Checchinato: Don't these small teams need a final approval?

Gianfranco Chimirri: Yes, they need it, but just at the end. Before this transformation, we needed huge research projects, big marketing investments, and a communication campaign. At the end of these steps, we launched the product hoping for success. Now it is different, we test the product immediately in a small market or in a specific. If our consumers do not like the product, the error does not have a big impact; investments have not been so big. We can change some features and try again without losing much money.

Francesca Checchinato: To sum up, technology and HR are at the heart of the digital transformation. What is the role of the digital transformation manager and the connection with the HR manager?

Gianfranco Chimirri: The Digital Transformation Officer (DTO) is the first one that has to create the new mindset and convey it to all of the company employees. The Digital Transformation Officer implements solutions and he/she is responsible for the digital capabilities inside the organization, mainly in the marketing, sales, and media area.

  Both the DTO and HR manager are part of the board. The DTO promotes the transformation from a technology point of view, but everybody has to be involved in the digital transformation, and the HR manager has to guarantee this last part, otherwise the digital transformation doesn't work properly. The DTO responds directly to the CEO. I think that in five years, more or less, this role will not exist anymore, because we will not need someone to help us transform the business, the transformation will be inside everyone.

Francesca Checchinato: Unilever is a global company. How can it deal with all of the countries' headquarters?

Gianfranco Chimirri: The model is the same in all of the countries, but nowadays the country subsidiaries have a completely different role compared to the past. Before the digital transformation, standardization was the key part of every activity; so, the countries had an executional role. The new model has overturned the organizational structure. Unilever has understood that a global consumer does not exist; a deep analysis about local consumers is needed, but only the local subsidiaries can understand them. For this reason, the local market starts to have more power in order to quickly innovate and satisfy customers' needs. Moreover, if a local headquarters launches an interesting offer, it could be implemented in other countries. It's like having 10,000 start-ups that work for the same company. We must be innovative and aggressive like a start-up with the advantage of being an MNE.

Of course, working in this way is more interesting. People know that their ideas count, and we do not have to just execute. At the same time, we have no alibi for success or failure. A lot of people criticized the old model, but for many, justifying the results based on the idea that they were a consequence of a global decision was a relief. Now we have to take responsibility ourselves.

Francesca Checchinato: You mentioned the role of Digital Transformation Officer, does it exist at a country level too?

Gianfranco Chimirri: Yes, the DTO exists both the local and global level. All local DTOs are part of a community.

Francesca Checchinato: What are the main challenges Unilever has to deal with?

Gianfranco Chimirri: We have to deal with new consumer behaviours and new channels, and we need to (1) understand our customers and (2) grow quickly in these new channels. New channels also mean new products and a new assortment, and this assortment evolves based on customer needs that we are now able to analyse.

This is also the reason why we adopted a new metric to evaluate our performance. We define it as the "quality of results". It means that the company has to perform well in the right channels. If we grow about 5% because we have double digit growth in retail, we are not satisfied. It means that we are not increasing our sales in the new channels, the channels of the future. Therefore, we do not achieve the quality of results, and in few years our competitors could win, and we would fail. Short-term metrics are important, but long-term metrics are the key for the company's sustainability.

Francesca Checchinato: And about the new consumers' behaviour, how do you understand your customers?

Gianfranco Chimirri: We know our customers thanks to the technology. Unilever is building its own database by using online customer registrations, data from store loyalty cards, and from social networks and other data. From a marketing point of view, this is a big shift from a mass market company and product driven marketing to a consumer and data-driven marketing. Consumer centricity has become a driver, and the company has begun to innovate and promote new products based on social media/websites and other data insights.

Then, thanks to the new start-up mindset, we can provide more solutions and adapt our product. When the digital transformation has been completed, we offer tailored products.

A mass-market company like Unilever is becoming a company with the ambition to have a one-to-one relationship with its customers.

Francesca Checchinato: Which kind of customization are you developing?

Gianfranco Chimirri: We are working on product platforms. We customize based both on channels and where our consumers live. Most of the products developed for e-commerce are not available in the GDO, and out-of-home products are different form home products. Delivery is personalized too, and special packs are developed.

Francesca Checchinato: So you claim that omnichannel and new consumers' behaviour are the main challenges. How does Unilever tackle them?

Gianfranco Chimirri: Omnichannel is a "must have" due to the new consumer behavioural patterns. The main attributes of products are the same in every channel, but packaging formats or solutions can be different.

The main point is to know our consumers. They search for different products based on channels. Now, thanks to the precise knowledge about our customers based on data and the possibility to push different products in different areas, we respond better to the customers' needs. Think about detergent. Consumers living in Northern Italy prefer a basic solution, people in the South want perfume; they want the smell of laundry diffusing all over the neighbourhood. Now we can promote Coccolino with fragrances in the South and without fragrances in the North. Before the digital transformation, we just made mass communication campaign and we had to choose between the two.

Speaking about Coccolino, we created Coccolino Moschino. It is more expensive, and it has special packaging and a high quality perfume. It works very well in the North, Milan in particular. This case is a typical example of our digital transformation product development. We were able to recognize the new trends of home detergents (that are becoming personal care like a deodorant), and we proposed a new product thanks to co-branding with a very famous fashion company. We sold it only within the right channels. If it works we win. Otherwise, we have just wasted a little money and testing on an idea.

Also, in some cases, co-branding or co-marketing strategies are part of the digital transformation revolution. Thanks to this strategy, the innovation is fast, and it is a win-win situation for both of the companies involved in the partnership. This openness is new for us. We are building partnerships with start-ups and freelancers too. Before this revolution, we only worked within our company. Now, we are connected with the external system; so we can quickly innovate and customize based on consumers' needs.

We have a lot of examples in the food market. We developed a balsamic vinegar with De Nigris (Calvè by De Nigris), dried fruit with Noberasco, and Kinder ice cream from a collaboration between Ferrero and us.

As well as the omnichannel, one of the main challenges is to be able to balance resources. The company has to monitor more variables, not only products and territories but also channels. Now, the online channel is small, but, as I said before, you have to look at the quality of the results. Moreover, the new channels are more profitable and there are no fees to enter. GDO considers prices; online is different, but the product should also be different.

Francesca Checchinato: It seems like the organization and marketing are facing the main and most radical changes. We have already discussed the organization, what about marketing?

Gianfranco Chimirri: Communication is completely different. We don't need to think about clusters based on demographics anymore. A few years ago, if we had to communicate the Cornetto brand, we analysed young people between 18 and 25 years old. Now, thanks to digital insights and the data-driven marketing approach, we combine information (who visits the Cornetto website or Facebook page and buys the product …) and create tailored communication. Nowadays, we really know who our customer is. We not only know their age but also if he/she likes travelling, where he/she spends holidays, his/her media diet, and so on. Thus, we can intercept him/her through the most effective touchpoints and convey the right information about our offer. It means that the marketers competences should be more related to data analysis in order to understand the insights.

Francesca Checchinato: How does Unilever deal with this requirement for new marketing competences?

Gianfranco Chimirri: We are investing in upskilling and reskilling because the majority of the roles are changing due to the digital transformation, especially marketing. We are helping people transition for the roles of the future in order to create physiological safety. We are creating opportunities for people to sharpen their skills. If we did not do that, we would have obsolete competencies within the company. Employability for all of our employees is at the centre of our program.

Due to automation, a lot of people have more time to work on added value activities, but they must have the competences to do that, and we must provide them.

Upskilling and reskilling must be a continuous process because required competences will be different (again) in two years. Nowadays, we select people by their learning mode and their ability to transform themselves. Are you able to become a different marketer every three years? If so, you are welcome in our teams. Attitudes become more important than expertise.

Marketers have completely changed their tasks and the way of working in the last years. Communication campaigns must be personalized and follow the customer journey. Working in marketing is difficult; this job is so different from the past, and everything has changed and it is still changing.

Francesca Checchinato: How do people embrace this new model caused by the digital transformation? Any opposition?

Gianfranco Chimirri: Yes, of course, it is normal to have opposition and fears in this situation. If you work on data entry for years and now a machine does it for you, it is not easy to accept. It means that you have time to provide insights from that data and became a sort of consultant. It is obvious that you need to be trained, otherwise you are afraid and not happy. It is an emotional pattern, so you need both upskilling and coaching for your employees. Coaching is important because change management is at the heart of digital transformation. Moreover, changes are quicker than human learning capacity; they have an exponential rate, so the burn out risk is high. We have to work on emotional intelligence to make people comfortable with the new situation.

Part of digital transformation investment should be related to people's well-being.

Francesca Checchinato: Any suggestions for companies in their first phases of digital transformation?

Gianfranco Chimirri: First of all, a high CEO commitment and a Digital Transformation Officer directly reporting to the CEO is an important message for employees.

Then you have to understand that digital transformation is not a technological change. Technology is just an enabler; you need the right culture and the right organization. You have to draw a new business model and not just to adapt the old one to the technology.

Francesca Checchinato: Best in class in the digital transformation?

Gianfranco Chimirri: The best is Microsoft. In Italy, Enel has to be mentioned and ING in the bank sector.

Discussion of the case and final remarks

Technology is changing the world, and mass market companies are suffering because their dimension is no longer a barrier for competitors. Therefore, a company must embrace digital technologies and start adopting them in an active way. If it uses the technologies to change its business model, it will lead again. It is not just a simple adaptation; it is a revolution. Analysing the case, we have to point out that technology is just an enabler of this new business model. Without an attitude and a cultural shift, digital technologies would just remain a tool.

The process that leads to the actual situation can count on two main elements: technology to understand consumers and respond to them and HR management to help people exploit technology's possibilities.

Companies must start collecting and analysing data so that they can know consumers and thus change their marketing approach from a mass-market strategy to a more one-to-one relationship. But, without motivated people, the new marketing approach will remain just a theoretical improvement.

Why does the Unilever digital transformation work?

In our opinion, there are five main elements:

- It started with a clear strategy.
- It was driven by the CEO.
- The DTO responds directly to the CEO.
- Changes were not marginal or incremental; they were disruptive for the company the traditional organization.
- HR was involved in the process.

Moreover, some solutions have also to be highlighted because they are part of the digital transformation success:

- Partnerships to provide meaningful products in quick time;
- Small interfunctional teams to cut the time-to-market and to be flexible;
- Coaching, upskilling, and reskilling to motivate people and guarantee employees' well-being.

Note

1. https://www.spglobal.com/marketintelligence/en/news-insights/latest-news-headlines/new-ceo-alan-jope-puts-digital-transformation-at-heart-of-unilever-s-strategy-49355732

Part 3

The digital transformation and business model innovation

Chapter 6

Digital transformation and business models

Tiziano Vescovi

Digital transformation leads to new business models

A fundamental restructuring of business processes associated with the adoption of digital technology implies the need to revise the views of the management of the organization. Management should be absolutely mobile, but one should not forget that behind any transformation, technology, or business process, there are people with their professionalism, involvement in the process, interest in the final result.

To use digital technologies, the company needs to abandon the old processes, rethink the content of the work, radically restructure the processes, and give the business a new format. That is, constantly learning new information, taking into account the current business context, and adapting the processes on the fly.

The process should begin with changing corporate culture. This point is quite crucial and difficult. Actually, it is interesting to notice that, if a company's culture remains the same, it is this culture that makes the choice of new resources (human, technical, financial) for digital transformation, and it is likely that culturally inappropriate choices could be taken.

Currently, most chief executive officers are concerned about understanding how the digital revolution is affecting and will continue to affect their firms, in light of the ongoing paradigm shift from an industrial to a more digital economy. In some industries such as retail, the profits of large dominant firms are declining, despite attempts to stem this decline and their brick-and-mortar business model is under threat (Björkdahl 2020).

Technological issues are not the first problem worrying business leaders engaged in digitalization efforts. The development and use of new digital technologies are prerequisites for digital transformation but are not sufficient for business success. Successful efforts require re-optimization to allow effective use of digital technologies and data, and creation and capture of value in new ways. In other words, digitalization includes more than digital technologies, it needs a new business model.

Nevertheless, there are many factors, such as the desirability of the products, that can affect a company's success as much or more than its digital capabilities. Therefore, no managers in traditional businesses should view digital as their guaranteed salvation. Digital is not just a thing that you can buy and plug into the organization to solve business problems. It is multifaceted and diffuse and does not just involve technology. Digital transformation is an ongoing process of changing the way doing business. Moreover, it is important to calibrate digital investments to the readiness of the industry – both customers and competitors. But when things are not going so well in the existing business, the call of a new business model can become more powerful than it should (Davenport and Westerman 2018). Amid the excitement and uncertainty of a new technological era, it can be very difficult to distinguish between investments you need to make ahead of the market and investments that must be in sync with market readiness. However, the speed of the transformation is changing the rules of the normal game. To wait for the right moment could mean to be late.

For example, the subscription economy grows five times faster than the S&P 500. Sixteen car manufacturers are now offering their cars on a subscription model. People living in big cities do not need to own vehicles; they are able to use them when they need them through these services. The subscription economy is coming, and companies need to adapt their business model (Chaniot 2019). For this reason, companies are seeking new ways to unlock value through new business models or reinventing existing ones to meet the ever-changing demands of customers and gaining a competitive advantage in the digital era (Von Leipzig et al., 2017, p. 518).

Digitalization can be seen as augmented generation, analysis, and use of data in order, on one hand, to increase the firm's internal efficiency, and on the other hand to grow the firm by adding value for customers through the change from analog to digital formats. This second aspect, being more strategic, pushes to new business models, because digital technology diffusion produces changes in customer behaviors and preferences, creates new players in competition, and offers new opportunities to execute and analyze operations. Failure to adapt business models to digitalizing business environments can have disastrous effects on incumbent firms (Lucas/Goh, 2009).

The degree of the digital transformation can concern the incremental as well as the radical change of a business model. Within digital transformation of business models, enablers or rather technologies (e.g., big data) are used to generate new applications or services. See Table 6.1.

Moreover, the generative and convergent characteristics of digital innovations incorporated in business models require increasing levels of organizational agility and absorptive capacity. The characteristics of digital innovations are "pulling" for specific organizational capabilities (Hanelt et al. 2018). Organizations seeking to deploy digital technologies to garner greater competitive advantages must also ensure their respective business models are aligned. See Figure 6.1.

Table 6.1 Digitalization of business models

Business model category	Short description	Examples
Ownership-based business model	Customer purchases a product and owns it	Purchasing a car
Usage-based business model	Customer purchase a certain usage period of a good	Car sharing, renting a car
Performance-based business model	Customer purchases a defined performance	Taxi, Uber services
Result-based business model	Customer purchase a defined end result	A-B mobility (moovel)
Freemium business model	Customer gets service for free (advertising etc.)	Google Maps

Based on Semples (2014); Cinquini et al (2013)

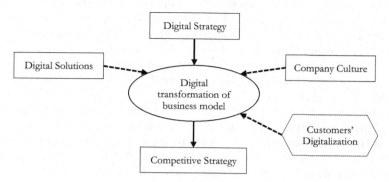

Figure 6.1 Main factors affecting the digital business model solution.

The digital transformation of business models concerns the individual elements of the business model, the entire business model, value chains, and the networking of different actors in a value network. The digital transformation serves to define the digital strategy more clearly within business models. It is based on an approach with a sequence of tasks and decisions that are logically and temporally related to each other (Schallmo, Williams, and Lohse 2019).

Different types of digital business models

The digital transformation of manufacturing firms does not necessarily involve a radical business model, a new product category, or response to a competing technology; it can take many shapes.

Westerman, Bonnet, and McAfee (2014), propose five archetypes of business model reinvention driven by digital technology: (i) reinventing the entire industry, as Airbnb did to the hotel market; (ii) substituting products and services, as Uber does with traditional rental cars; (iii) creating new digital businesses with the development of new products and services, as the banks developed in the last years; (iv) reconfiguring value delivery models, to provide more direct services to the customers, for example through e-commerce, and (v) rethinking value propositions.

In developing a new business model, the company can follow three key themes: (i) strategy-centric, (ii) customer-centric, (iii) organization-centric, and (iv) technology-centric to the fore (Loonam et al 2018).

Three key approaches can be followed to develop digital customer-centric approaches. The first is designing a customer experience from the outside-in, the second reaching and engaging customers and online communities, and the third blending a physical and digital (or virtual) customer experience. For example, creating a new "customer journey," where using digital technologies customers' expectations could be transformed by enabling them to connect with others and share their experience throughout their journey. Reaching and engaging customers and online communities is also an important step in developing a customer-centric perspective during digital transformation.

The organization-centric approaches include the importance of fostering a digital culture, focusing on the organization rather than the technology in delivering change, centered on four key themes: customer centricity, co-creation, market making, and digital culture. The technology-centric approach develops insights from data analytics, and relies across platforms, where digital technologies can seamlessly interact with one another. The value provided by underlying technologies is substantially dependent on the business model in which they are integrated.

It is clear that the importance of the strategic role of information systems in the digital era can no longer be restricted to automating and informing business processes but further extends to the transformation of organizations through innovating business models (Merali et al., 2012). According to Fichman et al. (2014), we define digital business models as a way how the enterprise creates and delivers value to customers, and then converts payments received to profits, which is enabled or embodied in information technology.

In categorizing the value, created through digital business models, the following three categories should be considered: (1) cost value (price transparency, consumption-based pricing, reverse auctions, buyer aggregation, rebates, and rewards), (2) experience value (customer choice, personalization, automation, lower latency, and any device any time), and (3) platform value (marketplaces, crowdsourcing, peer-to-peer, sharing economy, and data monetization) (Agaard 2019). Therefore, how to create value is the main challenge given by the digital economy.

Areas of business change

Digital transformation can create sophisticated products and services. Increasingly, manufacturing firms are integrating technologies in established products to make them more "intelligent". The data generated from products in use may increase performance or give rise to new functionalities. Hence, the integration of digital technologies allows the firm and its customers to collect valuable data on the product. These data allow products to be monitored, optimized, controlled, and sometimes operate autonomously. This is in line with manufacturing firm trends and reduces the distance between the firm and the end user while increasing business revenue streams. Companies create value for their customers through digitalization based on the Internet of Things and Artificial Intelligence (reinforcement learning neural networks). For example, automotive firms are connecting their vehicles to the cloud to enable new complementary downstream services. Several firms also use digitalization and connected products to allow direct links to end users. Digitalization has triggered a new relationship marketing strategy and allowed the development of innovative services based on direct sales to end users.

Companies can develop more integrated value chains, which increase the efficiency of various firm functions and enable better control over the operations. The sharing of information among systems and functions improves coordination, visualization, and planning of important processes. Manufacturing firms share data not only within the firm but also across firm boundaries with suppliers and customers.

In other words, digitalization has changed the game for incumbents in many industries by introducing competition and collaboration from firms outside the industry and competition with more asset-light business models.

Digital transformation initiatives focus on leveraging greater customer engagement, bringing further flexibility and agility to standardized, and centralized operational processes, and providing new strategic opportunities to organizations by reconfiguring business models. It created new products and services, and, in some cases, disrupting and reinventing entire value chains and industries (Westerman and Bonnet, 2014). According to Berman (2012, p. 17), "to succeed in digital transformation leading companies focus on two complimentary activities; reshaping customer value propositions and transforming their operations using digital technologies for greater customer interaction and collaboration." It highlights the importance of building across-platforms, where all customer touchpoints are reached.

As instances of digital innovations, digital business models exhibit several characteristics that differentiate them from traditional business model innovations. First, due to the generativity that is unleashed by the flexibility afforded by the digital technologies on which they build, these innovations are intentionally incomplete and, thus, dynamic and adaptable. Second, digital business models are characterized by convergence in various approaches, which stem from the openness afforded by digital technologies (Yoo et al., 2012). The impact of digital technologies is seen in the way they automate, extend, and transform business models – for automation, it refers to how a firm uses digital technologies to automate or enhance existing activities; extension refers to how a firm uses digital technologies to support new ways of conducting business; transformation refers to how digital technologies are used to enable new ways of conducting business to replace traditional ones (Li 2020).

Digital business models improve the interaction and continuous dialogue with customers. We can assume that the customers do not know what they want or need in two years. Thus, the value is the data that can be monitored through customers' interaction in exploring their "unknown" preferences and behaviors. Firms also need to create customer value in a more uncertain external environment. Customer usage patterns and customer behaviors are exploratory and formative and are accompanied by a lack of market knowledge and embryonic competition among new solutions. Another pathway compiles the development of service business models pursued by most of the service companies and also by the manufacturing companies. Here, the emphasis is on providing added services for the customer while using data to help optimize and tailor existing or new offerings to the specific customer usage and needs. Digital business models comprise network-based business models and business experimentations, relating to co-creating and co-developing common business models and business experiments. They constitute network-based connection with stakeholders and customers. In general, there is a trend among the firms toward becoming more service-oriented to realize value from digitalization.

Difficulties and problems in changing the business model

The first point is that most companies ignore or underestimate the role of corporate culture in the transformation process, thereby making it a failure. In order to avoid this problem, sometimes, in the process of transformation, companies create in the organizational structure a separate branch for a digital transformation manager, to whom they transfer the powers of strategic digital development, which is seen as part of the company's strategy, and not the general company's strategy itself.

Such a solution could be ineffective, because: (i) the old culture has chosen the manager for digital transformation and he/she will act consistently with it; (ii) the proposals of the digital transformation manager will be associated with the risks of reducing revenue from typical activities, increasing costs, etc., leading to general resistance.

In this situation, the digital transformation manager has to overlay the digital transformation on the existing system. Thus, subsequent development will make the company, at best, a hybrid, but not digital. This can lead to a sort of Frankenstein's monster, not analogic, nor digital. This is because manufacturing firms typically do not have previous experience with the concepts and solutions they want to develop, and they involve product features and customer segments that are uncertain. Additionally, many IT department members lack experience in building new business; they have an informatic culture more than a digital one.

Moreover, in digitalization efforts organized to develop new customer solutions, problems arise in relation to coordination with sales and marketing functions to launch new products and services. This can create a cultural conflict between traditional and digital approaches to the customers.

Smaller companies seem to lack the understanding of the importance of developing new business models or adopting already tested business models, well aligned with the demands of the digital era. They have a lower adoption rates and they adopt digital technologies in a more fragmented manner that is often restricted to certain functional areas such as moving vital assets to the cloud or use of social media in engaging with their customers (Nwaiwu 2018).

Considering the two alternatives connected to the development of a digital business model, i.e. cost-cutting/operational efficiency and growth/innovation, it appears quite easy to understand that the most difficulty relates to the growth/innovation side. Actually, in this second case there are:

- no established mechanisms to allocate capital needed by several functions;
- no established culture to develop and sell services or use data to build new business;
- no control on data generated from customer applications to identify growth options;
- uncertain market demand;
- dependence from partners in the value eco-system;
- new managerial processes to be implemented;
- implementation time difficult to be estimated.

What is best for the firm will depend also on the stability of the market, which indicates whether there is a need for a larger transformation for the firm to remain competitive. In a stable market, digitalization aimed at achieving operational efficiency because the growth prospects may be too small to justify sacrificing current efficiency and profits. In a turbulent market digitalization is asked to increase competitive position of the company.

The pathway to a digital business model

First of all, it is necessary to create, as in every big cultural change, appropriate motivators that will shift the focus of activity to experiments, eradicate the "fear of error," make decisions and responsibilities from the individual to the group, and encourage the exchange of knowledge and intellectual leadership of both the employees and the company. A new generation of leaders should also be identified. In a company that is serious about the digital transformation of business processes, succession planning requires the development of new vision and capabilities. This point could represent a high cultural obstacle that needs to be overcome.

The speed of a company's innovation processes has been changed from the past. Many companies, worked on a project for 18 months, then they put it on the market, and then 18 months after that they could discover that the project was not meeting customers' needs. Now digitalized companies co-design from day one, with their customers, dealers, and consumers, to make sure that they are satisfied by the solution proposed. So, speed is a crucial issue in the digital business model.

Aagaard et al. (2019) point five pathways for companies in driving business models and digital transformation:

1. improving customer interaction by initiating customer self-service through digital technologies, increasing the experience value for customers;
2. digitalizing the value chain to obtain data and to automate the production process, increasing cost value and enabling experience value;
3. producing data for business intelligence using real-time data obtained from products to increase external awareness;
4. developing service business models enabled through the increased usage of digital technologies, such as prototyping face-recognition for future payment options;
5. creating network-based digital experimentation, for cooperating with network actors to achieve a successful digital business model innovation by utilizing external capabilities and resources.

It seems clear that digitalization will entail ongoing complex organizational and managerial challenges. Its design is important because it has been shown to be a crucial enabler of dynamic capabilities and the capacity to sense changes, seize opportunities, and to transform a firm.

For example, Scania has reallocated the IT department's responsibilities among various firm functions, and every department in the firm includes data scientists. The result of these changes to the organizational structure is that all the firm functions are involved in the digital transformation and are less dependent on input from software developers and data scientists from other functions in their efforts to explore and develop new customer solutions (Björkdahl 2020).

Many of the digital firms cooperate outside the firms' traditional boundaries for the development of technological solutions. This means that firms increasingly need to be involved in more dynamic value constellations, where different actors from multiple industries contribute to firms' value creation, and where parts of the firms' value chain activities shift to new actors.

In order to develop a digital business model, managers should answer to some questions. For example:

• Why could digital technologies help to solve a company's problems and increase value for the customers?

- Where do digitalization creates the most value?
- What enables and what abilities are needed to support digitalization?
- How is the company approaching an operating model, culture, and processes to drive the transformation?

Frequently, business leaders invest in searching for growth in adjacent areas that have little overlap with the current business, rather than using digitalization to improve existing business to achieve growth. A strategic digital transformation strengthens the core business, making it costly and difficult for other firms to compete. In other words, the pursuit of new value propositions requires a transformational approach to customer engagement, where unmet customer needs can also be identified and met through digital transformation.

The keys to the development of a digital business model should consider, by mean of the digital technology, *personalization* of customer preferences, *asset sharing* with other organizations, *usage-based pricing*, and a *collaborative ecosystem* answering to the increase in demand of product and services, and *agility* adapting the company to the fast changing environment.

Conclusions

First of all, in defining a new business model for digitalization, there is no "one-size-fits-all solution," and firms need to have dynamic capabilities to achieve agility and address the business environment changes caused by digitalization itself.

The most important question to be addressed is why digitalization is important for a firm, that is, why and how digitalization might allow a firm to create and capture more value. Successful firms are starting by asking why digital technologies and data will help them to solve their main challenges in obtaining value for the company and its customers (e.g., customer experience, employee experience, and business decisions).

While organizations have always valued the importance of customer insight and opinion into respective product and service development, the issue is now challenging management to think very differently about their relationship with the customer. Instead of being casual onlookers, organizations seeking to implement effective digital technologies will need to ensure customers become active participants, where insights directly influence the product and service on offer. This obviously challenges the power and control held by many organizations and opens management to external influence, and indeed scrutiny.

A radical change is due to the need of a double direction integration of the company, both vertical and horizontal, of the business model. The typical pre-digital stand-alone business model may not lead to success in digital transformation.

Actually, digital transformation requires a change of the vision of the company, a change in the business culture of the company, a change of the relationship with the customers, according to the digital economy. There is a shift of importance from products to functions and activities and from products to services, where the products are part of the service.

References

Aagaard, A., Andersen, T.C., and Presser, M. (2019), "Driving business model innovation and digital transformation through Internet-of-Things," *ISPIM Innovation Conference*, Florence, Italy.

Berman, S. J. (2012), "Digital transformation: Opportunities to create new business models," *Strategy & Leadership*, 40(2).

Björkdahl, J. (2020), "Strategies for digitalization in manufacturing firms," *California Management Review* 2020, 62(4).

Chaniot, E. (2019) "Tools for transformation: Michelin's digital journey," *Research-Technology Management*, 62(6).

Cinquini, L. and Di Minin, A. (2013) *New business models and value creation: A service science perspective.* Basingstoke: Springer.

Davenport, T. H. and Westerman, G. (2018), "Why so many high-profile digital transformations fail," *Harvard Business Review*, March.

Fichman, R. G., Dos Santos, B. L., Zheng, Z. E. (2014), "Digital innovation as a fundamental and powerful concept in the information systems curriculum." *MIS Quarterly*, 38(2), 329–A15.

Hanelt, A., Leonhardt, D., Hildebrandt, B., Piccinini, E., and Kolbe, L. M. (2018), "Pushing and pulling – digital business model innovation and dynamic capabilities" *Journal of Competences, Strategy & Management*, 10, 55–78.

Li, F. (2020), "The digital transformation of business models in the creative industries: A holistic framework and emerging trends," *Technovation*, 92–93.

Loonam, J, Eaves, S, Kumar, V, and Parry, G. (2018) "Towards digital transformation: Lessons learned from traditional organization," *Strategic Change*, 27, 101–109.

Lucas, H. C. and Goh, J. M. (2009) "Disruptive technology: How Kodak missed the digital photography revolution," *Journal of Strategic Information Systems*, 18(1), 46–55.

Merali, Y., Papadopoulos, T., and Nadkarni, T. (2012), "Information systems strategy: Past, present, powerful concept in the information systems curriculum," *MIS Quarterly*, 38(2), 125–153.

Nwaiwu, F. (2018), "Analysis of emerging business models of companies in the era of the digital economy," *Journal of Sustainable Development*, 8(20), 18–27.

Schallmo, D., Williams, C., and Lohse, J. (2019) "Digital strategy: Integrated approach and generic options," *ISPIM Innovation Conference*, Florence, Italy.

Semples, Christophe and Hoffmann, Jonas (2013). *Sustainable innovation strategy: Creating value in a world of finite resources.* Basingstoke: Springer.

Von Leipzig, T., Gamp, M., Manz, D., Schöttle, K., Ohlhausen, P., Oosthuizen, G., and Von Leipzig, K. (2017). "Initialising customer-orientated digital transformation in enterprises". *Procedia Manufacturing*, 8, 517–524.

Westerman, G., Bonnet, D., and McAfee, A. (2014). *Leading digital: Turning technology into business Transformation.* Boston: Harvard Business Review Press.

Yoo, Y., Boland, R. J., Lyytinen, K., and Majchrzak, A. (2012), "Organizing for innovation in the digitized world," *Organization Science*, 23(5), 1398–1408.

Chapter 7

From disruptively digital to proudly analog

A holistic typology of digital transformation strategies

Zeljko Tekic and Dmitry Koroteev

Digital transformation is everywhere

Sixty years after the inception of the digital era and a decade after its development accelerated at incredible speed, business is dominated by intangible and knowledge-based assets, new products are developed in ever-shorter cycles, customers are well informed and globally connected, and venture capital is widely available. This new reality is populated by new global companies started by teams of two to five talented persons from anywhere in the world and with few financial resources, developed using new low-cost methodologies. The startups serving the needs of the digital native population have grown from zero to one (Thiel & Masters, 2014), creating new value, and then from one to one billion, scaling globally in just a couple of years. However, the new digital reality is not equally fertile for everyone; many established companies with long traditions and successful pasts have been struggling to change in a timely manner and have lost value; some even have disappeared. These companies were prevented from evolving by their knowledge and capabilities— built over years from handling problem identification and problem-solving tasks in the nondigital environment—and a reliance on old beliefs (Henderson & Clark, 1990). To increase chances for survival and success in the market, many companies across industries that failed to evolve started a major transformational change: trying to substantially integrate digital technology into their businesses. This organizational change is known as *digital transformation*.

Digital transformation is a multifaceted phenomenon in that it has different aspects/ implications for different companies. For one group of companies, it is about adopting new technologies, like the Internet of Things (Caro & Sadr, 2019); for another group, it is about using social media to engage with users to gain their feedback or even open new channels of sale (Kaplan & Haenlein, 2010); for some, it is about a completely new way of doing business (Crittenden, Crittenden, & Crittenden, 2019). These multiple layers are present on the other levels of observation as well. Some companies see digital transformation as a way to optimize processes and cut costs, while others view it as an opportunity to create new value by offering products and services that have never existed before; some companies see digital transformation as a change in a profile of people they employ, while others view it as a need to find and serve new customers. All these perspectives could be valid and correct.

Rooted in the importance and complexity of the phenomenon, interest in digital transformation is booming in the academic and practitioner communities. Scopus—one of the two largest abstract and citation databases—records 1,155 academic articles with the term 'digital transformation' in the title, abstract, or keywords written in the English language over the last 30 years (1989–2018)[1]. Of these results, 749 articles (82%) were published during the last 3 years (2016–2018), with 485 articles (42%) in 2018 alone. At the same time, as of January 10, 2019, a Google search returned more than 29 million hits for the same term, reflecting an ever-increasing number of consultancy reports, surveys, specialized websites, executive and educational courses, and case studies offered across the world.

However, too many diverse perspectives on digital transformation create problems understanding and evaluating strategic choices and their consequences on performance. In this article, we develop a typology of digital transformation strategies, reducing the complexity of the real world to a small number of richly defined types; this methodology follows examples from the literature in related fields (e.g., Paswan, D'Souza, & Zolfagharian, 2009; Taran, Boer, & Lindgren, 2015). By developing this typology, we aim to create useful heuristics and provide a systematic basis for comparison (Smith, 2002) in order to support strategic decision makers in understanding the big picture of digital transformation and in identifying and analyzing various options.

Taking the big picture

Bharadwaj, El Sawy, Pavlou, and Venkatraman (2013) define *digital business strategy* as an organizational strategy formulated and executed by leveraging digital resources to create differential value, and identify four key elements—scope, scale, speed, and source of business value creation/capture—that should guide thinking about it. However, it is observed that digital business strategy alone is not enough, and that firms typically need a standalone digital transformation strategy to help managers navigate through the transformation process (Hess, Matt, Benlian, & Wiesböck, 2016). The recently developed body of academic and practitioner literature is helpful here, but these works are dominantly focused on specific aspects—challenges, drivers, and failures of previous attempts of digital transformation (Ismail, Khater, & Zaki, 2017)—and/or certain industry sectors, and typically misaddressing the big picture (Hess et al., 2016).

Although strategy has been recognized as a driving force of digital transformation (Hess et al., 2016; Kane, Palmer, Phillips, Kiron, & Buckley, 2015; Matt, Hess, & Benlian, 2015), only a few recent contributions take a more holistic approach to the topic, each offering different dimensions for analyzing digital transformation. Matt et al. (2015) claim that, independent of the industry or firm, digital transformation strategies can be analyzed through four dimensions: use of technologies, changes in value creation, structural changes, and financial aspects. At the same time, one of the largest surveys among CEOs concludes that customer experience, operational processes, and business models are three pillars that should be focal points of digital transformation initiatives (Fitzgerald, Kruschwitz, Bonnet, & Welch, 2014). Westerman, Bonnet, and McAfee (2014) describe an additional three elements inside each of these three pillars, for a total of nine elements of digital transformation. Andriole (2017) outlines and discusses five myths about digital transformation.

The outcome is increased variability and diversity of topics covered, constructs used, and relationships between them. This translates into unclear and blurry understanding of the whole of digital transformation. The result is high uncertainty in decision-making, sometimes preventing successful identification and analysis of strategic options in digital transformation, their comparison, and evaluation of their consequences. In such circumstances, typologies, classifications, and taxonomies serve as useful tools in science and business practice to reduce the complexity of the real-world phenomenon (Meyer, Tsui, & Hinings, 1993; Smith, 2002). Hence, our efforts in this article are directed toward building a typology of digital transformation strategies based on a small number of determinants, aiming to support better understanding and strategic decision making.

Tastiest ingredient and master recipe for digital transformation

Digital technology is a business transformation enabler: a means for achieving strategic and powerful ends of digital transformation, not an end of digital transformation itself. Conversely, a business model is a digital transformation driver: a factor that causes transformation to succeed or not. A business model illustrates the logic of how to do business with new means. Digital technologies significantly accelerated experimentation with business models (Massa & Tucci, 2013), opening up new opportunities for organizing business activities. Although it may be true that "a better business model will beat a better idea or technology" (Chesbrough, 2007, p. 12), it is "trivial technology that screws up your business model" (Chesbrough, 2010, p. 358). Using the terminology of sweet delights, digital technology is the tastiest ingredient needed for making a digital transformation cake, while the business model is a master recipe for making the cake. If either of these two—the key ingredient or the recipe—is not optimal, the cake will not be worth serving. Based on this, we use digital technology and the business model as two distinctive dimensions for building our typology.

Digital technologies

The beginning of the digital era can be associated with the synergetic effect of two groundbreaking events at Bell Labs: (1) the invention of the transistor by William Shockley and colleagues in 1947 and (2) the pioneering work in information theory by Claude Shannon in 1948. Since then, progress in developing digital technologies has been systematic, with steadily increasing performance starting with the integration of digital logic circuits and continuing with the development of the microprocessor and computer, digital cellular phone, and internet. These and many other technological innovations have enabled successful and omnipresent representation of our analog reality through digital signals (i.e., encoded series of zeroes and ones). When it comes to manipulation and processing, digital signals have several important advantages over analog. Being more resistant to noise and easier to amplify, a digital signal can be replicated cheaply and more reliably, transmitted farther, more conveniently saved and retrieved, easily moved between media, and remotely accessed or distributed. In addition to many other factors, the properties of a digital signal enabled the creation of information and communication technologies, which significantly changed how the world looks and works—especially over the last 20 years or so.

At the end of the second decade of the twenty-first century, technologies like artificial intelligence, machine learning, big data analytics, cloud computing, the Internet of Things, and social media are the main enablers of digital business. They dilute borders between the world of physical (tangible) artifacts and cyber (intangible) space, making it possible for bits, atoms, and even cells to combine in new and interesting ways that previously were impossible.

Typically, these digital technologies do not displace and replace existing components; rather, they enhance and digitize them, turning them into data sources and enabling novel connections and recombinations (Iansiti & Lakhani, 2014). Digitization changes components, creating possibilities for new interactions and new links with other components in the surrounding, enabling the process of digital transformation. However, a component's digitization usually does not change the core design concept behind each component. Currently, the third wave of digital technologies (Case, 2017) moves from the internet and the traditional infrastructure and enters the highly regulated sectors of health, energy, transport, or finance.

The dramatic development and diffusion of digital technologies, paired with the potential to disrupt many fields of our lives, made them—individually and synergistically—the core tools and concepts of digital transformation. Companies from all over the world from low- and high-tech industries, as well as universities and government organizations, are investing in them with the goal of acquiring knowledge and building capabilities to implement the next generation of business solutions.

Business models

Creating and delivering value to a customer utilizing digital technologies is just half of the job. To be successful, companies must organize themselves to capture a viable portion of the value created. A business model is a systemic and holistic description of three key activities: how an organization creates, delivers, and captures value. Every organization uses a particular business model. Sometimes it is purposely designed, sometimes it spontaneously emerges, but it always exists. A business model defines a problem for an organization to solve and attempts to answer some/all of the following questions: for whom is the problem solved; what value is delivered; what technology is involved; who pays for the solution, and how much; how is the solution delivered; how will money be collected; and how should an organization be organized to best serve all of these activities. In this view, the business model is understood as a manageable device that mediates the value creation process between the technical and the social domains (Chesbrough & Rosenbloom, 2002).

Articulating the underlying business logic of a firm's go-to-market strategy (Teece, 2010), a business model must consider both individual customers and a complex network of exchange partners (Massa & Tucci, 2013). It must describe the position of the firm within the value network, including identification of potential complementors and competitors (Chesbrough & Rosenbloom, 2002). Describing aspects related to value creation, delivery, and capture, a business model also outlines the architecture of the firm. It prescribes how the firm should be organized and how it should shape its network of partners for creating, marketing, and delivering created value in order to generate profitable and sustainable revenue streams (Osterwalder, Pigneur, & Tucci, 2005). Thus, a business model is seen as the organizational and financial architecture of a business (Chesbrough & Rosenbloom, 2002; Teece, 2010).

Business model development is an iterative process. The right formula is rarely known up front, as it cannot be fully planned on paper. Why? Initial ideas about customers' needs, value proposition, product, delivery channels, number of customers, and other characteristics are just hypotheses—educated guesses, more or less—based on an entrepreneur's or manager's experience, education, and context (Blank, 2013; Chesbrough & Rosenbloom, 2002; Teece, 2010). In order to develop a business model, we need this initial set of hypotheses to provide a starting point for a discovery-driven process through which an organization tests and retests its initial assumptions and experiments with different approaches and prototype solutions, thus facilitating optimally quick and cheap learning and hypothesis validation (Blank, 2013; Massa & Tucci, 2013; Ries, 2011).

A good business model provides considerable value to the customer, enables the implementing organization to collect a viable portion of created value through revenues, and is hard to imitate (Teece, 2010). But building a good business model is not easy. Both processes—designing a new business model for a newly formed organization and reconfiguring an existing business model of an existing organization—are challenging, with specific issues and tasks (Massa & Tucci, 2013). They require creativity and a lot of hands-on field research work focused on customers, competitors, and suppliers, with the goal of extracting useful insights, information, and intelligence (Teece, 2010).

A holistic typology of digital transformation strategies

Based on our discussions, we posit that digital transformation strategies can be characterized in terms of two dimensions: (1) level of mastery of digital technologies relevant to the sector in which the company competes (high or low) and (2) level of business model readiness for digital operation (high or low). The results show four generic digital transformation strategies: disruptive, business model led, technology led, and proud to be analog (Figure 7.1).

Below, we describe these four types of digital transformation strategies. We pay particular attention to the following aspects of digital transformation: primary target of

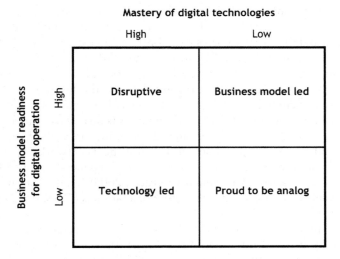

Figure 7.1 Typology of digital transformation strategies.

transformation, leadership style, importance of skills like creativity and entrepreneurial spirit among employees, risks and challenges faced in the process, and the consequences of failure. We also make improvement recommendations for each of the four types of digital transformation strategies and offer generic tactics for success. Table 7.1 summarizes the main characteristics of the four types of digital transformation strategies.

Type 1: disruptive digital transformation

Disruptive digital transformation is characterized by a high level of business model readiness for digital operation and a high level of mastery of digital technologies relevant to the sector in which the company competes. Companies that follow this strategy—so-called *disruptors*—are always newcomers to the industry/sector and typically are startups. Why? Because when it comes to disruption, startups have one significant advantage: they have nothing—no resources, no employees, no customers, no suppliers, no factories, no brands, no established routines or rules how to do business, no CEOs, no CFOs, no CDOs, no commitments, and no loans. Having nothing allows everything to be questioned and experimented with to find a fundamentally new way of doing business in the industry, to change the logic of an entire industry, and to become the standard for the next generation of products and companies. Startups do the magic of turning nothing into disruption by using a get-out-of-the-building approach (Blank, 2013), engaging in an intensive interaction with the environment (i.e., users, suppliers, stakeholders) to understand needs/problems/pains and how to serve/solve them in a sustainable way. As startups experiment, proving and disproving the hypothesis of its business model, the initial business model undergoes continuous change. Changes—especially failures—are natural for startups and are responded to by making small adjustments (iterations) or more fundamental alternations (pivots) with only one goal in mind: not simply to find a good business model but to find one that is repeatable and scalable (Blank, 2013).

If a business model has a significant degree of novelty or uniqueness—at least for the sector in which it is applied—the main ingredients for disruption are there. Digital technologies enable repeatability and scalability, and make possible (1) a significant degree of novelty or uniqueness by turning existing components into data sources, and (2) new connections in novel ways. Why are established companies less able to transform by disrupting? It is because of challenges idiosyncratic to existing organizations (Massa & Tucci, 2013), such as organizational inertia, management processes, and path-dependent constraints in general—which may not be easily resolved or ignored.

Disruption does not happen often, but when it does, it changes the landscape of the sector completely. Some of the most successful global companies (e.g., Uber, Amazon, Facebook) and winners on the big Chinese market (e.g., Tencent, Alibaba) are examples of companies that have succeeded with this type of digital transformation. Consider also the Russian Tinkoff Bank, which started with the founder's vision to build a 100% digital and branchless bank in 2006 and to become the world's largest fully online bank (Tinkoff, 2019). Tinkoff Bank disrupted the large Russian financial market—typically perceived as conservative, bureaucratic, and state controlled—by focusing on underserved digital native users and by utilizing digital technologies to solve their most common problems (e.g., long waiting lines, papyrology). Today, Tinkoff is one of the most profitable banks in the world and describes itself as a "tech company with a banking license" (Finance Disrupted, 2018), since 70% of HQ employees are IT specialists and most business processes are powered by machine learning and AI (Tinkoff, 2019).

The primary goal of the disruptive digital transformation strategy is to change the value proposition in well-established markets by altering how value is created, delivered, and/or appropriated, making existing products irrelevant. The starting point in this endeavor is identification of the right problem and the best user group to start with—the so-called *beachhead* market—in order to provide fast and quantifiable learning and an efficient bridge between tech-savvy early adopters and cautious mainstream consumers (Moore, 1991).

Companies that employ this type of digital transformation strategy start with a *founder's vision*—an idea of what is needed in the market and a dream of how to satisfy that need—and are fully led by the vision. The strength of the vision and the mission of the future company is what makes cofounders work for low or no salary; what brings early employees on board; and what helps in attracting initial funding, media attention, and first customers.

Typical risks/challenges include failing with experiments slowly and not pivoting ideas on time, scaling up any aspect of an organization too early, and building a functional organization before a business model is validated. Finally, careful management and timely supply of venture capital are needed to fuel disruption.

The consequences of failure under disruptive digital transformation are limited. Nothing will be hurt or lost, as nothing existed at the beginning. Co-founders will gain experience in learning from failure and venture capitalists always calculate risk before investing.

Our tactics for improvement: Fail fast and fail cheap. Fail as many times as needed. Failure is learning. Failure allows for great success.

Type 2: business model led digital transformation

Business model led digital transformation is characterized by a low level of mastery of digital technologies relevant to the sector in which the company competes and a high level of business model readiness for digital operation. Companies that use this generic type of transformation usually come from competitive milieus and are under pressure to transform in order to survive. Eroding profit margins and shrinking markets are major motivators to transform. In many situations, these companies need to change something, if not everything (Andriole, 2017), and they should do it quickly. The primary focus is on understanding new logic in doing business, frequently considering examples of similar successful companies in other geographical locations or in a similar sector (e.g., banks, insurance companies). Only once a new business model is understood do firms undergoing business model led digital transformation try to fill in tech gaps.

Companies of this type will predominantly reside in B2C sectors and emanate from industries with low barriers to entry (e.g., retail, telecom) or those that can be wholly digital (e.g., consumer financial services, media). For example, Philips Lighting and Cofely started selling light as a service on a pay-per-lux basis instead of selling installations, bulbs, and electricity, while Michelin moved from selling tires to offering a service that guarantees performance (World Economic Forum, 2019). Sberbank, the largest state-owned bank in Russia, is another example. Under the strong leadership of then-new CEO Herman Gref, Sberbank began its transformation from an old-fashioned, bureaucratic, Soviet-style bank into a modern international player. While facing internal inefficiencies and the arrival of new technologies, the bank was challenged by strong competitors from the West as well as by local contenders like Tinkoff Bank. After successfully completing the initial transformation and following best international practices, Sberbank is now looking to build an ecosystem outside of its core banking activities. Over the last few years, the bank stepped into different sectors (FinTech Futures, 2018), including: healthcare (DocDoc app, which links patients with doctors and clinics); real estate (domclick.ru, which manages all aspects of buying/renting a property); biometrics (invested in VisionLabs); cybersecurity (created its own company, BI.Zone); and e-commerce (created a joint venture with Yandex).

The primary objective of the digitalization efforts is exploration of new opportunities: finding and connecting new and enriching existing resources (i.e., products, people, platforms) and creating new schemes, especially when they lead to creating completely new products and opportunities to become a platform or a valuable add-on to a platform.

Companies that employ a business model led digital transformation strategy can successfully go through transformation if strong leadership is demonstrated from the top, primarily regarding empowering employees, resolving inertia, communicating new vision, and providing support in crises that may occur. In this scenario, entrepreneurial spirit is in high demand but typically not available. To fill the gap, companies scout for startups by running accelerator programs and hackathons, and organizing innovation labs. These activities are done in collaboration with a startup community and existing accelerators. The goal of these activities is not to search for a specific technology but rather to broadly scan for new, entrepreneurially oriented employees and for technologies that may contribute to the changing business model. The typical risk/challenge is to identify the useful parts of the strategy and to discard those that are not (i.e., useful to be reinforced, useless to be unlearned). Further challenges include dealing with employees' inertia and lack of support for changes to the way things have always been done; a lack of understanding regarding open innovation and the need to establish sustainable links with external partners; and making money out of data. In some cases, clients are not initially ready to switch to digital products and services, especially ones from the predigital era. Therefore, effort spent on educating such users is essential.

The consequences of failure under business model led digital transformation may be disastrous, especially if the transformation efforts started late. In competitive and disrupted industries, a limited window of transformational opportunity exists, and sometimes no second chance. In the words of Accenture CEO Pierre Nanterme (2016): "Digital is the main reason just over half of the companies on the *Fortune* 500 have disappeared since the year 2000." A majority of these companies would fit this generic transformation type.

> Our tactics for improvement: Take as much as possible from the disruptors. Copy, adjust, learn whatever is possible, and use your existing advantages (e.g., installed base, market share, brand, established channels/partnerships) to compete and fight back. Build capacity for open innovation as soon as possible. Plan and pursue massive educational activities/programs for both clients and employees, targeting and demonstrating the actual benefits of going and being digital.

Type 3: technology led digital transformation

Technology led digital transformation is characterized by a high level of mastery of digital technologies relevant to the sector in which a company competes and a low level of business model readiness for digital operation. Companies with these characteristics invest more eagerly in new technologies because they believe this pathway is less risky, more predictable, and easier to justify than experimenting with their own business models. This is truer for companies with previous success, especially if they are public (Andriole, 2017). They feel less affected and more cautious about change.

Many B2B and some B2C companies, particularly in industries perceived to have higher barriers to entry (e.g., higher education, legal services, healthcare) or that cannot digitalize their whole operation (e.g., oil/mineral extraction companies), are in this group. For example, heavy machinery producers such as Caterpillar, Komatsu, and GE equip their bulldozers, turbines, and other large engines with a number of different sensors to monitor temperature, vibrations, and other conditions; this allows the discovery of failure threats and preventive maintenance before collapse happens (Board of Innovation, 2019). In a similar manner, oil and gas companies like Russian giant Gazprom Neft profit from digital technologies. The company utilizes tools for pattern recognition in order to process seismic images and well logs, classifiers based on machine learning for optimization of directional drilling, and AI-aided tools for fast modeling of reservoir development. These and many other digital technologies enable Gazprom Neft to reduce costs and risks.

The primary target of the digitalization efforts is optimization and cost reduction: better exploitation of existing resources inside existing schemes using new technologies. The likely outcome is that a company will use individual digital technologies to solve discrete business problems (Kane et al., 2015). This outcome may offer positive results in the short term but, overall, the improvement is limited and marginal.

Leadership often adopts the concept of digital transformation but is very risk averse, and thus does not support it wholeheartedly. Reasons for that may differ, but typically center around not feeling the pressure of failure and the need to change. Change is not a natural state for humans or companies; it is risky, painful, time-consuming, and expensive (Andriole, 2017). Successful companies—especially market leaders and companies from industries with higher margins—need time and the fear of failing to begin transforming their business models. Companies that apply this generic type of transformation usually are characterized by mixed skillsets of employees and dominated by emotion-free, hard-skill professionals from the predigital era. Creative and entrepreneurial employees typically do not fit well and are underutilized. If put in the leadership role, entrepreneurs may be counterproductive by creating more digital initiatives than organizationally manageable and thus become a source of new crises.

The most frequent risk/challenge of this type of digital transformation is that initially positive results from using new technology for the singular purpose may incorrectly be interpreted as the promise of greater value, and seen as digital transformation itself. This can lead to greater investment in the same, only later to realize the result was a false positive. Other challenges are mainly associated with the conservative mindset of employees.

The consequences of failure under technology led digital transformation are neither disastrous nor negligible; they fall somewhere in between. This is especially true in the case of previously successful companies from publicly owned or monopolistic sectors. Backup Plan B, which must be started on time, entails changing gears and reconsidering the business model.

Our tactics for improvement: Allow and promote a bottom–up approach. Empower small groups of employees to conduct unofficial, unplanned-but-tolerated experiments that can result in identifying new pathways and more than just new technology-oriented digital transformation.

Type 4: proud to be analog

The proud to be analog strategy is characterized by a low level of business model readiness for digital operation and a low level of mastery of digital technologies relevant for the sector in which the company competes. The main characteristic of these companies is that their key products are valued by customers because they are analog: handmade, human inspected, and/or built exclusively or in very small batches. In contrast to disruptive digital transformation, there is no value in scaling up production and selling more of the same, no value in automation, and no value in replacing humans in production. The opposite is true: if any of these practices are employed, they will decrease the value of the company and put it in danger since consumers' needs for uniqueness and individuality—the primary reasons for the purchase of luxury brands—will not be met (Wilcox, Kim, & Sen, 2009). Companies of this type pinpoint analog, traditional, and handmade as main messages and unique sales points in a world of homogeneous competitors and use extreme differentiation as a competitive strategy.

Typically, companies of this type are privately owned and have a long tradition (e.g., Rolex, Rolls-Royce). They are exclusively B2C companies, with a business model that assumes niche clients will only accept robot-free and automation-minimal produced goods. If we think about companies that produce the finest, most expensive luxury watches, jewelry, chocolate, suits, shoes, porcelain, and cars, we get an idea about companies that fit this type. As Coco Chanel famously said: "Luxury is a necessity that begins where necessity ends." It is clear that luxury products—and especially extreme luxury products—are not built to satisfy our need, but our desire. They are not purchased to offer superb functionality or to fit cost/benefit ratios, but rather to be a status symbol for the purpose of emotional satisfaction.

The primary objective here is to identify parts of the business that could be digitalized and changed without jeopardizing the core of the business that has to stay analog. Typically, this will include communication channels and quality control of input materials. Because experience in buying—like packaging and handling these luxury products—matters to overall exclusivity and uniqueness perception, these companies will rarely sell online but instead try to help customers find the closest retailer. Digital efforts will focus on communicating brand image and building stories around it, as people do not decide to buy a Rolex watch, Crockett and Jones shoes, or a Fabergé egg because they have seen them advertised on television commercials. Companies that pursue this type of strategy try to modernize their brand identity by mixing their rich heritage and tradition with modern lifestyles and desires, and by telling their stories via social media to influence consumers' future purchase decisions.

Companies that employ the proud to be analog strategy are rooted in tradition (e.g., a business owned over the years by several generations of one family). They are small to medium size firms. Brand is the most valuable asset, and all innovation steps are undertaken with extreme caution. These characteristics favor risk-averse leadership style and selection of employees.

Since luxury goods sales will be significantly influenced by millennials by 2025 (D'Arpizio & Levato, 2017), the risks/challenges of this type of transformation revolve around the transition from pre-digital users to digital native users. Does the target audience understand the value proposition? Is it appealing to digital natives? How should efficient channels of communication be established?

The consequences of failure under proud to be analog are neither disastrous nor negligible; they fall somewhere in between. This is particularly true in the case of super luxury brands.

> Our tactics for improvement: Experiment through a separate entity. Partner with a digital native company to create luxury products for the digital native generation.

Toward more solid conceptual ground

In this chapter, we presented a typology of digital transformation strategies based on two critical dimensions: (1) usage of digital technologies and (2) readiness of a business model for digital operation. To the best of our knowledge, this is the first attempt to typify digital transformation strategies and provide a conceptual abstraction of the complex phenomena. The developed typology represents our effort to systematize and bring meaning into the booming field of digital transformation research and practice.

As with all typologies, this one is built using mental constructs rather than empirical cases. From a research perspective, the next logical step would be to test the robustness and generalizability of the framework. The proposed typology offers grounds for developing more theoretical approaches in understanding and explaining digital transformation as well as methods for measuring it. Future research efforts should consider that direction, too.

From a managerial perspective, our study and developed typology provide new insights and several important implications for managers engaged with digital transformation at strategic and operational levels. The typology creates useful heuristics and provides a systematic basis for comparison; thus, it helps managers to approach digital transformation more systematically. The four generic types differ in the primary motivation and target of transformation, leadership style, importance of skills like creativity and entrepreneurial spirit among employees, risks and challenges faced in the process, and the consequences of failure. Differentiation across these aspects should allow managers to circumvent an ad hoc manner of decision-making and to become more systematic and careful in identifying, estimating, and assessing the options in front of them. This, in turn, should lead to better resource management and the growth of new capabilities, creating a pathway to increased performance levels and higher returns of the upfront investment in digital transformation.

Table 7.1 Main characteristics of digital transformation strategies

Main characteristics	Types of digital transformation strategies			
	Disruptive	*Business model led*	*Technology led*	*Proud to be analog*
Primary target of transformation	Substantial change of the value proposition	Exploration of new opportunities	Optimization and cost reduction	Identification of parts of a business that could and should be digitalized
Leadership type	Vision led	Vision led	Risk avoidance led	Risk avoidance led
Creativity and entrepreneurial spirit among employees	Crucial for success and main fuel of the company	In high demand, but typically not available inside the company	Typically underutilized, sometimes even counterproductive	Not in high demand as all innovation steps are done with extreme cautions
Typical risks and challenges	Failing with experiments slowly; scaling up too early	Recognizing which parts of firm's knowledge base are useful and needed, and which are not	Using individual digital technologies to solve discrete business problems	Transiting from predigital generation of users to digital native users
Consequences of the failure	Minimal	Very high, may be fatal	Medium	Low to medium
Tactics for improvement	Fail fast and fail cheap	Copy from the disruptor as much and as quickly as possible	Allow and promote bottom-up approach in selected cases	Experiment through partnering with digital native companies
Companies pursuing this strategy	Dominantly startups from B2C sector	Dominantly from B2C sector (e.g., consumer financial and insurance services, retail, telecom, media)	Dominantly from B2B sector (e.g., oil and mineral extraction companies, heavy machinery, legal services, healthcare)	Exclusively from B2C sector (e.g., producers of the finest luxury watches, jewelry, suits, shoes, porcelain, cars)

Acknowledgments

Reprinted with permission from: Tekic, Z., & Koroteev, D. (2019). From disruptively digital to proudly analog: A holistic typology of digital transformation strategies. *Business Horizons*, 62(6), 683–693. All rights reserved: Copyright Elsevier, 2019.

Note

1   Excluding articles from the subject area of mathematics, in which digital transformation has a different meaning.

References

Andriole, S. J. (2017). Five myths about digital transformation. *MIT Sloan Management Review*, *58*(3), 20–22.

Bharadwaj, A., El Sawy, O., Pavlou, P., & Venkatraman, N. (2013). Digital business strategy: toward a next generation of insights. *MIS Quarterly*, *37*(2), 471–482.

Blank, S. (2013). *The four steps to the epiphany: Successful strategies for products that win* (2nd ed.). Pescadero, CA: K&S Ranch.

Board of Innovation. (2019). *Digital transformation examples*. Available at www.boardofinnovation. com/staff-picks/digital-transformation-examples/

Caro, F., & Sadr, R. (2019). The Internet of Things (IoT) in retail: Bridging supply and demand. *Business Horizons*, *62*(1), 47–54.

Case, S. (2017). *The third wave: An entrepreneur's vision of the future*. New York, NY: Simon & Schuster.

Chesbrough, H. (2007). Business model innovation: It's not just about technology anymore. *Strategy and Leadership*, *35*(6), 12–17.

Chesbrough, H., & Rosenbloom, R. S. (2002). The role of the business model in capturing value from innovation: Evidence from Xerox Corporation's technology spin-off companies. *Industrial and Corporate Change*, *11*(3), 529–555.

Chesbrough, H. W. (2010). Business model innovation: Opportunities and barriers. *Long Range Planning*, *43*(2/3), 354–363.

Crittenden, A. B., Crittenden, V. L., & Crittenden, W. F. (2019). The digitalization triumvirate: How incumbents survive. *Business Horizons*, *62*(2), 259–266.

D'Arpizio, C., & Levato, F. (2017, May 18). The millennial state of mind. *Bain and Company*. Available at www.bain.com/insights/the-millennial-state-of-mind/

Finance Disrupted. (2018). *Oliver Hughes, CEO, Tinkoff Bank*. Available at www.financedisrupted. com/oliver-hughes-ceo-tinkoff-bank

FinTech Futures. (2018, February 2). *Case study: Sberbank – tech them on*. Available at www. bankingtech.com/2018/02/case-study-sberbank-tech-them-on/

Fitzgerald, M., Kruschwitz, N., Bonnet, D., & Welch, M. (2014). Embracing digital technology: A new strategic imperative. *MIT Sloan Management Review*, *55*(2), 1–12.

Henderson, R. M., & Clark, K. B. (1990). Architectural innovation: The reconfiguration of existing product technologies and the failure of established firms. *Administrative Science Quarterly*, *35*(1), 9–30.

Hess, T., Matt, C., Benlian, A., & Wiesböck, F. (2016). Options for formulating a digital transformation strategy. *MIS Quarterly Executive*, *15*(2), 123–139.

Iansiti, M., & Lakhani, K. R. (2014). Digital ubiquity: How connections, sensors, and data are revolutionizing business. *Harvard Business Review*, *92*(11), 90–99.

Ismail, M. H., Khater, M., & Zaki, M. (2017). *Digital business transformation and strategy: What do we know so far?* [Working paper]. Available at https://cambridgeservicealliance.eng.cam.ac.uk/news/2017NovPaper

Kane, G. C., Palmer, D., Phillips, A. N., Kiron, D., & Buckley, N. (2015, July 14). Strategy, not technology, drives digital transformation. *MIT Sloan Management Review*. Available at https://sloanreview.mit.edu/projects/strategy-drives-digital-transformation/

Kaplan, A. M., & Haenlein, M. (2010). Users of the world, unite! The challenges and opportunities of social media. *Business Horizons*, *53*(1), 59–68.

Massa, L., & Tucci, C. L. (2013). Business model innovation. In M. Dodgson, D. M. Gann, & N. Phillips (Eds.), *The Oxford handbook of innovation management* (Vol. 20, pp. 420–441). Oxford, UK: University of Oxford.

Matt, C., Hess, T., & Benlian, A. (2015). Digital transformation strategies. *Business and Information Systems Engineering, 57*(5), 339–343.

Meyer, A. D., Tsui, A. S., & Hinings, C. R. (1993). Configurational approaches to organizational analysis. *The Academy of Management Journal, 36*(6), 1175–1195.

Moore, G. A. (1991). *Crossing the chasm: Marketing and selling high-tech products to mainstream customers.* New York, NY: Harper Business Essentials.

Nanterme, P. (2016, January 17). Digital disruption has only just begun. *World Economic Forum.* Available at www.weforum.org/agenda/2016/01/digital-disruption-has-only-just-begun/

Osterwalder, A., Pigneur, Y., & Tucci, C. L. (2005). Clarifying business models: Origins, present, and future of the concept. *Communications of the Association for Information Systems, 16*(1), 1–25.

Paswan, A., D'Souza, D., & Zolfagharian, M. A. (2009). Toward a contextually anchored service innovation typology. *Decision Sciences, 40*(3), 513–540.

Ries, E. (2011). *The lean startup: How today's entrepreneurs use continuous innovation to create radically successful businesses.* New York, NY: Crown Business.

Smith, K. B. (2002). Typologies, taxonomies, and the benefits of policy classification. *Policy Studies Journal, 30*(3), 379–395.

Taran, Y., Boer, H., & Lindgren, P. (2015). A business model innovation typology. *Decision Sciences, 46*(2), 301–331.

Teece, D. J. (2010). Business models, business strategy, and innovation. *Long Range Planning, 43*(2/3), 172–194.

Thiel, P. A., & Masters, B. (2014). *Zero to one: Notes on startups, or how to build the future.* New York, NY: Crown Business.

Tinkoff. (2019). *About Tinkoff.* Available at www.tinkoff.ru/eng/company-info/about-tinkoff/

Westerman, G., Bonnet, D., & McAfee, A. (2014, January 7). The nine elements of digital transformation. *MIT Sloan Management Review.* Available at https://sloanreview.mit.edu/article/the-nine-elements-of-digital-transformation/

Wilcox, K., Kim, H. M., & Sen, S. (2009). Why do consumers buy counterfeit luxury brands? *Journal of Marketing Research, 46*(2), 247–259.

World Economic Forum. (2019). *Digital transformation case studies.* Available at http://reports.weforum.org/digital-transformation/go-to-the-case-studies/

Chapter 8

Digital transformation, the Holy Grail and the disruption of business models

Interview with Michael Nilles, Chief Digital and Information Officer, Henkel

Andreas Hinterhuber and Michael Nilles

In this interview Andreas Hinterhuber and Michael Nilles, CDIO of Henkel, discuss the game-changing opportunities that the digital transformation opens to companies that embed a digital core into their business models. Michael Nilles sees digital transformation as the Holy Grail: a force that is not easy to find, not easy to capture and that has the potential to dramatically improve the customer experience. In B2C, the Holy Grail for Henkel's Beauty Care business is for instance the beauty tech ecosystem, a series of connected devices harnessing big data augmented reality allowing to create meaningful, personalized and direct relationships with consumers. In its B2B business, the Holy Grail is Henkel's digital twin along the digital thread. Henkel builds a digital twin along the entire value chain of the customer – starting with the initial customer request via deployment, until and after sales service – in order to sell outcomes to customers. In B2B the digital twin enables outcome-based servitization. Digital technologies are thus more than enabling technologies: digital technologies allow you to create fundamentally new, disruptive business models. The digital transformation needs small, agile teams with end-to-end responsibility for project delivery, start-up mentality and customer obsession. The final aspect that this contribution discusses are metrics of the digital transformation such as digital efficiency/productivity, digital revenues and digital growth (net of cannibalization).

Andreas Hinterhuber: How do you interpret "digital transformation"?

Michael Nilles: Digital transformation is like the quest of the Holy Grail – different for each business. I use the term Holy Grail purposefully since the Holy Grail is not so easy to find and not so easy to capture. We are constantly asking the question: what is it that is moving the needle in terms of long-term impact of business model disruption?

Take our business unit Beauty Care for example, a business with globally about €4 billion revenues. The market dynamics are changing rapidly: consumers now demand personal, meaningful, and direct relationships with brands. We see new business models emerging, such as subscription-based businesses. In B2C, the Holy Grail of digitalization – the force that is moving the needle in terms of disruption – is our beauty tech ecosystem. We build a digital ecosystem around the consumer.

Andreas Hinterhuber: So digital transformation is the use of technology to create customer value that could not have been created in the absence of technology?

Michael Nilles: Some executives and academics (Porter, 2001) see digital transformation as enabling technology – I disagree. Digital transformation goes far beyond technological enablement. Digital technology, for sure, is also an enabling technology that leads to efficiency improvements in supply chain or other functions. Digital transformation fundamentally improves existing business models or allows to totally disrupt existing business models.

The point is: digital transformation will lead to the disruption of business models, either by us or by competitors, and – paraphrasing Steve Jobs – I would rather disrupt my own business than having a competitor do it. This is why we invest in start-ups since some of these companies will one day disrupt current business models.

One key aspect of digital transformation is building digital business models. One example is StreetBees, a Henkel dx portfolio company that uses digital technology to create a meaningful relationship between consumers and brands by allowing consumers to directly give insights into their buying and usage behaviors. Streetbees encourages consumers to record their decisions, emotions and attitudes in the "brick-and-mortar world" with their own words, using video, photo and text on their digital devices. This is an approach that is really in contrast to pure, traditional market research.

This is one example where the digital transformation leads to disruptive opportunities that simply would not have been possible before.

Another great example is eSalon, a D2C business focused on individually customized hair coloration for at home application. The consumer can shop a truly personalized product as part of a subscription model or as a one-time purchase. This, again, is a digital business model, where technology is deeply embedded in the business model. This close link between digital transformation and digital business models is also reflected in McKinsey's growth framework.

Take the three horizons of McKinsey's growth framework (Baghai, Coley, & White, 2000). In horizon 1 – *extend and defend core businesses* – digital technology is an enabling technology. In horizon 2 – *build emerging businesses* – and in horizon 3 – *create genuinely new businesses* – digital technology is deeply embedded in the business model.

Andreas Hinterhuber: I appreciate the link between McKinsey's three horizons of growth framework and the role of technology in the context of the digital transformation. In horizon 1 digital is an enabling technology; in horizons 2 and 3 digital technologies are the building blocks for disruptive new business models. Great insight. Let us now explore a world that I know better than the B2C world: B2B. B2B, Adhesive Technologies, is Henkel's largest business unit with over €9 billion sales.

Michael Nilles: In B2B, the digital transformation is essentially built on data. In our Adhesives Technologies business unit, we are global market leader for one reason: we understand customer needs, we have the right technologies and we know how to apply them to create custom solutions. We are a know-how company.

So, what is the Holy Grail of the digital transformation in B2B? It is the so-called digital twin along the digital thread. We build a digital twin along the entire value chain, from the original customer request, to production, supply chain, deployment, and after sales service.

Andreas Hinterhuber: I am quite interested in pricing and value-based pricing in particular (Hinterhuber & Snelgrove, 2017). How do you define key business outcomes, or KPIs if you prefer, which you guarantee to your customers? Once they are defined, how does a pricing model look like?

Michael Nilles: For me a good example are industrial goods companies which transform from a pure product business to a service business and ultimately to data-driven outcome based service models. You typically move within such a service transformation from a service delivery model based on break/fix to a service model based on outcomes: you are for instance guaranteeing a certain uptime of your machinery. Customer uptime is the key factor in the price the customer pays, with agreed-upon penalties for severe underperformance. I had the chance to drive such a transformation myself.

Andreas Hinterhuber: This is a good example of value-based pricing.

Michael Nilles: You should consider that such a pricing model is the result of a long experience in the product business in the first place. You typically build on a very long history and experience in the product business. Starting with a simple break/fix model, you then shift towards a company offering services. And building on that you offer service contracts based on human experience. You then offer service contracts based on human experience. For a data-driven outcome based business model you have to offer digitalized services (e.g. 24/7 remote monitoring, preventive maintenance, digital twins). The experience, the evolution of the service delivery model then have to go along with the development of digital capabilities allowing you to deploy digital technologies.

Andreas Hinterhuber: Which company is, in your view, best-in-class in digital transformation? From which company or which industry are you currently learning?

Michael Nilles: Except for born-digital companies, there are, as of today, few product-based or service-based companies that have mastered the digital transformation. GE under Jeff Immelt was an early leader (Immelt, 2017). Immelt's strategy of building an industrial internet and an open platform, Predix, was a pioneering and bold move at the right time. GE's five pillars of the digital transformation still capture the main challenges quite well (Talya & Mattox, 2016). (See Figure 8.1).

As we nowadays know, the strategy of GE was great, but the company ultimately had to deal with other issues than the digital transformation.

There are few lighthouses that could serve as examples on how to get the implementation right. This book is an ambitious answer to this very important question at the right time. There are, this is clear, no easy answers on how to get the digital transformation right.

There is, however, one fundamentally important shift that I would like to point out: a move within product architecture towards a software-driven stack. Take Tesla: One thing that Tesla did extremely well was the development of an own digital stack enabling the

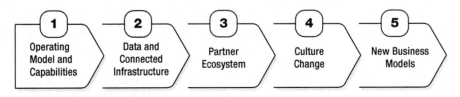

Figure 8.1 The five pillars of the digital transformation (Talya & Mattox, 2016).

company to update the software – and thus the car – over the air (OTA). An OTA software update is almost like getting a new car – over the air! This is what industrial businesses can learn from Tesla: industrial companies could complement their traditional product architecture with a software-based architecture with OTA updates. OTA update capability is then a key value driver and many machinery companies are struggling with this challenge.

The second thing that Tesla is doing extremely well, is the D2C approach. Tesla knows its real customers and is very well managing the relationship with them along various touchpoints.

Andreas Hinterhuber: Let us now turn to organizational aspects of the digital transformation. Which capabilities, what type of structure are relevant to implement the digital transformation?

Michael Nilles: Structural changes are required to implement the digital transformation successfully. Henkel changed its own digital structure recently – in July 2020: Henkel dx is the name of our new organizational unit now combines digital capabilities, traditional IT, new ventures and business expertise under one roof. Our mission is "creating the next": we are truly committed to creating the next big opportunity for Henkel as well as more operational things like the next level of better services. In our dx compass we have identified five cultural drivers that are crucial for our digital transformation (see Figure 8.2).

Business and customer obsession: Every day, without exception, we think and live business, customer and consumer first. Our passion is to create high impact and value through digital innovation. We believe in the power of digitalization.

Proud to be tech: Our heartbeat is technology. We love software, data and analytics. We take pride in building tech to solve the toughest business challenges. We are passionate about exploring emerging technologies and leveraging their full potential.

Business domain expertise: We are trusted experts to our businesses. We speak their language and deeply understand business models, products, processes and the dynamics of our markets. We never stop improving ourselves.

BizDevOps: Together we create, develop and operate everything end-to-end. By bringing together our business domain and tech experts, we leverage our unity of effort. This enables our dedicated and interdisciplinary teams to deliver a better and faster outcome.

Start-up mentality: We are bold, persistent and committed to growth. We believe in the power of open ecosystems. Driven by the market, we test & learn, we build fast – in weeks, not months – and are always ready to adapt.

Andreas Hinterhuber: Innovation is key in the context of the digital transformation. We know little about the origins of innovation in the context of the digital transformation. There is one recent piece of research that suggests that the digitization of existing products and services is mainly driven top-down, whereas the creation of radically new digital products and services is mainly a bottom-up approach (Chanias, Myers, & Hess, 2019). What is your take on this?

Michael Nilles: In my opinion, this view is too simple and probably incorrect. As a prerequisite for innovation, companies need to change their mindset about innovation from

FIVE CULTURAL DRIVERS TO EMBED DIGITAL INTO OUR DNA

Business & Customer Obsession Every day, without exception, we think and live business, customer and consumer first. Our passion is to create high impact and value through digital innovation. We believe in the power of digitalization.

Business Domain Expertise
We are trusted experts to our businesses. We speak their language and deeply understand business models, products, processes and the dynamics of our markets. We never stop improving ourselves.

Proud to be Tech
Our heartbeat is technology. We love software, data & analytics. We take pride in building tech to solve the toughest business challenges. We are passionate about exploring emerging technologies and leveraging their full potential.

BizDevOps
Together we create, develop and operate everything end-to-end. By bringing together our business domain and tech experts, we leverage our unity of effort. This enables our dedicated and interdisciplinary teams to deliver a better and faster outcome.

Start-up Mentality
We are bold, persistent and committed to growth. We believe in the power of open ecosystems. Driven by the market, we test & learn, we build fast - in weeks, not months - and are always ready to adapt.

OUR AMBITION

BECOMING DIGITAL BUSINESS LEADER IN OUR INDUSTRIES

Digital applied is our world. We see economies, societies, customer and consumer behavior massively changing and thereby reshaping our markets. Radically focusing towards individual customer and consumer needs, we build digital and data-driven business models – to optimize, transform and disrupt, turning threats into pportunities. Consequently, we aim to play at the digital forefront of our industries.

DRIVEN BY OUR MISSION

CREATING THE NEXT

We are creative minds. Our dedication to innovation is endless. From ideation to business impact. What we have achieved today, won't be enough tomorrow. We explore the impossible, push boundaries to make it happen. At all levels and without limits. Applying data and digital, from continuous optimization to the next big thing. Together, we are CREATING THE NEXT.

Figure 8.2 The cultural drivers of the digital transformation.

viewing innovation as a closed-system approach to viewing innovation as an open, collaborative approach where intellectual property (IP) protection is not the only question anymore. Key to managing innovation successfully in the context of the digital transformation is thus a mindset change towards open innovation aimed at producing solutions that can be connected to an ecosystem. This requires a true transformation of the traditional innovation mindset.

In terms of origins of innovation, companies need to provide alternative pathways for innovative ideas. They need to develop a roadmap for digital innovation outlining key activities and initiatives. I see three alternative pathways:

Strategic innovation, aimed at producing disruptive new business models, is top down. Getting these big bets right is difficult and risky.

The digital transformation
and the three horizons of growth

	Horizon 1	Horizon 2	Horizon 3
	Extend and defend core businesses	Build emerging businesses	Create genuinely new businesses
Role of digital technologies	Enabling technologies	Creation of new, disruptive business models	Creation of new, disruptive business models
Key metrics	Digital efficiency/ productivity	Digital revenues Digital growth (net of cannibalization)	Digital revenues Digital growth (net of cannibalization)

Figure 8.3 The three horizons of growth and key performance indicators of the digital transformation.

Open innovation, by contrast, is diffuse and bottom-up: we co-innovate with other companies, with start-ups and with academics – we get a lot of impulses on data-models related to innovation from academics and the contribution of academics to corporate innovation should not be underestimated.

Incubation, finally, is also bottom-up. On the one side, via corporate venturing activities we invest in start-up companies not only through acquisitions but also through minority shareholdings; we see the latter as viable scouting model to get insights into emerging trends and cutting-edge technologies. On the other side, we have internal incubation activities in the form of a digital innovation fund. The logic for investing internally, i.e., in ideas coming from Henkel employees around the globe, is quite similar to the logic of investing externally: people come with ideas, they get money with literally zero bureaucracy and we apply the metrics of any brave venture capitalist expecting that out of ten ideas, eight will fail. For our internal incubation we have announced digital Innovation Hubs in Berlin, Shanghai and Silicon Valley with the objective of providing a platform to commercialize the one or two successful ideas rapidly on a global scale.

In sum, innovation in the context of the digital transformation takes three distinct formats: strategic innovation from the top aimed at producing disruptive new business models, open innovation and incubation, bottom up, that can both produce incremental and radical innovation.

Andreas Hinterhuber: Thank you. A final question concerns metrics. Which key performance indicators (KPIs) do you use to keep your digital transformation on track?

Michael Nilles: The most important digital KPIs are digital revenues, digital growth (net of cannibalization) and digital efficiency/productivity.

Digital revenues and digital growth are closely linked to transforming and disrupting the business model, digital productivity/efficiency KPIs are closely linked to optimizing the current business model. There is therefore a clear correlation between strategy – disruption versus improvement of the business model – and the KPIs that we are using (see Figure 8.3).

Andreas Hinterhuber: Michael, I truly thank you for the privilege of having this first-rate exchange of thoughts on the digital transformation.

Michael Nilles: Thank you. I likewise enjoyed our exchange of thoughts.

Acknowledgments

Reprinted with permission from: Hinterhuber, A., & Nilles, M. (2021). Digital transformation, the Holy Grail and the disruption of business models, *Business Horizons*, 64. All rights reserved. Copyright Elsevier, 2021. DOI: doi.org/10.1016/j.bushor.2021.02.042.

References

Baghai, M., Coley, S., & White, D. (2000). *The alchemy of growth*. New York, NY: Basic Books.

Chanias, S., Myers, M. D., & Hess, T. (2019). Digital transformation strategy making in pre-digital organizations: The case of a financial services provider. *Journal of Strategic Information Systems, 28*(1), 17–33.

Hinterhuber, A., & Snelgrove, T. (Eds.). (2017). *Value first, then price: Quantifying value in business markets from the perspective of both buyers and sellers*. Milton Park, UK: Routledge.

Immelt, J. (2017). How I remade GE. *Harvard Business Review, 95*(5), 42–51.

Porter, M. (2001). Strategy and the internet. *Harvard Business Review, 79*(3), 63–78.

Talya, A., & Mattox, M. (2016). *GE's digital industrial transformation playbook*. San Ramon, CA: GE.

Töytäri, P., & Rajala, R. (2015). Value-based selling: An organizational capability perspective. *Industrial Marketing Management, 45*, 101–112.

Chapter 9

How artificial intelligence and the digital transformation change business and society

An interview with venture capitalist Vinod Khosla

Ajay Agrawal and Vinod Khosla

In this interview Vinod Khosla, one of the world's most successful technology entrepreneurs, talks to Ajay Agrawal, founder of Creative Destruction Lab, about the AI-generated opportunities and dangers that lie ahead, and how to prepare for them. This interview explores the role of academics in shaping the role of education in a world characterized by an increasing ability of machines to replace jobs formerly occupied by humans. This discussion anticipates the coming age of hyper-innovation. Technology, Vinod Khosla, reminds us, does not rule—it serves those masters that are able to use technology to better serve business and society.

Ajay Agrawal: In a 2017 essay (Khosla, 2017), you wrote that AI might improve metrics like GDP growth and productivity, but, at the same time, it may worsen less visible metrics such as income disparity. Are you still concerned about that?

Vinod Khosla: Even more so. Without a doubt, AI is the most important technology we have seen in a very long time. Some people even refer to it as 'the last technology' because it will likely be responsible for all of the technologies that follow. As such, it presents massive potential for contributing to society. Having said that, where we get to will depend on the path we take.

It's great to talk about creative destruction if you're the one doing the disrupting; but if you're the one being destroyed, it isn't much fun. Disruption will always be unpleasant for someone, and in the coming years, it will take its toll on jobs. The core issue is that 'efficiency' in the business world generally means reducing costs, which results in replacing lower-wage, less-skilled workers with far fewer well-paid, highly skilled people. Because of this, I do worry that the machine learning revolution will lead to increasing income disparity—and that disparity beyond a certain point could lead to social unrest.

On the positive side, the jobs we all covet are jobs that we would do even if we didn't get paid to do them, and that is the long-term potential of AI: to eliminate the need for unsatisfying work. However, long before we get there, we will have to go through the dynamics of shifting from today's economy into its next iteration. The path ahead will be extremely uneven, and, as a result, society may push back on these technologies.

Ajay Agrawal: You have said that lower-skilled jobs like truck drivers and food services are actually less at risk from automation than radiologists and oncologists. Please explain.

Vinod Khosla: I do think that the higher-skilled, more knowledge-based jobs will be the easiest ones to replace with AI. For example, radiologists are toast. That just should not be a job anymore. I would go even further and say that any radiologist who plans to practise ten years from now will be practising out of their own arrogance, because they will be misdiagnosing patients much more frequently than AI-driven systems. Likewise, oncology is going to be much easier to automate than a factory worker, because a factory worker has much more dimensionality. Just within the Khosla Ventures portfolio alone, entrepreneurs are already trying to use machine learning to replace human judgement in many areas, from financial services to farming to law and cardiology. And our portfolio represents just a tiny fraction of the efforts underway around machine learning. With less and less need for human labour and judgement, labour will be increasingly devalued relative to capital—and even more so relative to ideas and machine learning technology.

Ajay Agrawal: What is the timeline for rolling out these technologies across industries?

Vinod Khosla: Most people might agree that AI taking over radiology is a done deal—that there is nothing a radiologist can do that AI can't, and that AI is both cheaper and more accurate. But the fact is, only about 1 per cent of radiologists are currently using these technologies. There is a significant time period for rollout that is very important to con-sider in looking at all of this. My bet is that we are ten years away before AI is commonly accepted in radiology, and maybe 15 years from the point where we say, 'We don't really need oncologists anymore'.

This is actually good news for some people. When it happens, your primary care phys-ician will be in the best position to look after you, because just about all specialty expertise will be provided by AI. The average patient has seven ongoing health conditions, yet each specialist has no idea about the others, nor any contact with them. In the future, the role of a GP will be to provide integrative care. As a result, medical schools should start focusing on recruiting and training individuals with high emotional intelligence and empathy, because their future will largely involve managing patients rather than deter-mining medical interventions.

Ajay Agrawal: You believe the rollout for AI will vary between industries. How long before entire assembly lines are robotic?

Vinod Khosla: I think we are five to ten years away from being able to completely replace human workers on assembly lines. And by the way, this represents what is probably the lar-gest market in the world. One trillion dollars would be an understatement, and there may be some important indirect effects. For example, if you build an assembly line robot that doesn't need any programming to do a task like 'assemble an iPhone'—if it can just learn by watching a dozen or so examples and can then perform better than a human worker—that could result in an inversion of the supply chain. All of those manufacturing jobs that moved to China might actually come back to the West, because of course, it is much more cost-effective to have your manufacturing done locally. Between assembly line robotics and 3D printing, we could see a complete inversion of the supply chain. In fact, 40 years from now, there might be no need for jobs of any type. If that happens, governments will have to turn their attention away from job creation and figure out how to help people find meaning in their lives.

Ajay Agrawal: How far off is the income-disparity crisis that you have warned of?

Vinod Khosla: There are many variables involved in income disparity—policy being one of the larger ones. But in terms of AI as a contributor, I believe that 20 or 30 years out, it will be the biggest variable. While the future looks promising in terms of increased productivity and abundance, as indicated, the process of getting there raises all sorts of questions about the changing nature of work.

Clearly, there are also significant implications for education. I suspect that if and when software systems exceed the capability of the average—and eventually the smartest— humans in judgment and skill, the avenue of personal growth through education that has traditionally been open for career advancement may be closed. I gave a talk at the National Bureau of Economic Research meeting recently, and former Harvard President Larry Summers came up to me afterwards and said, 'You just blew my solution for countering the effects of AI!' Most people don't realize that education is not the solution here.

Ajay Agrawal: Is the solution to slow down technological change in order to preserve jobs?

Vinod Khosla: Definitely not, but we do need to address the issue of income disparity. The easiest answer seems to be what economist Thomas Piketty has advocated for—some form of income redistribution. I suspect that will be a necessary component. We also need to look at our capitalist system, which is filled with arbitrary policies in favour of either labour or capital. When you allow certain partnership tax structures or you don't provide a tax credit, you are advantaging certain kinds of activities and disadvantaging others.

We need to look at that very closely and make some changes. For example, giving an R&D tax credit back to companies favours innovation, whereas giving favourable depreciation is a bias towards capital instead of labour. Keep in mind that large corporations tend to shape most rules and regulations—at least in the US—so many of these biases have been engineered into today's economy. But the cost of labour and the cost of capital can be effectively altered by some simple changes in rules, regulations and laws. More and significant manipulation will be required to achieve reasonable income disparity goals.

Social mobility is a tougher goal to engineer into society's rules. I suspect the situation will become even more complex as traditional economic arguments of labour versus capital are upended by a new factor many economists don't adequately credit: the economy of ideas driven by entrepreneurial energy and knowledge. This factor may become a more important driver of the economy than either labour or capital. Of course, all of this is mere speculation. The future is nearly impossible to predict.

Ajay Agrawal: Are there things leaders should be doing that they aren't currently doing?

Vinod Khosla: Despite all of the dramatic benefits it offers, there isn't nearly enough investment in AI. But that is only a matter of time; it will happen. I think what would help the most right now is broad social acceptance and adoption. We need to condition people for the consequences of AI and think carefully about how to roll it out so that the most disadvantaged in society are not disproportionately affected. Indeed, the hope is that they will be positively affected.

Either way, as indicated earlier, we are going to need a version of capitalism that is focused on more than just efficient production—one that places greater prioritization on the less desirable side effects of capitalism. We need to adjust the playing field in terms of how the system works. Hopefully, some of this work can be initiated in the academic world.

I mentioned universal basic income, and I think that could make the adoption of new technologies much easier. At one point, Bill Gates talked about a robot tax—placing a tax on every robot adopted by an organization. In the environmental movement, as carbon reduction has become a widespread goal, people have talked about carbon taxes that disproportionately benefit the bottom quartile of society. I don't think we are studying these kinds of mechanisms enough.

Ajay Agrawal: To what extent should we be worried about the influence of machine intelligence on the sphere of public opinion?

Vinod Khosla: Of course, this is already happening. The 2016 US presidential election was one example, and Brexit has been the most visible example to date—because in that case, it actually changed the outcome. Cyberwarfare powered by AI is now one of the most powerful weapons for launching what traditionally would have been very visible attacks on other countries. Developing a better air force or bigger nuclear bombs had much more transparency. As a result, you could actually have agreements between nations: 'You promise not to develop X, we promise not to develop it, and we will both be able to verify that over time'. With AI, there is no verifiability. The battle between nations can now be conducted silently and without any transparency. That is a danger that I worry about more immediately.

On an individual scale, we already have the ability to hack into any human mind if we have enough interaction with the person. That is what is happening on Facebook when somebody sells you a pair of jeans that you didn't think you wanted: with machine learning, they are reverse-engineering a narrow part of your mind, making you do something they want you to do. A simplistic view of this type of activity is, 'We're just selling more stuff to more people'; but the more dangerous lens is, 'We're getting you to behave in a specific way'. This is a real danger, and I don't know if there is an easy solution.

Ajay Agrawal: In terms of creating AGI—artificial general intelligence—who do you suspect will get there first?

Vinod Khosla: We were the only venture investor in OpenAI, the for-profit company whose stated aim is to promote and develop friendly AI in such a way as to benefit humanity as a whole. As an investor, I like situations where the chances of success are less but the impact could be extremely consequential. That is the only $50 million dollar cheque we have ever written, so we believe that it is one strong possibility. Vicarious is a small company working on AGI for assembly line robotics. DeepMind is clearly doing some stunning work and there are some interesting efforts underway within Google. There's also a lot going on in China, which has a clear national focus on winning the AGI race. I am generally optimistic that one of these technologies will win in a really big way—but I believe more than one of them will win in creating AI that does more and more of the economically valuable human functions that we need done.

Ajay Agrawal: Whose responsibility is it to protect everyday citizens from AI's negative effects? Is it up to governments?

Vinod Khosla: Like most people, I hate government solutions because they are engineered to narrowly benefit some citizens but not others. Creating effective government policy is great to talk about theoretically, but effective policy is very rare. There may be other solutions that could change the playing field. We talked about capitalism before, and it's a philosophy we all buy into. Historically, why has capitalism been important? Because it increases economic efficiency. You just have to compare North Korea with South Korea to see the difference economic efficiency can make. Having said that, we are moving into an era where efficiency and productivity will no longer be major variables in success. The emerging form of capitalism is more about generating demand and making you want things you didn't know you wanted than it is about producing the things that you need.

As we consider changes to the system, we need to be very careful. In complex systems in general—whether it be the global economy or software code with millions of lines—there are always holes or bugs, and as a result, such systems bring with them the possibility of unintended effects. This danger exists with AI, too. Despite this, I am generally optimistic that we can contain the negative effects. If you think about it, every single powerful technology in the history of the world has been a valuable tool that can be used for either good or bad. The danger always exists, but that is not a reason to slow down progress.

Ajay Agrawal: You believe that some venture capitalists actually do damage to the companies they invest in. How so?

Vinod Khosla: Just because you invest in a company and sit on its board doesn't give you the right to put pressure on or advise an entrepreneur. Anyone who has done it knows that starting a company is a painful, difficult process, and unless you have empathy for the entrepreneur from having experienced it yourself, I don't believe you have the right to pressure them. That is one rule that I have followed for the past 30 years. I have never voted against an entrepreneur, and I would advise entrepreneurs to only listen to individuals with concrete experience as input to their thinking.

Ajay Agrawal: You also believe that technology can—and should—help to reinvent societal infrastructure. How so?

Vinod Khosla: Today, 700 million people—the top 10 per cent of the world's population—enjoy a rich lifestyle in terms of environment, healthcare, housing, food and education. The other 90 per cent wants what we have, and technology is the only bridge to making this a reality. We need a 10× in resource utilization multiplication—not a 10× in the number of doctors, or buildings or cars. Technology has the potential to achieve food goals, reshape cities, cure disease, mitigate climate change and enhance human capability.

The future is not knowable, but it is 'inventible'. I believe the amount of innovation we will see over the next ten years will explode by 10× over what we've had. That's because as axes of innovation increase, possibilities for solutions increase. And at the same time, the cost of experimentation is going down. That's why I believe we are at the very beginning of a hyper cycle of innovation. The tools available now can be combined in an endless number of ways to innovate: AI and data collection, 3D printing, quantum computing, robotics, social networking, genomics, dematerialization, to name just a few. When you combine medical imaging with AI; when you combine 3D printing with new material science, you get very different things. And this won't be a temporary cycle—I think we are entering a permanent hyper-innovation cycle.

Ajay Agrawal: Any parting advice for entrepreneurs, AI-focused and otherwise?

Vinod Khosla: First, keep at it, because most major disruptions are non-institutional, and the world really needs you and your ideas. Second, in my experience, the more money you raise, the less likely you are to succeed. That's because when there is little money, it forces you to think about the problem much harder. If you get a lot of money, you tend to start executing without analysing the problem sufficiently. You have to be hyper-efficient with your dollars to be more creative with the problem. And third, always remember, technology doesn't rule: It serves. We get to decide its goals.

Acknowledgements

Reprinted with permission from: Agrawal, A. (2020), Thought leader interview – Vinod Khosla, *Rotman Management Magazine*, Winter 2020, 14–18. All rights reserved: Copyright Rotman School of Management, University of Toronto.

Reference

Khosla, V. (2017). AI: Scary for the Right Reasons, online article (13 September), Medium.com, accessed 2 July 2020.

Chapter 10

The digital company culture
Interview with Luca Ferrari, CEO, Bending Spoons

Tiziano Vescovi and Luca Ferrari

Bending Spoons started in 2013 in Copenhagen, Denmark, as a collaboration between five young friends with expertise in business, technology, and design.

They saw an opportunity in developing mobile apps and, since then, have created some very popular ones indeed. These include, among others, 30 Day Fitness (the world's most downloaded workout app in 2019) and Splice, one of the leading video editors in the US.

In 2014, Bending Spoons relocated to Milan, looking to build a pro tech team with a strong chance of sustained and potentially exponential success. The company's performance has grown dramatically, with revenues of approximately $7M in 2017, $45M in 2018, and $90M in 2019.

Their apps average more than 200,000 daily downloads, and they have more than 12.5 million monthly active users (including 5.6 million in the USA and 4.6 million in Europe), enjoying around 300 million downloads in total. To date, Bending Spoons has received more than 40,000 job applications, with less than 1 percent of applicants being selected to join the company. Recently, Bending Spoons designed and developed Immuni—the app selected by the Italian government to fight the COVID-19 pandemic.

Luca Ferrari is the CEO of Bending Spoons.

Summary

In this interview Luca Ferrari, CEO of Bending Spoons, deepens the strategic way of thinking of a company born digital. It emerges that being digital is not about adopting software, which remains just a tool, but is a way of finding solutions to customer problems using, in large part, digital solutions. The business model of a digital company such as Bending Spoons is based on two key aspects: partnership, the network of valuable connections, and key resources, above all represented by people. In a digital company, people are more important than technology. The surprising changing trends of digital products are also highlighted.

Tiziano Vescovi: What is the vision and the mission of the company?

Luca Ferrari: Our goal, our dream, is to create the best technology company ever. A company that creates value for society and, at the same time, offers excellent jobs. A company that creates products of the highest quality, succeeding from a commercial point of view. Ideally, we would like to develop individual products that are truly revolutionary for the end user.

Founders are always less interested in what the specific product being built is. Rather, we're excited by the potential of technology in general.

We identified the world of mobile services as an interesting market. Even a decade ago, it was large and constantly growing, with low barriers to entry. It was an attractive situation for a startup with little capital available.

Over time, we have developed several product strategies, trying to balance investments in three key activities:

1. Exploitation. Activities aimed at maximizing the success of a product (for example, in terms of profitability or market share).
2. Exploration. The systematic study of the market and its opportunities.
3. Optionality. Activities aimed at establishing the right conditions for the success of the two previous phases.

Tiziano Vescovi: Was there a strategic reason behind the decision to move from Copenhagen to Milan?

Luca Ferrari: We moved Bending Spoons from Copenhagen (where we founded it) to Italy because we thought there was enormous untapped potential here. Italy is a large country with an excellent school and university system, but it's hindered by a pessimistic outlook in younger and older generations alike. In reality, the infrastructure is good—although there's room for improvement—and many things can still be achieved despite the bureaucratic problems. In the new digital business landscape, intellectual and passionate individuals are vital, and here there are many. The tax system is not the most important consideration if the business is successful and of great value. It can be crucial if the business creates little unit value, as can be the case for certain large companies in mature markets. In such industries, margins may be low, and optimization and efficiency are the target.

Tiziano Vescovi: How do you develop strategy and planning in a digital company?

Luca Ferrari: As an entrepreneur, I don't believe in making extremely detailed plans and looking at numbers constantly. Actually, I think that excessive attention to economic details can prove a distraction from what makes the real difference in quality: macro ideas, macro changes. Growing up, the company has taken on many different projects—often minor. Now, we're focusing on fewer, bigger things. In the beginning, it was more pragmatic not to be the best at one thing, but to be good enough at several different things. That's how we reached the economic threshold where we could sustain costs and grow. The more we grew, the more we focused on the activities where we were most successful, abandoning those that were marginal or too difficult to scale. Today's strategy goes in the direction of making fewer different types of apps, focusing instead where we're among the very best: the health and wellness and video editing markets. In the mobile world, these two represent vast markets. In the health and wellness sector, we also see development potential outside of pure digital: from personalized coaching to hardware products.

Tiziano Vescovi: Why aren't analog companies in the health and wellness sector (for example) strong digital players?

Luca Ferrari: Most big analog players have minimal or low-quality digital presence. Digital transformation requires a different culture that isn't present in analog companies, precisely because of the business's structural aspects. These are two corporate cultures that require two different mindsets. We've noticed that it's much simpler for a digital company to move into developing analog products than the reverse. Analog suppliers are easier to source and manage. For example, fitness apps evolve as a result of direct contact with real personal trainers. Everyone uses the same available hardware (exercise bike, treadmill, etc.). However, hardware companies have difficulty offering software products for two main reasons. First, they often don't understand what they need because it requires a skill set so different from what they have available. Second, analog companies find it challenging to attract the necessary talent. The best software designers typically want to work in digital companies where they find others of their level, rather than being relegated to departments in continuous conflict with other parts of the company.

So, the big tech companies expand into analog markets more easily and effectively than the other way around.

Tiziano Vescovi: Is Bending Spoons a digital company? What features does a digital company have?

Luca Ferrari: In my opinion, a digital company has the following characteristics:

- It mainly produces digital products.
- It maintains a high level of ubiquity and penetration of software in business processes and operations. We make software and use software, so we're digital. By contrast, a restaurant is not digital.
- Data are central to decision-making and daily business management. For example, we use statistical and machine learning models that process the data from our apps. Through data, we derive the user behavior, from which we derive the value of the communication channels and content, and the financial value of the customers themselves.

Tiziano Vescovi: Can you comment on the process that leads to the current digital strategy? I.e., the scope of the strategic analysis (internal/external), any alternative options considered, the decision criteria when selecting the current digital strategy, any goals, etc.

Luca Ferrari: Having a digital strategy doesn't require using software. Software is one tool, one tactic, one more option. Every six months, we meet with a group of innovative entrepreneurs strongly immersed in digital, but none of us speak of digital innovation as such. Innovation is nothing but the consequence of searching for solutions to the customer's problems. We need to ask ourselves how to achieve the goals we've set ourselves. What are the problems we want to solve? Then we use the technologies that allow us to solve these problems.

The main strategic drivers are the same as for analog firms. I have seen many companies that would like to digitize for the sake of digitizing. They make it a goal instead of a tool. So, they end up building apps of poor quality and no value. With this approach, you risk losing any positive attitude towards digital, because the result is frustration. Then, you blame digital by saying it doesn't work, instead of attributing failure to a flawed approach.

Therefore, it's necessary to have very clear business objectives. Make sure that you have a clear overview of the spectrum of possibilities offered by technology. If you don't, then have someone in management who excels in this field. This person will be able to identify the correct solutions and the most suitable suppliers. However, technological solutions should only be used in terms that help achieve business goals. The approach must be to use digital technology to do a few crucial things very well, not to do a lot of marginal things in a mediocre way just to feel digital.

Tiziano Vescovi: Which company is, in your view, the best-in-class in digital transformation? What have you learned from their initiatives?

Luca Ferrari: There are several companies that we look to as a model, for different reasons. Regarding the quality of the products, in my opinion, Apple, Amazon, and Tesla are top class. This is because of their ability to serve customers. For example, at every management meeting, Jeff Bezos used to put an empty chair and say that the client was sitting there.

As for the organization of work and the management of human resources, Facebook and Google can probably be considered the best for the work environment they have managed to create.

If we look at innovative ideas with a strong social impact, I would say Airbnb. The company has created a revolutionary product, involving a large population in the business, both as supplier and consumer, and improving life for both. It's an excellent example of analog-digital integration. And to mention a lesser known, but very interesting and increasingly successful company, WeRoad sells trips for young people. It focuses on somewhat adventurous groups, providing itineraries, guides, and coherent travel companions. In doing so, it integrates social, analog, and digital.

Tiziano Vescovi: Digital transformation and business model innovation: Can you describe the key aspects of your business model?

Luca Ferrari: We consider two aspects—partnerships and key resources. Our main partner is Apple. On the one hand, to have Apple as a partner is extremely beneficial, because it's a continuous incentive to improve performance. On the other hand, it also represents a significant constraint, as the App Store is by far our main distribution channel.

The key resources in our company are, above all, people and apps. In the real estate business, you need good properties with an attractive location. In our business, the equivalent scenario is to have the perfect app, well-positioned and well-named so that it can be found easily. The value of our team and their skills are worth 2/3 of the company. People generate the company's potential and development, and they're responsible for the quality of the proposed solutions.

Tiziano Vescovi: Organizing the digital company—what are the key points to be developed and controlled? Any comments on pros and cons of these options?

Luca Ferrari: The attractiveness of the business is of fundamental importance for technological talent. To make quality software, you don't need just software workers. You need people with rare characteristics, excellent in understanding and using data. It's a labor market sector where demand is far greater than supply—and this gap is growing. Demand continues to explode, yet universities and most countries are unable to respond by forming skills at a sufficiently rapid rate. In this area, passion and the necessary skills are essential. It's not enough to just "do the job."

The company's success is directly linked to its ability to attract and retain people with these skills, passions, and talents.

Significantly, technology companies develop products that don't have strictly economic justifications behind them. Some are used to motivate talent. This explains, for example, the choice of many companies to develop open source. Such decisions aren't always motivated by business concerns—they're often made to satisfy the desire of their software engineers.

Improving the company's attractiveness to talent must be the first thought of every manager of a technology company. I'm convinced that it's much better to have ten real talents than a thousand who are mediocre. But the competition for those ten talents is very high and international.

There's another aspect, too. A company needs to know how to attract a minimum critical mass of talented people since two or three would be isolated and would soon leave. For a top talent, being able to work with others of similar quality is extremely stimulating and motivating, while being in a place where you cannot compare yourself with your peers is frustrating. People need to be in love with what they do—if they can't do things the right way, then they'll leave. That's why, for example, managers of computer engineers must themselves be computer engineers.

Tiziano Vescovi: Are digital products changing? What are the main trends?

Luca Ferrari: First of all, the quality of the software is incomparable when you think back ten years. The improvements have been enormous. We have gone from digital utilities (e.g., weather, word count, translator, etc.) to content-rich digital services such as Netflix, Spotify, and others. This phase is now in its maturity, and we're entering the next phase, where what was analog becomes digitized. Think of Revolut, the bank where everything is done digitally. Think of Amazon, which transforms physical retail products into a largely digital system. Think of fitness, where the personal trainer interacts through digital software and no longer needs to be in the same room. The physical product doesn't become totally digital, of course, but an interface is created that reduces or even minimizes the need for the physical part of the products. Classic brick and mortar businesses are reinvented with a large digital component, to reduce costs and minimize customer effort, like with Uber.

Tiziano Vescovi: Pricing: How do you price digital products and services? What are the approaches behind understanding, measuring, and quantifying the customer's willingness to pay? What new revenue streams does the digital transformation enable?

Luca Ferrari: In the digital world, it's much easier to test prices—primarily, because the answers are immediate and easily measured. Furthermore, the marginal costs are practically non-existent. That lets us establish a price that optimizes based on demand alone.

For example, 100,000 users are offered a price of $2, and another 100,000 get a price of $3. The reaction obtained over the medium term is projected and the most profitable price of the two is set. So, you test the perception of value. It wouldn't be easy to do this in the analog world because the points of sale, test areas, and customer segments, etc. should be changed. The comparison wouldn't be correct. It would also take a long time and be visible to competitors. In the digital world, such tests can be done privately and globally in a few hours.

Tiziano Vescovi: Implementation: What are the main obstacles to a successful digital company?

Luca Ferrari: The most difficult thing is to find products that generate large volumes with good margins.

Time also represents a significant stressor in digital. Everything has accelerated compared to analog. As I have already mentioned, a big constraint is the need to market the products through the big players, such as Apple or Google. They direct and control the market. They do it with integrated logic, not only on apps, where they ask for 30 percent mark-up, but also on smartphones and tablets. Sometimes, these big players remove a developer's app from the sales platform because they think it's no longer in line with their strategies or because of third-party pressure. Another critical aspect is the acquisition of talent. This is very difficult for small, semi-recognized companies, especially at an international level. It's about developing a remarkable capacity for persuasion and involvement. Accessing the capital required to grow is another constraint—bank funding is always difficult for a startup. Another ongoing challenge is never to let the determination to reach the goal waver. This involves overcoming the fear of failing, of having to lay off people. For a startup, there is also the crazy effort of the huge workload needed in the early years, which can make you say, 'Enough, it's not worth it! I'm going to work for a company instead!'

Tiziano Vescovi: Advice for companies at the beginning of the journey: What lessons have you learned in the context of digital transformation? What advice would you give a company at the beginning of its journey?

Luca Ferrari: Find some very intelligent people, even if they're not necessarily super experts in digital technologies. I wouldn't give too much weight to experience. Technologies change so quickly that experience is secondary. In fact, it can sometimes be a hindrance, because it's based on the rules of a reality that no longer exists or, at least, that's ending. Smart people who are capable of understanding and acting quickly are required. Finally, realistic business objectives that don't require too much complexity to manage in the short term, and where technology can solve tangible and important problems in the immediate term, must be identified.

Chapter 11

Consulting for digital transformation

Interview with Giuseppe Folonari, European Head of Business Strategy, AKQA

Tiziano Vescovi and Giuseppe Folonari

The company

AKQA is a digital agency that specializes in creating digital services and products. It employs 2,200 staff globally with 29 studios. The turnover is about 250 million US$. AKQA is part of WPP, the biggest communication, advertising and public relations group in the world. AKQA Venice is the main office in Italy.

Giuseppe Folonari is the European Head of Business Strategy at AKQA.

Tiziano Vescovi: How do you interpret "digital transformation," from the consulting company point of view?

Giuseppe Folonari: In my opinion, digital transformation follows three converging lines:

1. Provide the company with technology enabling tools (CRM, corporate sites, data analysis technologies, new service technologies)
2. Update business models based on the possibilities that technology offers constantly, to respond to new dynamic competitive scenarios
3. Building experience in the digital age, conveyed through digital touch points, but also offline based on digital expectations (for example in relation to speed of response and personalization)

This convergence was separate concerning business management, the use of technology and the relationship with the customer, but now it is integrated.

Digital transformation is the combination of these factors: the ability of a company to adapt its business model so that it is consistent with strategic objectives, which responds to customer needs, by acquiring digital technological infrastructure.

Tiziano Vescovi: Can you comment on the process that leads to the current digital strategy?

Giuseppe Folonari: My reflection starts from the observation of the existing. Large companies are at the forefront of digital transformation compared to medium-small ones, which are not currently posing the problem significantly or strategically.

The possible balance must be identified considering what the company may be able to do and the safety of the company from a digital point of view. This means doing an external analysis between what the market expects and an internal analysis; often this is an underestimated aspect, to verify that the digital transformation process is consistent with the values of the company.

We are emerging from a market phase in which there has been a great deal of interest in experimentation, perhaps sometimes a bit of an end in itself. Experimentation creates interesting and different models, but not always defensible or proprietary, ending up remaining an experiment. Obviously, it must be said that in digital transformation there will always be a certain level of experimentation. The way to minimize the degree of failure of these experiments is to verify that there is not only a technological opportunity to be equipped with. Sometimes it happens just for fashion, such as "doing something" with machine learning, virtual reality etc. because a competitor has used them. The return on investment of the digital solution, the competitive value it allows to achieve and its defense against imitations or copies should be assessed in depth. This mainly affects the way in which the company and customers apply and use digital solutions.

Tiziano Vescovi: Which company is, in your view, best-in-class in digital transformation? What do you learn from their initiatives?

Giuseppe Folonari: The best companies are those that find the right balance between trust in the digital transformation process. They are processes that must bring a certain result in an uncertain context, and the ability to make decisions in an uncertain context. If you expect everything to be clear, which is also impossible, you never stop waiting and therefore never make a decision. On the other hand, if you decide too quickly, many innovative processes are killed in the cradle. So the companies that get the best results are the ones that have sufficient means to be patient and have the courage to make decisions.

Digital transformation does not work in those companies where there is a bulimia of strategies and the expectation of immediate return.

Excellent examples are Siemens and A2A. They placed the managers of the digital transformation project in a separate unit from the company, in a situation of freedom within the borders, keeping them connected to the company, but with the possibility of creating a new culture. This prevents company-established routines from slowing down or sometimes destroying innovation. The organizational choice is to create a pilot. The people of the company involved in the digital transformation working group should have been employed for a relatively short time, so as not to be nailed to past behavior. The people involved in the working group are the evangelists that the company puts in its body to spread the DT.

Tiziano Vescovi: Digital transformation and business model innovation: How will digital technologies enable the company to adapt and change the business model?

Giuseppe Folonari: Digital transformation is active in a logic of simplification and transparency. In B2B2C models there is a sort of small Copernican revolution. There have been numerous innovations in consumer models, very evident, logical and important, and this has been a very studied and documented phenomenon. In the complex value creation chains, such as the one of Siemens, where there is an enormous number of interlocutors and participants, the business model changes. But this also happens in relatively simpler contexts, such as that of the tire, the real innovation of the business model takes place when it is possible to create a positive relationship with the tire dealer. There they introduce innovations such as sharing information as a transactional lever. The company requests data from the tire dealer to have knowledge of what the end customer wants, who is the customer of the tire repairer. At this point the business model is modified to understand what can be given in exchange for such data, what services, which help for the tire repairer business.

B2B2C has a lot to do with digital transformation because this makes previously unthinkable types and quantities of services possible, both for B2B and B2C. Very often, B2C companies, especially in the case of small companies such as installers, craftsmen, etc., find it very difficult to find solutions to satisfy the growing demand for services from their customers. They need the support of the supplier, which in general is a larger and more structured company. B2B2C requires that DT and DT favor B2B2C in a circular way.

Tiziano Vescovi: Digital products – is the digital offer a complement or substitute to traditional products?

Giuseppe Folonari: There are some products that go from analog to digital, but the most significant aspect is given by the combination of the two. The services mainly take on a digital character. The combination is actually the most important challenge. Consumers expect more and more digital service parts related to the physical product.

Digital has shifted the evaluation criteria in purchasing decisions. The offer of good-enough products is growing, sold online in a simple way for the connected consumer, offline stores are subsequently opened; this happens for example for mattresses, fashion and accessories. These products can be easily developed and managed by digital operators. This evidently poses a threat to companies who remain only in the physical tradition, so the drive for digital transformation is growing.

Tiziano Vescovi: Is DT developed with network logic? I intend to ask if the company must become digital because its partners, suppliers, distributors, consumers, etc. are, and if it must interact with them, it is forced to use digital.

Giuseppe Folonari: Certainly, there is a theme of networking and mutual incentives to do so. There is a network effect. This, on the other hand, questions the decision to be the first mover because there are high costs (education, technological development, information) and long times to recover them, waiting for the value chain to become digital. This is why digital transformation is easier where there are great player leaders, who can bear a long period of return on investment. Subsequently, the existence of multiple digital operators reduces these costs, creates partnerships, increases the convenience to participate. Once the critical point has been reached, there should be a mass digitization for the network economies that derive from it.

Tiziano Vescovi: As a concept, does DT push towards partnership?

Giuseppe Folonari: On the one hand it produces a push, on the other it represents an incompressible reality, because it is no longer possible to reason on a logic of geographical defense or sectoral boundaries. This leads to the search for competitive advantages over specialization skills, towards partnership and aggregation between companies, favoring systemic logic. For example, Blue Apron is a company that supplies home kits of high-quality food ingredients and recipes to prepare them, a kind of Ikea food. Airbnb Experience brings the experience of visiting and knowing the place with an expert local person, who is not a professional guide. This experience can also include food. Blue Apron asked

Airbnb for the best recipes created in their Experiences to be included in the Blue Apron menus, and offering them to Airbnb, expanding the service, working on the concept of the common target, on which they build a digital-based partnership (information and service).

Tiziano Vescovi: Does DT increase the value and, therefore, can allow price increases or, on the contrary, does it produce price competition? Is the digitized product sold at a higher or lower price? What does the customer expect?

Giuseppe Folonari: The digital perception of the consumer creates an expectation of obtaining a lower price, but there can also be a greater willingness to pay for an adequate service. The reference point is Amazon. Amazon built its success upon low prices; on the other hand, there are statistics on the incredible frequency of purchase of Prime customers. Therefore, Amazon works on the frequency of purchase. Move the price assessment from the single purchase to the overall experience and therefore on the customer's lifetime value. The evaluation of online metrics focuses not so much on the average receipt, but on the customer lifetime value. This is an aspect on which Amazon is extremely aggressive and competitive. The shift to the customer's lifetime value amplifies the possibilities for companies. However, the definition of pricing always remains tied to the perception of value. Nevertheless, there is always the ambiguity linked to the perception of digital as a free product. There is therefore a conflict between the perception of greater value that pushes the customer to greater availability at a high price, and the belief that digital does not cost. Obviously, DT does not act only on the consumer or on the final product, but even and a lot on the process, therefore it is not connected to a digital object, but to a redesign of the business model, so the price is an indirect result. Sometimes the results of the DT process are not even read as digital by the end consumer. This generates perceived value and is not connected to digital gratuity perception.

Tiziano Vescovi: How is the sharing of data held by a company, often considered a strategic resource to be protected, judged? Positive or negative?

Giuseppe Folonari: Sharing is acceptable if it enriches the data itself. If it allows the company more knowledge, sharing makes sense. The business model relating to the sale of data, by selling the data in my possession to those who pay an affordable price, is becoming less popular, increasingly being replaced by enriching exchange.

Tiziano Vescovi: Communication and branding in DT, what changes?

Giuseppe Folonari: Change is radical. There is no distance between what is communicated and what is reflected in the service and product. This necessarily requires very strong coordination and cohesion between the departments. So, communication, brand management, distribution and sales must work in close symbiosis. A great effort of multidisciplinarity is necessary, an attitude of the type "I deal with sales and that's it" or "I make communication, someone else has to sell …" is no longer acceptable. Integration and multidisciplinarity have always been true to a certain extent, but now they are absolutely fundamental and critical. The market and the consumer would immediately perceive discrepancies between departments. People enter the shop with their smartphone in hand, saying "something else is written here than what you say." Often the information reached the consumer before the seller.

In the past, communication and brands followed a top-down process, from the company (which decided what to say and do) to the customer. Now customers want to experience the brand and check its promises (e.g. speed of response). The brand passes from a storytelling model to an experience model. From a one-voice brand, now we are at the many-voices brand.

Now the brand manager no longer has to think only about what to say but has to understand how others can talk about the brand. In some cases, three out of four people, before making a certain choice, listen to the suggestion of an influencer. This obviously substantially influences the selection process. Moreover, employee messages about the brand also have a very high impact. The brand partners, when describing or telling it, produce a very high impact, so we have a multiplicity of storytellers. There is a control problem, but a multivoice communication model can be set up, creating campaigns composed of many actors, such as customers, employees, suppliers, distributors, providing the concepts and communication production tools, offering different images of the brand, consistent but not the same. The identity of the brand is important but not as unique as it could have been previously.

Tiziano Vescovi: What metrics can be used to measure DT?

Giuseppe Folonari: In the overall context, where experience, service, product / service combination and communication change, a new measurement problem arises.

Traditional ratios (sell-in / sell-out etc.) continue to be relevant, but these indicators do not keep track of the customer's "temperature," nor if his/her expectations are changing. From this point of view, new measurement ratios are needed to integrate existing ones with indicators of more immaterial phenomena. AKQA has developed the Customer Experience Impact (CX Impact), which is based on a questionnaire to measure the perceived quality of the customer experience and demonstrate a statistically significant relationship between this index and some business indicators, such as the intention to purchase, word of mouth, brand equity (credibility, transparency, etc.). Progress in DT (digital literacy, predisposition to digital innovation) is measured with speed and autonomy, i.e. how much people feel empowered to move independently and quickly, and with the change of work processes. The latter aspect requires an organization close to the logic of digital businesses: small teams, with very clear objectives and a sufficient degree of autonomy. An evolution is therefore also an evolution of the organization chart; in this case we cannot speak exactly of a metric but rather of an indicator. The cultural aspect is not measured with a specific metric but there are other indicators: the adoption of digital tools, in this case the number is evaluated; digital literacy, number of people who have attended courses; exposure to external innovations, how many people have been exposed to innovative methodologies.

Tiziano Vescovi: Which are the main obstacles to the application of the DT?

Giuseppe Folonari: There are hindering factors both internal, the existing organization, and external, customers, consumers, suppliers, sales force, distributors, etc.

The real culprits are almost always internal factors, external conditions are generally an enabler. For example, the performance of market and profit results, i.e. the focus on quarterly deadlines, can lead to a distraction from digital objectives and a re-focusing on routine systems, in order to achieve budget results in a business emergency. The concentration on DT is therefore postponed to subsequent times. This aspect is not the first killer of the DT, but it is an endogenous condition that can influence a lot: it is a catalyst for internal obstacles.

The real root causes are due to an internal cultural nature, about how extensive a DT process should be. Very often DT is approached following a single guideline (business process, technology, customer experience): "We spent a lot of money on CRM and now nobody uses it." In the end, the reason for the crisis ends up being the lack of continuity of the transformation process, which stops at only one pillar of the palace.

Tiziano Vescovi: What are the necessary skills and competencies for operators in marketing, sales management, etc. to be an engine or at least a digital transformation actor?

Giuseppe Folonari: The ability to maintain their specialization understanding the complexity and interconnection between the parts necessary for the DT. It is about developing a set of DT skills, both vertical and horizontal.

It is not possible for marketing people not to have sensitivity on numbers and, vice versa, the numbers are not enough to understand the situation of the markets. A transversal study or specialized professional path is preferable. Then you need to develop familiarity with digital processes.

Tiziano Vescovi: Are there areas where it makes no sense to talk about DT?

Giuseppe Folonari: It makes sense to talk about it on all business areas and economic sectors, albeit in different time periods. There are some sectors where the process is already very advanced, for example the media, communications, tourism, hospitality. There are others where DT is coming, traditional services. Then there are others, such as, for example, mechanics and steel, where the business model continues to remain the same, is optimized, fine-tuned, since in those industries DT has a lot of impact on process technologies rather than the relationship with customers.

Tiziano Vescovi: What is the first advice to give to a company that wants to go digital?

Giuseppe Folonari: The awareness of a need for cultural change. Often there is a misalignment between management, who above all feels the need for DT and who is concerned about the general business situation, and some internal situations, where there is already DT or in any case a lot of technological modernity. So, it is important to identify innovation ambassadors, understand where they are, protect them, building on them the development of people, patents, innovations, products, that is, the future of the company.

DT is strongly driven by market change, so knowledge, information gathering, market awareness are fundamental. The question to ask is: how has customer life been digitized?

Tiziano Vescovi: How do you measure the impact of DT on profit and ROI?

Giuseppe Folonari: DT's processes penalize profitability in the short term because they require large investments that affect company accounts. They only start to become profitable after 3–5 years. If the manager's performance is measured at the end of a year, the situation becomes critical for any investment in the medium term. It is easier for listed companies that DT is made convenient even before the results, because it is much more expendable in year-end and prospective reports, as signs of innovation and future value.

Tiziano Vescovi: Are there other aspects to be taken in account in DT?

Giuseppe Folonari: Often there is difficulty in finding the right partner to accompany the company in the DT journey, especially if it is a SME. It is difficult to select it, since there is still considerable confusion and fragmentation in the type of ideal partner; must it be technological? Must it be strategic consulting? Marketing consulting? The consultant should act with great responsibility and not only with sales objectives. Increasingly, the companies build a network of DT suppliers, where a common, shared and systematic work method is still often lacking.

Tiziano Vescovi: Who is the key reference person in the company?

Giuseppe Folonari: The key reference person is increasingly the chief executive officer or the entrepreneur, i.e. top management. They are not the company's IT systems managers, who do not always have right digital skills and digital work logic. Top management commitment is paramount.

Chapter 12

L'Oréal digital consumer operating system

Axel Adida

Introduction

COVID-19 turned 2020 into a period of instant digitization, an acceleration challenge for most businesses and decision makers. Confined at home, office workers have fallen back on video conferences (2019: 10 million daily Zoom participants; April 2020: 300 million). With restricted access to physical retail, consumers have unabashedly embraced eCommerce (2019: 16% penetration in the US; April 2020: 27% penetration).

Numerous digitization frameworks, organizations, and transformation methods are found in scholarly and business literature. However, the pace of digitization of consumer behaviours remains a challenge for large-scale organizations, with a vast workforce, diverse brand portfolio, and global footprint. Every day, press articles invite marketers to reconsider how much and how deep digital is affecting people's longstanding habits and tastes: "have Millennial killed the doorbell?" questions the *Daily Mail*; "generation Z feels intimidated by full stops" remarks *The Telegraph*.

Lessons can be learned from L'Oréal digital acceleration (2014–2019), which instilled marketing and sales with a fresh digital consumer operating system that paved the way for business growth (2013: 23 billion euros revenue; 30 billion euros revenue), and resulted in sustained competitive positioning (#1 beauty group) and business resilience.

This chapter shares my personal experience of L'Oréal's digital acceleration (2014–2019) and outlines (1) Consumer-centricity, focusing attention on eCommerce, personalized consumer relationships, and brand love. (2) People powered, with digital expertise and upskilling enabling change. (3) Systematic pilot–proof–publicize– replicate approach to scale fast. (4) Technology and data factories, delivering lasting economies of scale.

A particular highlight is given to advertising, a core business driver for fast-moving consumer goods to promote brands and products; this chapter specifically articulates the consumer shift from traditional broadcast media (TV/press) towards digital precision media (social / video networks), and the specific precision advertising transformation challenges for the organization and its creative / media agency ecosystem.

2010s' rapid beauty consumer digitization

In 2014, L'Oréal embarked on a digital acceleration journey. Beauty eCommerce sales were still nascent; yet, consumers were showing signs of mass digitization.

History of digital beauty innovation

Following its "seize what begins" motto, L'Oréal had constantly innovated in the digital space when the internet was in its infancy. Digital consumer connections were being fostered: Lancôme rolled out its first website in 1995, rapidly followed by 35 global brands (i.e. L'Oréal Paris, Maybelline, La Roche Posay, Kerastase …) and 70 countries. As YouTube and Facebook were rising to prominence, digital content was being developed. Circa 10% of media budgets were being shifted from TV and press to digital media.

New digital experiences were being developed that defined personalized consumer relationships in the beauty category: in 2014, L'Oréal Paris Make Up Genius reached the App Store shelves, making virtual try-on accessible to consumers, followed powered with artificial intelligence and augmented reality.

Consumer digitization at mass scale: smartphones, digital platforms

Smartphones were increasing digital access at an unprecedented speed: US smartphone penetration had grown from 17% in 2009 to 75% in 2014. Suddenly, it became common place to see younger customers holding smartphones in store to check prices, verify ratings and reviews, before proceeding with a purchase.

Video and social networks reached a viewership never reached before by any media in any country during the broadcast era (TV / press / radio): each month, 700 million beauty videos were being watched on YouTube. Daily, 10% of the world population was connecting to Facebook (890 million daily average users in December 2014).

Unconventional new normal: new contents, new talents, new brands

Unconventional new content and ads, suited for 6–8-inch smartphone screens, were reshaping consumer branding and communications.

New talent proved particularly gifted at attracting millions of viewers. Michelle Phan's beauty tips resonated with young and daring beauty addicts: "Barbie transformation tutorial" or "Lady Gaga Poker Face tutorial" targeted those with a passion for beauty.

New social savvy brands, such as Anastasia Beverly Hill, Huda Beauty, or NYX, also proved particularly gifted at gaining consumer attention and preference. Rapidly, ColourPop and Kylie cosmetics would follow a similar playbook.

Lower barriers to entry for DNVBs

Rising digital native vertical brands (DNVBs), operating social networks and eCommerce expertly, were now challenging leading heritage brands, founded during the TV / press and mass retail era.

In the past, quality beauty products could be sourced from independent manufacturers. Still, upfront TV deals had favoured the bigger brands and media budgets. And large teams had proven essential to benefit from the scale of national retailers.

Digital was contributing to lowering traditional barriers to entry in the beauty industry: digital media auctions now favoured the savvier brands; direct to consumer eCommerce could be kickstarted in a Silicon Valley garage, with minimal staff and operations.

Consumer-centricity: eCommerce, personalized relationships, brand love

Pointing to the magnetic north

L'Oréal has chosen to universalize its brands: a core identity, with diverse local expressions reflecting differences in beauty needs, traditions, and desires. With a distributed organization, each team is in charge to deliver business results, whilst contacts with corporate teams are regular but infrequent.

To achieve change at scale and at speed, the strategy formulation needed to provide a simple magnetic north for every team member that was memorable to mobilize people, easily sharable to onboard new team members and agency partners, and applicable everywhere to foster entity-centre conversations. Persistent over time to maintain organizational focus. And, above all, self-evident from a consumer-centric perspective.

New beauty consumer digital operating system

During an interview, a young consumer pointed out she had been born in a digital world; hence she was equipped with a digital operating system to discover new brands and purchase new products.

eCommerce fit well with her laptop usage at home and smartphone usage on the go: "I occasionally go to Sephora, I shop more regularly on Sephora.com, for novelties, or discount sites, for prices". And most of the brand content she saw was on digital media and social networks: "it has been a while I have not seen your TV ads: the TV is rarely switched on, an only turned on cable news". DNVBs piqued her curiosity: "my latest beauty buys are from brands I discovered on Facebook and YouTube". Her loved brands and desired product were often recommended by other users: "I Googled new brands, their product user reviews convinced me they would work for my particular skin issues".

Horizon: eCommerce, personalized consumer relationships, love brands

Hence, the digital strategy shared internally and externally reflected this consumer centric perspective: eCommerce, personalized consumer relationships, and love brands.

Over a five-year time-span, L'Oréal Group's annual reporting has echoed the Group's digital achievements, in line with the initial consumer-centric strategy (see Figure 12.1).

People powered: digital expertise and upskilling

Inflow of senior leadership veterans from GAFAM, agencies and start-ups

In a bold move, L'Oréal gave digital a voice at the Executive Committee, nominating Lubomira Rochet as chief digital officer.

With a mandate to drive the digital acceleration across the L'Oréal Group, veterans from GAFAM, agencies and start-ups joined the senior leadership.

To sustain change, a CDO team was set up, with the right blend of strategy, delivery, and support experience: COO (digital operations – myself), CINO (digital innovation), commerce lead, brand lead, CHRO, CFO joined the team. Rapidly, brand (and divisional)

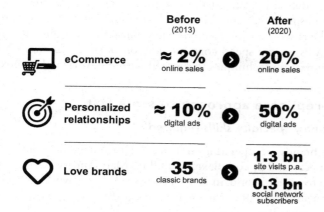

	Before (2013)		**After** (2020)
eCommerce	≈ **2%** online sales	❯	**20%** online sales
Personalized relationships	≈ **10%** digital ads	❯	**50%** digital ads
Love brands	**35** classic brands	❯	**1.3 bn** site visits p.a. **0.3 bn** social network subscribers

Figure 12.1 L'Oréal Group annual reporting.

CDOs, and country (and regional) CMOs added digital expertise and changed management skills across the organization.

Enabling marketing and sales organizations: 2000 experts, 33,000 upskilling

With centrally strategized and locally executed operations, having the critical mass of digital skills within each team is essential to assess local market conditions, implement digital marketing autonomously, and on occasion request CDO team guidance and assistance.

Digital is a learned trade for every professional; still, time is needed to attain proficiency. A total of 2000 digital expert nominations cut short the time required for each brand and country to have a critical mass of digital skills, from content development to web site management.

Digital marketing is at the core of marketing and sales challenges, hence bound to be on every job description. Hence, 33,000 marketers and salespeople followed a formal digital upskilling, highlighting the new techniques required to succeed in eCommerce, personalized consumer relationships, and brand love.

Lines and roles were bound to blur over time: acquired digital talent would develop beauty category expertise, and upskilled beauty marketers and salespeople would grow digitally.

Digital first management rituals

Salespeople often say that "retail is detail": following the consumer journey to prevent any mishap, focusing on every marketing and sales metrics to ensure operations profitability.

With a similar culture where "digital is detail", the CDO team provided frameworks and material to ensure that all management rituals started with digital from country visits to brands reviews.

Meetings would start with a digital portrait of the consumer landscape, L'Oréal and competitor business KPIs; the consumers and insight would be highlighted through eStore visits. By 2019, country presidents would enquire on ad viewability, brand presidents on social engagement, marketing VPs on site speed consequences on organic searches, sales VPs on bundling and text optimization techniques to succeed with the Amazon buy box.

Pilot–proof–publicize–replicate approach to scale fast

Software has eaten FMCG: new specialty skills required

Fast-moving consumer goods brands are operating in a more demanding technological environment that has grown in size and in complexity. Fulfilling Marc Andreessen's 2011 prediction, software has proven to eat the world. In 2020, a large share of the world's largest companies are technology companies; 483 unicorns are private companies with valuations above $1bn, dominantly in the technology sector.

New skills are required to operate successfully in today's environment. Analytics experts tag websites and apps, eMerchandisers handle data flows with eRetail partners, data engineers and scientists assemble and analyse data, precision media experts buy social networks ads, and digital communications deliver engaging short / mobile messages, and excel at influencer outreach.

New skills are required to meet rising consumer expectations. New sales channels need to have fewer glitches (i.e. pixel perfect product detail page, searchability, easy buys and returns). Personalized consumer relationships are expected everywhere (i.e. programmatic, Google walled garden, Facebook/Instagram walled garden, etc.). Conversations on social networks need to be personal (i.e. answers to Instagram comments). Products need to be searchable on Google, Amazon or Sephora (i.e. search engine optimized).

New competitive advantage: critical mass of skills on the frontline

In the digital space, small brands can emerge with great talent and small operations. Still, large brands have the critical mass to attain technology prowess, which they can turn into a new competitive advantage

Specialty skill sets are critical to win the algorithmic competition, for brands to top Google result, for products to appear on Amazon's buy box, or to command the lowest cost per contact or per acquisition. In the TV/press and mass distribution era, competitor market shares were easily accessible (with a fee disbursed to panel organizations). In the digital space, data engineers are required make sense of patchy data, monitoring competitor brand performance on Amazon, or highlighting success with specific consumer segments and tribes.

A critical mass of business allows a critical mass of specialty skills for each country and for each brand. Beauty, as with a number of FMCG categories, requires a strong local flavour to flourish: products, communications, commercials, on/off trade promotions and activations – and in country (or regional hubs) are necessary. Typically, precision advertising experts are needed to optimize the media ROI (return on investment) optimization funnel, to deliver favourable cost per effective contact: right media mix (digital vs traditional, right touchpoints by category), right buying (fraud, viewability, auctions), right creative (launch pack, checklist, what works by platform). As every digital platform imposes its own software / self-service for data management and ad buying, bigger operations can advertise across the board – small operations are constrained to specialize.

Scale-up method: pilot–proof–publicize–replicate

Home growing digital specialty skills in each of L'Oréal's 35 brands × 70 countries would have equated to operating over 2000 companies, achieving limited if any economies of scale. Pilot–proof–publicize–replicate has proven a successful method to scale-up, allowing the distribution of new methods and techniques across the organization, and leveraging emulation to spread the knowledge across L'Oréal.

In a limited set of pilot countries and brands, local marketers were offered the opportunity to lead a pilot of new marketing activations, with CDO team support and funding. Proof points were made: what worked – what did not work. CDO team guidelines were documented with Golden Rules and How-To supporting material. Local marketers would then become pilot champions, and publicize the results across the Group (papers, seminars, upskilling). Other teams eager to boost their own results would call-in for assistance to scale up proven recipes locally (insights, guidance, assistance). Wash–rinse–repeat: each year a new set of pilots is chosen to bring digital operations to the next level, while the proven business practices piloted the previous year are being scaled.

Relying on local entrepreneurship to deliver results fostered organizational agility: it relied on the traditional team emulation across the marketing and sales communities to spread results fast. It proved economic, with local teams uphauled with central resources at first, resulting in straightforward guidelines that could be deployed with limited central support.

Relying on collaboration to scale up pilot learnings, the method also fostered new ways of working, and a sense of community across L'Oréal experts and agency professionals. Over time, seminars and calls led to similar working habits, and raised the ability of the organization to learn.

Case study: precision advertising

2017: digital media attention surpasses TV attention

By 2017, US consumers watched more digital than classical media (out of 12 hours and 7 minutes media time per day), and more on their smartphones than on personal computers. Older consumers' TV attention remained high, but younger generations shifted massively their attention to smartphones and online TV.

Additionally, multitasking was depleting attention to TV commercials: in front of the TV, consumers kept browsing their smartphone – with peak usage during TV commercials.

By 2019, digital platforms further dominated the digital advertising space: 2.5 billion users on Facebook, 2 billion on YouTube, 1 billion on Instagram, half a billion on TikTok. With a new normal in the media landscape:

TV era	Digital media era
TVC 30 seconds	New digital short formats
4:04 daily watch time	5:50 daily watch time (>50% mobile)
Yearly commit, volume discounts	Personalized to consumer interest
Traditional buys (phone, meet)	Technologic buys (auction platform)
Decreasing ROI (price up, attention down)	ROI not guaranteed (price for attention)

Media ROI digital challenge

TV ROI has long been predictable. TV ad space price is fixed beforehand. Media plans are well mastered by media agencies everywhere (reach, frequency). Obviously, the cost per GRP could vary between channels, notably because of the audience mix (age, interests, etc.). Yet the impact on sales could be attributed to product market fit and creative quality.

Digital media ROI is not as predictable as TV ads were: price is not fixed (digital ad space is auctioned, ineffective ads pay more), consumer response is not certain (many ads are skippable), reach and frequency are not automatic (targeted advertising is the new normal: consumer data is key to reach the right socio demographic, tastes, interest, and intent to buy). Lack of buying prowess results in media space waste; lack of creative congruence results in attention waste. For similar products, competitive brands may have to incur very different costs per useful contact.

The price of digital ads varies. Google sponsored keywords prices vary depending on the category – cars, makeup or shampoo command different prices; broad and precise keywords are also priced. Programmatic, Facebook or YouTube auctions foster important price variations: the cost per a few seconds of attention can greatly vary: 2 seconds of attention on Facebook, 6 seconds and attention on YouTube, and a sponsored link on Google for a query showing intent to buy may command very different prices. Additionally, platforms favour engaging ads to occupy limited ad space: low performing ads may incur restrictions in reach, or a majored price point.

Consumer response to digital ads greatly varies. In front of a TV, consumers were used to paying minimal attention: only a few changed channels or muted the sound during commercials. In front of a smartphone, consumer attention is not guaranteed: ads are skippable. Consumers can skip the content with their thumb: a lesser quality ad may not be heard at all. TV programmes that lasted 30 minutes with the product and pack shots at the very end, rarely drove strong brand and product memorization, since most users on Facebook or YouTube skipped beforehand.

Pilots to develop precision advertising prowess

Precision advertising prowess is decisive to deliver digital media ROI: lack of buying expertise, limited data-driven optimization, and light supervision may result in marketing dollars being wasted. Digital ROI is absolutely predictable – if experts are on deck, analytics deliver relevant insights and benchmarks, and if the client–media agency relationships are learning relationships, they are fuelled with data-based decision making.

Batches of precision advertising pilots have been led since 2014, on select brand and countries to adjust to the new media landscape. Pilots provide proof points on simple precision advertising activation mechanisms. Proof points are then summarized in precision advertising guidelines, KPIs and ROI measurement, for roll out to other brands and geographies.

Which platforms are more effective for each category? Which deliver the best ROI? What is the winning TV vs digital mix for each category and country? Which formats should be produced for all new product campaigns? How to internalize successfully part of media buying? What are the right tactics for search? Which first, second or third party data delivers a significant boost of consumer relevance?

A yearly cycle of pilots and scale-ups constantly empowers the community and raises the bar of operational excellence. Each year, new pilots are being launched, and precision marketing operations are raised to a higher level. Each year, as guidelines (strategies, tactics, KPIs benchmark, and ROI measures) from past year pilots are being scaled, the whole organization is further empowered.

Precision advertising transformation results

As the precision advertising programme unfolded, the following results were achieved (see Figure 12.2).

Technology factories and economies of scale

Shared group-wide digital capabilities achieved economies of scale (i.e. code reuse, duplicate production avoidance, time to market) and greater agility (i.e. with the breadth of required features, most countries are subscale to develop technologies on their own).

KPI monitoring

A 360-degree digital cockpit monitoring 20,000 data points (i.e. Google Analytics, Facebook Power Editor) was set up to measure the scale, efficiency, and effectiveness of digital activations. The performance of L'Oréal brands was compared to past benchmarks (i.e. cost per ad view per category), and to competitors with panels (i.e. Similarweb, Clavis insight). Marketing and salespeople can monitor in real time aggregate KPIs and detailed

Broadcast different brand propositions	>	Systematic targeting Personalization at scale
TV native 30" TVC	>	Digital-native 6" formats
Maximize SoV Minimize GRP	>	Data-driven marketing
Media & comms agencies lead	>	Brand direct agencies digital acceleration
Digital experts lead digital marketing	>	Marketing and media teams are fully digital

Figure 12.2 Precision advertising programme results.

metric commerce (e.g. ratings, buy box), media (e.g. cost per 6-second views), and brand (e.g. Share of UGC).

Web, eCommerce and services

A website factory systematized technology reuse for the 2000-plus L'Oréal properties, based on SiteCore and Demandware technologies. A service factory facilitated the roll-out of technological apps across countries and brands (i.e. Modiface cosmetics and hair try on).

eCommerce and digital contents backbone

A backbone for eCommerce and digital communications was deployed (product information management system – Salsify, digital asset management system – OpenText). In a fast-paced environment, images and texts are developed once, and shared with brands in countries for adaptation, and with sales in countries for eMerchandising. Thus, duplication of marketing material was reduced, and exhausting email one-to-one communications were limited: after a new product launch call, marketers worldwide would know where to find the right brand material for localization.

References

Axel Adida, interview (2018, June). Quand le digital transforme la publicité de L'Oréal. Available at https://bfmbusiness.bfmtv.com/mediaplayer/video/business-transformation-quand-le-digital-transforme-la-publicite-de-l-oreal-1106-1082269.html.

Jean-Paul Agon, interview (2014, September). Google ne peut pas tuer L'Oréal, au contraiire. *L'Usine Digitale*. Available at www.usine-digitale.fr/article/jean-paul-agon-google-ne-peut-pas-tuer-l-oreal-au-contraire.N285733.

Comscore (2015). US Smartphone Subscriber Market Share. Comscore reports. Available at www.comscore.com/Insights/Rankings/comScore-Reports-December-2014-US-Smartphone-Subscriber-Market-Share?cs_edgescape_cc=FR.

François Dalle (2001). L'Aventure L'Oréal. Odile Jacob.

Marc Duquesnoy, interview (2018, December). L'Oréal fait rayonner 40 marques à l'international. Hub Institute. Available at https://hubinstitute.com/2018/hubday/transformation/replay-MarcDuquesnoy-LOreal-VesnaCosich-Sprinklr-social-media-manager.

eMarketer (2017, 2019). Average time spent per day with major media by US adults. Available at www.emarketer.com/chart/206481/average-time-spent-per-day-with-major-media-by-us-adults-2017-hrsmins.

Nico Grant (2020, August). Zoom crash leaves students and businesses scrambling on crucial day. *Washington Post*. Available at www.washingtonpost.com/business/on-small-business/ zoom-crash-leaves-students-businesses-scrambling-on-crucial-day /2020/08/24/1e4b68fc-e64a-11ea-bf44-0d31c85838a5_story.html.

Helena Horton (2020, August). Generation Z finds full stops intimidating, experts find. *The Telegraph*. Available at: www.telegraph.co.uk/news/2020/08/23/generation-z-find-full-stops-intimidating-experts-say/.

Influencer Marketing Hub (2020). Number of subscribers per platform (Google, Facebook, Instagram, etc.). Available at influencermarketinghub.com.

L'Oréal (2020). Portrait of the Group, its culture and its brands. Available at loreal.com.

L'Oréal Finance (2015–2019). Annual reports. L'Oréal Finance. Digital expertise reports available at www.loreal-finance.com/fr/rapport-activite-2015/digital, www.loreal-finance.com/fr/rapport-activite-2016/digital, www.loreal-finance.com/fr/rapport-annuel-2017/digital, www.loreal-finance.com/fr/rapport-annuel-2018/digital-4-3/, www.loreal-finance.com/en/annual-report-2019/digital-4-4-0/.

L'Oréal Finance (2020). First Quarter 2020 Sales. L'Oréal Finance. Available at www.loreal-finance.com/eng/news-release/first-quarter-2020-sales.

Emil Protalinski (2014, January). Facebook passes 1.23 billion monthly active users, 945 million mobile users, 757 million daily users. The Next Web. Available at https://thenextweb.com/facebook/2014/01/29/facebook-passes-1-23-billion-monthly-active-users-945-million-mobile-users-757-million-daily-users.

Hayley Richardson (2019, June). Have millennials killed the doorbell? *Daily Mail*. Available at www.dailymail.co.uk/femail/article-7107379/Have-Milennials-killed-doorbell-People-admit-theyre-scared-answer-aggressive-ring.html.

Lubomira Rochet, interview (2015, May). Quand Lubomira Rochet, la CDO de l'Oréal, détaille son plan d'action. Petit Web. Available at www.petitweb.fr/entreprise/quand-lubomira-rochet-la-cdo-de-loreal-detaille-son-plan-daction/.

Lubomira Rochet and CDO team, presentation (2017, November). Capital Market Days in Berlin. L'Oréal Finance. Available at www.loreal-finance.com/system/files/2019-08/CMD_2017_L_Rochet.pdf.

The ShawSpring Partners Team (2020, April). Accelerating Digital adoption: US eCommerce penetration 2009–April 2020. Analyst newsletter. Available at www.jaguaranalytics.com/wp-content/uploads/2020/05/ShawString-Quarterly-Letter.pdf.

The Pixability team (2014). The Social Video Beauty Ecosystem 2014. Available at www.pixability.com/insights-reports/beauty-youtube-2014/.

Sarah Vizard (2018). How L'Oréal drives marketing effectiveness and media neutrality. Marketingweek. Available at https://marketingweek.com/loreal-drives-marketing-effectiveness/

Tom Warren (2020, April). Zoom grows to 300 meeting participants despite security backlash. The Verge. Available at www.theverge.com/2020/4/23/21232401/zoom-300-million-users-growth-coronavirus-pandemic-security-privacy-concerns-response. www.theverge.com/2020/4/23/21232401/zoom-300-million-users-growth-coronavirus-pandemic-security-privacy-concerns-response.

Chapter 13

Internal start-ups as a driving force in the digitalization of traditional businesses

Interview with Jörg Hellwig, Chief Digital Officer, Lanxess

Andreas Hinterhuber and Jörg Hellwig

In this interview Andreas Hinterhuber and Jörg Hellwig, CDO of Lanxess, discuss the role of internal, digital start-ups as a driving force for the digital transformation of traditional businesses. Jörg Hellwig played a key role in establishing CheMondis, a digital start-up that today is the largest marketplace for chemical products in the Western world. Digital, internal start-ups allow traditional, analogue companies to understand, absorb, develop, and diffuse the digital capabilities required to compete successfully in the digital future. Internal start-ups are a driving force allowing companies to pilot large-scale organizational transformation on a small scale. The culture of the start-up – fast communication and decision-making – is a force that potentially can positively change the culture of the established business. Jörg Hellwig stresses the importance of building a platform supporting the move from selling products to selling services or solutions. Platforms, by nature, are environments in which transactions are executed allowing the platform provider to add additional services that then can be profitably sold. The digital transformation has direct repercussions on the nature of work and required capabilities. Big data analytics capabilities are a key differentiator: Lanxess has hired dozens of data scientists to internalize capabilities in data analytics. Digital technologies strongly influence innovation processes, both in terms of limiting the number of low-value targets as well as in terms of allowing an early parametrization of desirable innovation outcomes.

Andreas Hinterhuber: How do you interpret "digital transformation"?

Jörg Hellwig: Digital transformation is about combining new, digital technologies with existing core competencies to improve internal processes and to leverage new business opportunities.

Andreas Hinterhuber: Can you describe your approach to digitalization?

Jörg Hellwig: It is my firm belief that digital transformation needs to be treated as a business and change initiative, not as an IT project. As Chief Digital Officer (CDO), I am a business executive and former VP of a business unit, not an IT manager. Together with my team, I am reporting directly to our CEO Matthias Zachert, which demonstrates his full support. This allowed us to avoid typical mistakes, such as rigid deadlines (e.g. "results within 12 months") or tight resource constraints (e.g. "max. 10 people").

Another key insight is that we cannot implement the digital transformation sequentially: we digitalize our company concurrently and do everything at the same time. To do this we need to leverage new technology: in this case, we started scanning the market for technology, found partners, ran use cases, and then scaled the technology up.

We have a small, central team led by the CDO and our other team members are the 14,300 people working for Lanxess globally. We have created a digital driver community and run pilot projects with managers in our operating units. We do not pull away the best people from their operating responsibilities but leave them in their current functions. We thus treat operating executives as multipliers; we coach them, and we run pilot projects together with them.

Many of our competitors are taking a different approach: they pull their best people into large, central, digital teams doing only digital – this keeps digital activities away from operating activities and this cannot be good in the long run.

Andreas Hinterhuber: So you created a centre-led team of operating managers running digital projects. What were some of the external manifestations of the digital transformation?

Jörg Hellwig: One of our first initiatives was to create a web-shop for our products. After three or four weeks, I came back to our CEO asserting "We are on the wrong path", since I realized that we were trying to force our customers to come to us, that we were trying just to translate analogue process to digital.

So we changed our approach: we founded an independent software company serving as a digital marketplace – CheMondis – 100% owned by Lanxess, but selling not only our own products, but all products a customer may need. One thing is key here: CheMondis is fully independent, applying the culture and working conditions of a software company. Currently, more than 2500 companies are registered on the platform and about 50,000 products are listed. Lanxess represents less than 10% of the products on the CheMondis platform.

CheMondis went live in November 2018 and today it is the largest digital marketplace for chemical products in the Western world.

Andreas Hinterhuber: CheMondis must have enabled you to gain new business, but for sure it also cannibalized existing business from Lanxess?

Jörg Hellwig: CheMondis was for sure a disruptive approach for Lanxess. We lost business, less because of price, mostly because of superior availability and information of CheMondis. The new company is a disruptive force, not only for Lanxess, but also for competitors. We see the creation of this new venture as major milestone in our digital transformation, principally for two reasons: it has delivered valuable information about future market changes, which more than covers the cannibalization effects. In other words, it allows us to look into the future.

The new start-up allows us, on a small scale, to understand the future capabilities that all companies need to master in order to be digitally savvy. It reminds us about the importance of being platform ready, it stresses the importance to collect, update, recombine, and intelligently deploy data. Compare this with how most companies today deal with data: they have ERP systems, but then use Excel for production planning; they have data, but do not use it properly.

CheMondis shows us what the future can look like and is a constant trigger for our need to change. We will need to become data-driven in the future, and this approach allows us to understand the tangible benefits of being data-driven already today. This is our input for change: It is really cool to bring the digital world to Lanxess.

Andreas Hinterhuber: In other words: an important part of your digital transformation was building a platform – CheMondis.

Jörg Hellwig: Very well said – it became a cornerstone of our efforts. A platform only works as an open platform without limitations regarding competitors. The best example is Amazon, where competing sellers are listed side-by-side to Amazon. Convenience is key: customers shop around but end up with Amazon because the site is so convenient and so easy to use. The word "convenience", of course, takes on a different connotation in B2B, but not so much. In B2B, we do not need repetitive work done by people on different sides of the buyer–seller relationship. Our customers have started to demand from us the ability to connect their systems to our systems – and this was the initial idea when we started CheMondis.

We had the products and we wanted to create a software platform that we could put in the middle to serve the industry. CheMondis really is a huge application programming interface (API) connector for the chemical industry: somebody has the product, somebody else needs the product, CheMondis connects the two dots and the business is done on our platform to which then the CheMondis team add services, such as financial, logistic, insurance services, and so on.

The data that CheMondis collects as a result can then be used to serve the industry better: for a small company without a professional procurement department CheMondis can act as procurement office, procuring a catalogue of products at a price below the one the company would have been able to obtain on its own. In this way, CheMondis creates value for the industry and by doing so creates value for itself.

The independence of the platform is important. There is no lifeline between Lanxess and CheMondis. Employees working for CheMondis have a contract with CheMondis and there is no connection, no side contact, with Lanxess. Employees from Lanxess who wanted to work for CheMondis had to quit their jobs and apply for a position there. CheMondis is disruptive, also for our human resource department. Today, we are proud of CheMondis. Our baby is now in elementary school and will be in college sometimes soon.

Andreas Hinterhuber: I sense a potential conflict between your subsidiary and the parent company. How do you deal with established customers purchasing, presumably at lower prices, from your subsidiary, but then demanding after-sales service from the parent company?

Jörg Hellwig: I get this question frequently from sales and marketing executives around the world. My view is this: the assumption that there is conflict is a very old-fashioned view. We do not take anything away, we offer options and then let the customer decide where she or he wants to purchase and where she or he wants to obtain customer service.

CheMondis offers unprecedented convenience: we have 24/7 hour access, we have chat-bots with 80%–90% success in answering customer questions, we have a repository of answers to customer questions because many of these questions are the same, etc.

We do not say that a service done by a specialist is not required anymore: personal service will perhaps always be required, but digital tools facilitate service delivery. We equip our sales people with augmented reality/virtual reality (AR/VR) headsets, and this allows specialists who normally would have to travel in person to deliver customer service remotely and instantly. Previously, we should remember, customers would have to wait at least a few weeks until our specialists find a timeslot to physically travel to customer locations to solve any issues. Digital technologies bring speed and excellence to services.

Andreas Hinterhuber: So digital technologies allow you to create value that you could not have created before.

Jörg Hellwig: Yes. The idea is to create customer addictiveness via technology. We add service in such a way the customer says "You treat me better", regardless of technology.

At Lanxess we use data analytics to improve the information available to customers. The most frequent questions we receive are shipment-related questions: "Where is my container?" – our customer service reps would start to hunt the container down, calling production, logistic providers, until, finally, they locate that thing somewhere. We now have created a central repository of information – combining own data with data from our carriers – and an "At risk shipment alert" where the first things our service reps see when they log on their PCs or tablets in the morning are the five or ten shipments that are at risk of being late. Our service reps find that the machine has done the work for them, i.e. information collection, and that they can now work to prevent the delay or inform customers proactively.

Andreas Hinterhuber: An output of the digital transformation is the intelligent use of data.

Jörg Hellwig: The starting situation for most companies is unstructured data kept in silos. At a very early stage, we started a collaboration with Palantir, a software company specializing in Big Data and analytics. We created a data-lake, combining the massive amount of internal data with data from suppliers, logistic partners, and customers. With the help of Palantir we analyse this data: our vision is to become a data-enabled company.

We have Palantir employees sitting in our offices helping us to make sense of the data that we have: as a highly automated company, we generate petabytes of data in our production processes; in the past, this data was deleted when the hard drive was full. Now, with the help of artificial intelligence (AI) we use data from the data-lake and we build data models that help us to improve energy efficiency, capacity utilization, or on-time performance. We will realize our vision of becoming a data-driven company by making data-driven decisions. Data-driven models allow us to achieve small, but noticeable improvements, let us say 3%–5%, in key business outcomes. Over time these improvements accumulate.

Andreas Hinterhuber: You probably build hypothesis that you test with data, a bit like we academics in our best scientific publications.

Jörg Hellwig: This is exactly the point. Data and theory allow us to build hypotheses. We then use data to test hypotheses and implement improvement actions. We are building up our own expertise in this area: I recruited 15 data scientists, including an astrophysicist – we never had this type of talent working for our company. These people love data and they learn from Palantir. Our data scientists go out to the production plants, work with our managers, and sometimes they come back with solutions to problems we did not even know we had. This works when our data scientists look at the entire value chain, from sourcing to final customer deployment.

The chemical industry is complex, machines are so much better in analysing large datasets than humans; when we have the data on our tables, then we let human creativity, expertise and intelligence kick in to make much better decision.

The digitalization helps us to preserve information that would be lost if the person with that data left the company. The digitalization also helps to eliminate boring jobs.

Digitalization is frequently seen as a job killer. I do not use the digitalization to eliminate jobs, but digitalization ultimately means that every job is going to change. Not surprisingly, many colleagues are willing to help me as CDO to put their knowledge into a system and then they say. "Look, I help you to get rid of my current, boring job, which I have been doing for years; I would like to do something else, but then please train me because I want to stay with Lanxess since Lanxess is a great company." So this is the deal: we jointly have discussions, enrol them in certification courses on data analytics, for example, and help these associates to get more interesting, intellectually stimulating jobs. We also have these types of discussions with the works councils and the labour unions. With this new knowledge, these colleagues can create more value for Lanxess and they have much more interesting jobs than before.

Andreas Hinterhuber: A huge topic, both for the academic literature as well as for practicing managers, is innovation. How has the digital transformation affected your ability to innovate?

Jörg Hellwig: We are trying to accomplish an important thing here: we want to change the chemical industry for good. The digital transformation is a never-ending journey. The first years were about understanding data, structuring data. Now the digital transformation is about machine learning and AI.

Innovation in the chemical industry is largely a trial-and-error process. We may have an idea, go to our labs, and conduct hundred trials, five hundred trials, to create a product. If the first trial works, everybody is celebrating. So you could say: there must be a better way of doing this!

And although we have worked like this for decades, there is probably a better way: So we teamed up with a company from Silicon Valley, take the data from prior innovation projects, combine this with a powerful AI engine with one objective: we want to limit the number of trials that we need for successful innovations. We figured out that AI indeed can help us in telling us which trails to eliminate. AI furthermore helps us to define certain product parameters ex-ante, instead of finding these parameters after hundreds of trial-and-error iterations. Take recyclability and the circular economy, huge topics now and in the future. AI is helping us in product development to bring products to the market that have recyclability built in. The automotive industry is a very substantial end market for many of our products: having total recyclability built into the DNA of our products right from the start, the ability to use cars at the end of their lifecycle as input for raw materials for our new products, will change our industry for good. Using AI for innovation is a huge step forward. It is not that we have not done this before, but now digital technologies have opened up new possibilities.

Andreas Hinterhuber: AI speeds up innovation because it identifies less promising targets early and reliably.

Jörg Hellwig: Software alone is stupid, but if you add domain knowledge, like our chemistry expertise, then this combination is valuable. With the help of AI we will in the future conduct perhaps only 10% of the trials that we would have done if we had worked without this data. In addition, AI, as outlined, can help us to come up with radically better new products. We now have new tools that we did not have before.

Andreas Hinterhuber: Finally, metrics. Which metrics or key performance indicators do you use to make sure that your digitalization is on track?

Jörg Hellwig: For our traditional projects we build business cases, calculations showing 4–5 years payback time before we move ahead. For digital projects, this is very difficult – there are too many unknowns. We invest in software, but we do not really know if this money will ever come back. Matthias Zachert, our CEO, fully supports the digitalization. He says: "I do not know if this initiative will be successful or not, but I do know that if we do not do this, then I can be 100% sure that in 2–3 years we will fall behind".

Digitalization leads to productivity and efficiency improvements. We are able to record the positive effect in production capacity utilization, and many other metrics. Digitalization also leads to a different approach to the market: we are faster, customers value us more.

We do not use tangible calculations to document the impact, but observe the intangible benefits of the digital transformation: less pushback on prices because we deliver more value to customers, resulting in improved customer loyalty, faster time to market, or higher new product success rates.

Finally, we do not want to put too much effort in tracking the return on digital investments. Matthias Zachert says: "I do not need 15 accountants trying to calculate the benefits of the digitalization". We are doing the right thing, it feels good, and we have established a new way of working together. Nowadays, people from different business units share ideas, we work much faster than we did before, we work in sprints, have scrum meetings and have learnt a lot from the digital world, where this new way of working together – speed – originated. We learn a lot from CheMondis. Meetings, for example, used to last for hours. We now meet for 30 minutes, we move out tables and chairs, people do not bring their laptops, speak out and really contribute in a different way. These are all small steps, but they have a major impact on our business, also on our ability to innovate.

Andreas Hinterhuber: Mr Hellwig, I truly thank you for the privilege of having this first-rate exchange of thoughts on the digital transformation.

Jörg Hellwig: Thank you. I likewise enjoyed our exchange of thoughts.

Part 4

Digital transformation and customer value creation

Chapter 14

Digital transformation and consumer behaviour

How the analysis of consumer data reshapes the marketing approach

Francesca Checchinato

Introduction

Digital, and especially mobile technology, has fundamentally changed the way people consume media, develop brand preferences and choose products. The rise of social media has facilitated customer-to-firm communications as well as customer-to-customer interactions, affecting traditional value-creation structures and relationships (Reinartz et al., 2019). Since the aim of marketing is to satisfy clients in order to generate profits, new technologies and their digital adoption greatly impact marketing structures and strategies.

Digital transformation affects the entire marketing ecosystem (Quinton & Simkin, 2017), and thus scholars suggest that a paradigm shift has occurred, in which digital marketing is a very important part. Digital marketing involves computer-mediated environments that allow consumers, firms and other stakeholders to create, interact and access digital content (Hoffman & Novak, 1996), contributing to enhance not only customer satisfaction and retention but also management complexity.

Even if the firm acknowledges the potentiality of digital technology for developing its marketing strategy, it is sometimes not able to strategically embrace it. Competencies as well as an organisational culture supporting this transformation are needed and should be developed, especially in the case of traditional companies.

One of the main shifts that differs from the past is related to the relationship with prospects and customers. According to Berman (2012, p. 16), 'leading companies focus on two complementary activities: reshaping customer value propositions and transforming their operations using digital technologies for greater customer interaction and collaboration'. New tools such as website and social network profiles have been implemented to answer the call for interaction, but the key point seems to be to understand the customer journey and develop marketing strategies accordingly.

This chapter analyses how companies deal with the new consumer behaviour, since this issue represents one of the main challenges of the digital transformation. First, we briefly analyse how digital technologies impact consumer behaviour and customer experience. Then, we describe the data-driven approach and how companies can exploit this data to define their marketing strategies. Finally, we analyse the impact of this new approach on the main marketing levers.

Technological developments and their impact on consumer behaviour

Technological developments are transforming consumer behaviour and customer experiences (Lemon, 2016; Van Doorn et al., 2017). Internet of Things, augmented reality, virtual reality, mixed reality, virtual assistants, chatbots and robots, which are typically powered by artificial intelligence, are transforming the customer experience (Hoyer et al., 2020). This dramatically affects company marketing strategies. Three elements must be considered to understand why companies need to embrace a digital transformation:

1. In the last years, the amount of consumer data available to companies is huge.
2. The sources and amount of information about companies available online are increasing day by day.
3. There is a shift in customer expectations and thus in the company/brand–consumer relationship.

As Erevelles et al. (2016) highlighted, technology has turned consumers into generators of data. An increasing number of consumers direct their attention away from traditional media to embrace online media, a context in which almost all movements are trackable, and thus contribute to the proliferation of data about users. Moreover, COVID-19 is an enforced change for consumers to adapt to digital and this contributes to accelerate this ongoing phenomenon. There are approximately 3.8 billion active social media users in the world (WeAreSocial Report, 2020), with a year-on-year growth of about 9.2%. The effect is that consumers share a great amount of personal information on social network sites, revealing not only their demographics but also their interests, passions, favourite locations and so on. They are also engaged in conversations about companies and brands, and this provides other key information for companies. The reasons consumers share such information online have been deeply analysed and classified in the literature and refer to self-centred motivation or community-related motivation and/or are seen as a social function for an ego-defensive reason or for reputation building (Daugherty et al., 2008; Hsu & Lin, 2008; Chang & Chuang, 2011; Ghaisani et al., 2017). Moreover, the adoption of a wearable system as well as other tolls related to Internet of Things is expected to increase in the next years (Kalantari, 2017). Data such as heart rate, skin conductivity and skin temperature are potentially available for firms.

All these changes represent a huge shift for company information systems, since marketers need to listen and monitor this information to draw user paths and create or define the so-called customer journeys (Demmers et al., 2020), and then personalise their offerings accordingly.

As far as the second key point is concerned, it is well known that the number of brand–consumer touchpoints is proliferating due to the digital transformation. Content about brands/companies is created both by firms (the so-called market or products generated content MGC/FGC [market- or firm-generated contents] [Checchinato et al., 2015; Pehlivan et al., 2011]) and consumers (the so-called UGC [user-generated contents] [Cheong & Morrison, 2008; Smith et al., 2012]). Available content about brands and companies is often widespread because information exchange among consumers and between consumers and brands is increasing. Consumers share their opinions about products, their feelings and attitudes about their product or service experiences, produce how-to videos, etc. Other consumers read this information, which makes them more aware of companies and brands. They collect information about product attributes, pricing and other factual information and use it in their purchasing decisions. Then they share new information about products in the post purchase phase, in a loop that firms can take advantage of in this consumer-driven marketing (Court et al., 2009).

The concept of information asymmetry, which occurs when sellers have more information on products than consumers and thus can use this extra information to their advantage, has changed. Now consumers can access information provided by the company as well as third parties; however, they have to cope with another problem: an excess of information from a source that is not always trusted and a time-consuming search for the right information.

This new landscape changes the way consumers research information and buy products. Thus, we come to the third key point we highlighted: digital technologies have transformed customer expectations and the relationship between the consumer and organisation or brand. The Internet, mobile devices and social media have revolutionised consumer behaviour, enabling them to research and shop anytime and anywhere; thus, they are experiencing and expecting no sleeping services by companies. They have very high expectations, not only related to shopping anytime, anywhere but also concerning the relationship with companies (Holbrook & Hulbert, 2002). They expect to be able to interact with companies and engage with them in a personalised way. Therefore, the relationship with the customer must be implemented and nurtured along the customer journey. This new marketing approach needs data to be properly developed, to produce higher outcomes or just to survive, especially when industry competition is high and customer preferences rapidly change (Germann et al., 2013).

Consumer data and data-driven marketing

If the consumer behaviour has been affected by the development of new technological tools, customer-oriented companies cannot manage their businesses as they used to do just a few years ago. This is even more true if digital technologies are relevant in the sector in which they compete. As previously discussed, due to digital technologies, a huge amount of data is available that comes from different sources. Scholars and practitioners used to call this 'big data'. The main features, which also represent its challenges, are its volume, variety and velocity as well as veracity and value in collecting data (Erevelles et al., 2016).

To truly understand consumer behaviour and create value within a customer experience, firms require a comprehensive view of the customer experience itself over both the digital and physical realms (Bolton et al., 2018). This is a challenge, especially for traditional companies. Born-digital companies are fundamentally different from bricks-and-mortar firms because they already possess digital capabilities within the organisation; they do not have to internalise them (Tippmann & Coviello, 2020), and they start their business collecting such data about their prospects and real customers. One of the problems that traditional companies face is related to the analysis of this data. Even if they can understand the potential value, technology must be updated. In fact, traditional software is not capable of analysing big data (Bharadwaj et al., 2013). Thus, they need to establish a new platform that is capable of storing and analysing the data (Erevelles et al., 2016). Moreover, to develop real consumer insight, firms need to extract structured, semi-structured and unstructured data (Erevelles et al., 2016). This requires a different organisation and the implementation of a data intelligence system integrating all kinds of data coming from the main company's touchpoints (the website, social networks pages, loyalty card, point of sales interactions, etc.) and/or bought, if necessary, from other sources.

As far as the marketing is concerned, so-called data-driven marketing, the use of marketing analytics to make decisions must be developed. Marketing analytics 'involves collection, management, and analysis—descriptive, diagnostic, predictive, and prescriptive—of data to obtain insights into marketing performance, maximise the effectiveness of instruments of marketing control, and optimise firms' return on investment (ROI)' (Wedel & Kannan, 2016, p. 98). Information flow and transactions have been greatly facilitated by Internet development. If firms are able to develop both technological and analytics capabilities, their ability to engage in one-to-one marketing, create personalised content and target the right customers with the right marketing mix increases (Shah & Murthi, in press) as does their performance. The process of digital transformation that starts as a marketing need enables new ways of value creation because it allows the fulfilment of long-standing consumer needs in unprecedented ways (Reinartz et al., 2019).

The use of data-driven marketing also becomes necessary for "physical" companies because they compete with the digital firms and face the risk of losing their market position if they are unable to understand, intercept and satisfy their customers.

New technologies help companies to enhance the value of their services and outperform their competitors, but if the value is just on the technology, it is not durable. For example, automatically generated status mails or the delivery status check can provide an extra value-adding service to the customer, but it is an easy replicable value that competitors can copy: in a short time, these services became a basic offer. Companies must analyse their customer journeys and optimise their offers accordingly, considering all the marketing mix elements.

Davenport (2006) highlighted three key attributes to help companies to proficiently adopt analytics: (1) the use of predictive models instead of basic statistics to identify the most profitable customers and how marketing-related activities (but not only marketing) impact financial performance; (2) the systematic approach, in which all functions are involved in the analytics and share common platforms. In traditional companies, 'business intelligence' is generally managed by departments, each of which controls their own data warehouses and trains their own people, and thus companies lose time and the chance to have a comprehensive picture of the effect of their actions; (3) the use of senior manager advocates. Adopting analytics with a systematic approach requires changes in culture, processes, behaviour and skills and so it requires leadership support.

Since marketing data are often required by multiple departments, starting the acquisition and elaboration of consumer data helps not only the marketing department but also other departments, such as production, sales or finance. Internal functions need to start working together and networking activities among department employees should be fostered. The use of predictive models and the systematic approach allows marketers to determine the success of marketing initiatives by incorporating economic and finance data and thus develop marketing accountability that, in turn, will help in future decisions. For example, most of the leading hotel chains, such as Marriot International, have integrated real-time data from a variety of sources (e.g. economic factors, events and weather forecasts) with key internal metrics (e.g. average daily rate, cancellation and occupancy, reservation behaviour) to predict customer behaviour, convey messages to them and adjust pricing strategy accordingly. In a short time, the majority of hospitality businesses have created an owned online presence and/or a presence in some touristic portals (online travel agencies) because they need this to intercept customers, to be visible. The next step to survive will be to use data to optimise performance, like the big chains, to compete with them. Of course, the system will be less complex, but their decisions about prices, rooms services, promotions, etc. should be driven by data.

Therefore, the ability to collect and analyse customer data shared by all of a company's departments and the capacity to integrate this data with other external and internal data sources are the basis to exploit these data and properly affect the marketing strategy. As Tabrizi et al. (2019) stated, if the goal of DT is to improve customer satisfaction and intimacy, then any effort must be preceded by a diagnostic phase with in-depth input from customers. Google, Amazon and other digital leaders have eclipsed competitors with powerful new business models that derive from an ability to exploit data.

As a consequence, new capabilities and skills are required by marketers as well as a strong effort from a company's board of directors to help people become aware of the importance of this shift.

From consumer data to the marketing strategy

Digital technologies impact on the customer–company relationship

According to the IDG report (2019), the top objective of firms' digital business strategies is to meet customer experience expectations. To reach this aim, companies embrace a per-sonalisation/contextualisation of customer interactions, a real-time capture of customer feedback and interactions, and an omnichannel customer experience strategy.

The starting point is the relationship between consumer and company along the cus-tomer journey, the path of the customer from the first encounter with the brand to the post purchase and repurchase phase. The rise of social media has facilitated customer-to-firm communications as well as customer-to-customer interactions (Lamberton & Stephen, 2016), thereby affecting the traditional value-creation structures and relationships.

The development of this relationship is extremely important because it is well known that it affects firm performance (Izquierdo et al., 2005). Contrary to advertising, building a relationship fosters sales indirectly and increases loyalty in the long run. The adoption of digital technologies encourages relationship building because they make it easier to per-sonalise services and messages, as well as to monitor consumer feedback.

Even if the product is not digitizable, firms can use technology to improve its value and transform the intangible part as much as possible to increase the customer experience. For example, Rosso Pomodoro, an Italian pizza chain, has enhanced the user experience thanks to digitalisation. Customers can book tables and order their meals before entering the restaurant. This affects not only customer experience—less time spent waiting for waiter services and the meal—but also profitability—higher table rotation, leads gener-ation and more knowledge about customers.

Digitalisation can also be added to traditional products to collect data about the use of the product, thanks to the 'Internet of Things': products are able to communicate with each other. This data can then be used by firms to suggest the right products among its assortment. This is the case of the new L'Oreal smart hairbrush[1] that analyses hair type and recommends products accordingly.

Finally, digitalisation can be adopted to assist consumers, to help them in product choice in order to avoid frustration. Driven by data, Nike introduced a new fit scanning technology able to map customers' feet. More than 60% of people are wearing the wrong shoe size and this causes dissatisfaction. After acquiring computer vision company Invertex, Nike decided to develop a new scanning solution that uses a proprietary combination of computer vision, data science, machine learning, artificial intelligence and recommendation algorithms. In this case, through its app, it created a personalised communication flow to better serve its customer without changing assortment.

The path is clear: more and more companies are shifting their focus from off-line to online communication and distribution, whether they are selling goods or services.

As written in the 2019 Burberry Annual report, companies are adopting a social-first strategy[2]:

> Over the year, we re-imagined the content and cadence of our social media presence, with an increased frequency of posts, tailored social media campaigns and a deeper focus on products. […]. As a result of this social-first strategy, we have seen significant improvements across reach, engagement and volume of followers.

Moreover, digital channels improve customer excitement, anchoring them to their online communication and maintaining the relationship: Burberry releases limited-edition product drops on the 17th of each month and unveils them on the company's Instagram and WeChat profiles. Customers only have a restricted time period to buy the items (24 hours); thus they continue to pay attention to Burberry social pages to avoid the loss of some news.

Thanks to the combination of digital technologies and customer data, companies can also intervene in relationship reinforcement with their VIP clients, convey messages to them, give them the chance to buy or book products before the official release date or generate a unique experience. This is the case of the Lamborghini Unica application that is available only for owners and, for example, allows them to have access to exclusive events around the world and to record memorable trips and store them.

To sum up, digital transformation can extend the customer experience and create a direct customer–company relationship because of the fast interaction and because digital is a key enabler of personalisation in the different steps of the customer journey (pre, during and post purchase).

Effects on marketing strategy[3]

This new context and the development of one-to-one relationships mainly affects how businesses manage their channels to reach consumers as far as communication and distribution are concerned. This does not mean that other marketing elements are not affected by this transformation, but the impact in some way depends on the decision related to distribution or communication channels. For example, the increase of co-creation and product personalisation are related to a shift in how companies produce their goods, but to make them possible and scalable, companies must implement a one-to-one way of collecting consumer desires and then distributing the final product.

The transformation of communication channels is challenging because it requires new skills and a new systematic and strategic approach (Royle & Laing, 2014). Communication is one of the main realms affected by technology developments, not only because the advent of the Internet and social media revolutionises a firm's communication mix but also because of the uncountable opportunities to customise and use marketing automation to optimise results.

As we have already discussed the impact of the Internet on consumer behaviour, we now want to highlight how firms are facing this new landscape, the main challenges and how to exploit the web and other linked technologies.

Digital advertising is probably the main representation of the new communication landscape, and it is increasingly adopted by brands. Ad spending in the digital advertising market worldwide is projected to reach US$384.96 million in 2020 (based on eMarketers data, or $355.784 million based on Statista.com data), becoming the dominant ad medium in the United Kingdom, China, Norway, Canada, the United States and the Netherlands. Despite traditional advertising, digital advertising is a way to more precisely reach and engage with consumers, thanks to the available data, and perform better than traditional advertising. Personalisation is one of the main effects of the digital transformation and it is one of the main opportunities of digital advertising; De Keyzer et al. (2015) demonstrated that personalised advertising on social networks leads to more positive consumer responses than non-personalised advertising.

To personalise and optimise advertising investments, the adoption and the integration of data management platforms or other similar technologies is suggested for leading firms that invest a huge amount of money in this instrument. These platforms help to follow the dynamism of markets in real time. This dynamism is often a challenge for traditional companies because of their background. They have to look at digital company communication strategies to learn how to exploit digital media in their own communication mix because it is not just a translation of offline content to online content: new technologies, a systemic approach and new skills are required. The traditional communication model—primarily based on company created touchpoints—is no longer effective; companies need to monitor information about their products or brands and respond quickly. Depending on the market, they also need to create content that fits timely events. This requires a big effort to practise the so-called real time marketing, the practice to align firm 'messages with public events happening in the moment' (Willemsen et al., 2018).

Digital marketing services are booming, but due to the rapid change in the rules, platforms and the high technical knowledge required for each one, firms have to face a fragmentation of function that requires specialised people for every digital section (SEO, analytics, social media management, advertising, etc.). The development of analytical skills, technical skills and effective performance metrics becomes crucial, and it is no longer possible to internally hire people for all of them. Companies need to collaborate with external specialists, often more than one, to optimise their performance. This sheds light on the need to change the way in which firms work: firms should open the door and the 'make or buy' dynamism increases and outsourcing is necessary. Networking and collaborations with more partners are becoming a building block to succeed.

Moreover, delivering greater value for customers and conveying customised and interesting messages are considered the best ways to succeed in this new landscape. To do so, firms should join consumer conversations beyond the product itself, enter into consumer daily life and support them along the customer journey towards the right touchpoints. To achieve these results, companies use both a combination of chatbot and human interactions among instant messaging applications (e.g. Facebook Messenger,

WhatsApp, Telegram and WeChat; Lo Presti et al., 2019) and discussion forums or chat robots. These new approaches have produced disruptions to traditional product-centred business models (Lo Presti et al., 2019) and require the adoption of marketing automation. Automation needs new content organisation and a digital thinking approach. As highlighted by Järvinen et al. (2020), marketing automation revolutionises organisational structures, processes and the culture of how marketing is conducted. Again, it is not a simple or incremental change but is a shift in the communication approach.

Another key element of transformation in the communication approach is the opportunity to control the communication activities performance and quickly change messages or channels accordingly. In fact, firms can test and modify messages to convey to consumers in real time and this can open the door to a 'try and learn' mindset. The digital environment creates endless opportunities for real-world experiments through A/B testing and other means.

Another challenge to face is the (potential) global reach of firm communications. As customers across the globe are directly exposed to brand messages, they expect to have the same consistent message and experience across all their touchpoints. This puts intense pressure on firms that must control the consistency of their messages both on different channels and in different countries. Moreover, they must control and quickly answer customer requests and/or reviews in public spaces in order to prevent uncontrollable sharing of fake information.

As far as the distribution is concerned, the increase in the variety of channel formats and the progression from single to omnichannel marketing has made distribution trickier to manage for marketers (Ailawadi & Farris, 2017). As consumers start purchasing online, firms face new challenges to manage a combination of many types of channels to match how customers want to search, buy and return. One of the main challenges is related to logistics and, for some aspects, to financial accounting. Most of the firms are clicks and mortars manufacturers with huge store networks both direct and independent. To follow consumers along their purchase journey and to make the online–offline strategy possible (Checchinato et al., 2017), businesses must reconfigure their systems, integrating all the actors of the chain in order to better fit the journey. The wider the integration (e.g. not only direct stores but also wholesales), the more complete the digital transformation.

Again, to succeed companies need to implement effective data-driven decisions, thus the first step in managing omnichannel distribution is to find the specific metrics that facilitate these decisions. Different channels compete with and complement one another; thus, a portfolio of metrics should be developed (Ailawadi & Farris, 2017) and then monitored over time.

Moreover, manufacturers must figure out how to measure the contributions of different channels along the path to purchase and reward them in order to avoid conflicts between suppliers and their retailers and to engage and motivate retailers to embrace the omnichannel approach. The lack of data on the role played by channels and media in the consumer purchase funnel contributes to the attribution of the conversion to the last touch channel before purchase. Nowadays, with the availability of path data related to all the different touch points along the customer journey, companies can implement so-called attribution modelling, that 'practitioner defines as the science of using advanced analytics to allocate appropriate credit for a desired customer action to each marketing touch point across all online and off-line channels' (Moffett et al., 2014; Kannan et al., 2018). One of the main challenges is the quality of the data that, following the classification provided by Bradlow et al. (2017, p. 83), refers to 'three primary groups, namely, (1) traditional enterprise data capture; (2) customer identity, characteristics, social graph, and profile data capture; and (3) location-based data capture'. By incorporating this kind of data in the business intelligence system, companies must move to estimate basic descriptive models of household paths to estimate predictive models for each customer behaviour.

Lastly, another challenge affecting firm distribution strategies is related to price consistency across both online and offline channels (Ye et al., 2018). Companies need to monitor prices, to not lack control over this marketing lever and thus allow it to affect their brand image.

Conclusions

Changes in technology and in consumer behaviour have always been the key drivers of business strategy. Despite the past revolution, what is different nowadays is the velocity, interwovenness and, therefore, complexity of these elements (Kung, 2008).

Firms need to understand that this is a completely new landscape and there is no chance to come back to the old paradigm, especially after the COVID-19 pandemic that accelerated the digitalisation process both for firms and consumers. Digitalisation is strategic to survive and better serve consumers and thus should be embraced in a strategic way. Firms that continue to view digital marketing as a tactical opportunity rather than considering digital as a new approach to marketing will fail. This does not mean that digital transformation should follow the same rule or process for each company nor that each company must adopt the same level of digitalisation; but each company should start analysing the customer journey (how long it is, on which occasion it starts, where consumers search for information, what the triggers are, etc.) and ask itself how digital transformation could provide a better customer experience and avoid drop-offs from the journey due to missing information or customer support or other relationship issues. Companies must start to learn customer needs, values, frustrations and practices.

Approaching digital transformation can be an opportunity to analyse the value proposition (and the firm's main processes): if the firm identity is not clear and the unique selling proposition is poor in the physical environment, the digital transformation can be used as an opportunity to start rethinking and digitalise accordingly. It will take some time, but it is necessary to start with the right blueprint and train employees or to collaborate with external experts to implement digitalisation. It is crucial to stop thinking about making everything internal: the market dynamism requires outsourcing or, if possible or strategic, the acquisition of partners.

Moreover, the start of digital transformation is just the first step of a long trip. Companies should continue to analyse the situation and adopt the best and most efficient solutions, considering time and available resources. Anticipating rather than responding to changes in customer expectations has become a strategic imperative for firms, thus data-driven decisions and changes in previous decisions should be the rule.

Notes

1   www.cnbc.com/2017/01/04/loreals-smart-brush-listens-to-hair-recommends-luxury-tr eatments.html
2   www.burberryplc.com/content/dam/burberry/corporate/oar/documents/Burberry_201819-Annual-Report.pdf [Accessed 03.30.2020]
3   Some of the ideas provided in this paragraph come from a conversation about digital transformation with Marco Ziero, CMO of Moca Interactive.

References

Ailawadi, K. L., & Farris, P. W. (2017). Managing multi-and omni-channel distribution: Metrics and research directions. *Journal of Retailing, 93*(1). https://doi.org/10.1016/j.jretai.2016.12.003

Berman, S. J. (2012). Digital transformation: Opportunities to create new business models. *Strategy and Leadership, 40*(2), 16–24. https://doi.org/10.1108/10878571211209314

Bharadwaj, A., El Sawy, O. A., Pavlou, P. A., & Venkatraman, N. (2013). Digital business strategy: Toward a next generation of insights. *MIS Quarterly 37*(2), 471–482.

Bolton, R. N., Mccoll-Kennedy, J. R., Cheung, L., Gallan, A., & Witell, L. (2018). Customer experience challenges: Bringing together digital, physical and social realms. *Journal of Service Management, 29*(5), 776–808. https://doi.org/10.1108/JOSM-04-2018-0113

Bradlow, E. T., Gangwar, M., Kopalle, P., & Voleti, S. (2017). The role of big data and predictive analytics in retailing. *Journal of Retailing, 93*(1), 79–95. https://doi.org/10.1016/j.jretai.2016.12.004

Chang, H. H., & Chuang, S. (2011). Social capital and individual motivations on knowledge sharing: Participant involvement as a moderator. *Information & Management, 48*(1), 9–18.

Checchinato, F., Disegna, M., & Gazzola, P. (2015). Content and feedback analysis of YouTube videos: Football clubs and fans as brand communities. *Journal of Creative Communications, 10*(1), 71–88. https://doi.org/10.1177/0973258615569954

Checchinato, F., Hu, L., Perri, A., & Vescovi, T. (2017). Leveraging domestic and foreign learning to develop marketing capabilities: The case of the Chinese company Goodbaby. *International Journal of Emerging Markets, 12*(3), 637–655. doi: 10.1108/IJoEM-04-2015-0060

Cheong, H. J., & Morrison, M. A. (2008). Consumers' reliance on product information and recommendations found in UGC. *Journal of Interactive Advertising, 8*(2), 1–29.

Court D., Elzinga D., Mulder S., & Vetvik O. J. (2009). The consumer decision journey. *McKinsey Quarterly,* July.

Daugherty, T., Matthew S., Eastin, & Bright, L. (2008). Exploring Consumer Motivations for Creating User-Generated Content. *Journal of Interactive Advertising, 8*(2), 16–25.

Davenport, T. H. (2006). Competing on analytics. *Harvard Business Review,* January, 1–9.

De Keyzer, F., Dens, N., & De Pelsmacker, P. (2015). Is this for me? How consumers respond to personalized advertising on social network Sites. *Journal of Interactive Advertising, 15*(2), 124–134. https://doi.org/10.1080/15252019.2015.1082450

Demmers, J., Weltevreden, J. W. J., & van Dolen, W. M. (2020). Consumer engagement with brand posts on social media in consecutive stages of the customer journey. *International Journal of Electronic Commerce, 24*(1), 53–77. https://doi.org/10.1080/10864415.2019.1683701

Erevelles, S., Fukawa, N., & Swayne, L. (2016). Big data consumer analytics and the transformation of marketing. *Journal of Business Research, 69*(2), 897–904. https://doi.org/10.1016/J.JBUSRES.2015.07.001

Ghaisani, A. P., Handayani, P. W., & Munajat, Q. (2017). Users' motivation in sharing information on social media. *Procedia Computer Science, 124,* 530–535. https://doi.org/10.1016/j.procs.2017.12.186

Hoffman, D. and Novak, T. (1996). Marketing in hypermedia computer-mediated environments: Conceptual foundations. *Journal of Marketing, 60,* 50–68.

Holbrook, M. B., & Hulbert, J. M. (2002). Elegy on the death of marketing – never send to know why we have come to bury marketing but to ask what you can do for your country churchyard. *European Journal of Marketing, 36*, 706–732.

Hoyer, W. D. Kroschkeb, M, Schmitt, B, Kraumed, K., & Shankare, V. (2020). Transforming the customer experience through new technologies. *Journal of Interactive Marketing*, 51, 57–71. doi: 10.1016/j.intmar.2020.04.001.

Hsu, C., & Lin, J. C. (2008). Acceptance of blog usage: The roles of technology acceptance, social influence and knowledge sharing motivation. *Information & Management, 45*(1), 65 –74.

IDG Communications (2018). State of digital business transformation. Retrieved from www.idg. com/tools-for-marketers/2018-state-of-digital-business-transformation-white-paper/

Izquierdo, C. C., Gutierrez Cillàn, J., & Gutierrez, S. S. M. (2005). The impact of customer relationship marketing on the firm performance: A Spanish case. *Journal of Service Marketing, 19*(4), 234–244. https://doi.org/10.1108/08876040510605262

Järvinen, J. M., Tarkiainen, A., & Tobon, J. (2020). Effectual and causal reasoning in the adoption of marketing automation. *Industrial Marketing Management, 86*(March 2019), 212–222. https://doi.org/10.1016/j.indmarman.2019.12.008

Kalantari, M. (2017). Consumers' adoption of wearable technologies: Literature review, synthesis, and future research agenda. *International Journal of Technology Marketing, 12*(3), 274–307. https://doi.org/10.1504/IJTMKT.2017.10008634

Kannan, P. K., Reinartz, W., & Verhoef, P. C. (2018). The path to purchase and attribution modeling: Introduction to special section. *International Journal of Research in Marketing, 33*(3), 449–456. https://doi.org/10.1016/j.ijresmar.2016.07.001

Kung, L. (2008). *Strategic management in the media: Theory to practice*. London: Sage

Lemon, K. N. (2016). The art of creating attractive consumer experiences at the right time: Skills marketers will need to survive and thrive. *GfK Marketing Intelligence Review, 8*(2), 44–49.

Lo Presti, L., Maggiore, G., & Marino, V. (2019). Mobile chat servitization in the customer journey: From social capability to social suitability for customers. *The TQM Journal* https://doi.org/10.1108/TQM-10-2019-0241

Moffett, T., Pilecki, M., & McAdams, R. (2014). The forrester wave: Cross-channel attribution providers, Q4 2014. November. www.forrester.com/ report/The+Forrester+Wave+CrossChannel+Attribution+Providers+Q4+2014/-/E-RES115221.

Pehlivan, E., Sarican, F., & Berthon, P. (2011). Mining messages: Exploring consumer response to consumer- vs. firm-generated ads. *Journal of Consumer Behaviour, 10*(6), 313–321. https://doi.org/10.1002/cb.379

Quinton, S., & Simkin, L. (2017). The digital journey: Reflected learnings and emerging challenges. *International Journal of Management Reviews, 19*(4), 455–472. https://doi.org/10.1111/ijmr.12104

Reinartz, W., Wiegand, N., & Imschloss, M. (2019). The impact of digital transformation on the retailing value chain. *International Journal of Research in Marketing*. https://doi.org/10.1016/j.ijresmar.2018.12.002

Royle, J., & Laing, A. (2014). The digital marketing skills gap: Developing a digital marketer model for the communication industries. *International Journal of Information Management, 34*, 65–73.

Shah, D., & Murthi, B. P. S. (n.d.). Marketing in a data-driven digital world: Implications for the role and scope of marketing. *Journal of Business Research*. https://doi.org/10.1016/j.jbusres.2020.06.062

Smith, A. N., Fischer, E., & Yongjian, C. (2012). How does brand-related user-generated content differ across YouTube, Facebook, and Twitter? *Journal of Interactive Marketing, 26*(2), 102–113. https://doi.org/http://dx.doi.org/10.1016/j.intmar.2012.01.002

Tabrizi, B., Lam, E., Girard, K., & Irvin, V. (2019). Digital transformation is not about technology. *Harvard Business Review, 13*.

Tippmann, E., & Coviello, N. (2020). Born digitals: Thoughts on their internationalization and a research agenda. *Journal of International Business Studies, 51*, 11–22. https://doi.org/10.1057/s41267-019-00290-0

Van Doorn, J., Mende, M., Noble, S. M., Hulland, J., Ostrom, A. L., Grewal, D., & Petersen, J. A. (2017). Domo arigato Mr Roboto: Emergence of automated social presence in organizational frontlines and customers' service experiences. *Journal of Service Research*, *20*(1), 43–58.

We Are Social, Digital (2020). Retrieved from https://wearesocial.com/digital-2020

Wedel, M., & Kannan, P. K. (2016). Marketing analytics for data-rich environments. *Journal of Marketing*, *80*(6), 97–121. https://doi.org/10.1509/jm.15.0413

Willemsen, L. M., Komala, M., Kamphuis, A., & van der Veen, G. (2018). Let's get real (time!) The potential of real-time marketing to catalyze the sharing of brand messages. *International Journal of Advertising*, *37*(5), 828–848.

Ye, Y., Lau, K. H. and Teo, L. K. Y. (2018). Drivers and barriers of omni-channel retailing in China. *International Journal of Retail & Distribution Management*, *46*(7), 657–689. doi: 10.1108/ IJRDM-04-2017-0062.

Chapter 15

Digital transformation and the salesforce

Personal observations, warnings, and recommendations

Frank V. Cespedes

What is digital transformation?

Initially, digital transformation referred *primarily* to a process of turning paper documents into digital copies for easier storage and searching. But over the past few years, it's become a much broader and conveniently vague term, now regularly used to refer to artificial intelligence (AI; itself an elastic term of art), big data, virtual online communications, internet of things (IoT) enabled devices, and almost any application of software or computing technology to a business process.

I understand how this happens. In the cross-functional battle for one's share of a limited budget, managers hitch their pitch to the current "big thing." But that does not help decision-making and resource allocation.

The economist Fritz Machlup (1963) once wrote an essay about "weaselwords," by which he meant "words concealing voids of thought, used to avoid commitment ... which destroy the force of a statement as a weasel ruins an egg by sucking out its content." Machlup was talking about how economists often use words like "structure" or "system" in lieu of empirical cause-and-effect linkages. His point was that using these words is often just a way of sounding smart or up-to-date, and it ultimately confuses analyses and actions. "Digital transformation" runs the risk of being a current example of Machlup's point, and that can have real costs for an organization. Vague initiatives that do not differentiate the choices and alternatives cannot be tested and contested by managers as market conditions change. And both the market conditions and animating technologies change constantly in the spaces often labeled as "digital," so clarity here is more than a semantics issue.

My working interpretation of the term "digital transformation" is that it generally refers to how data collection, algorithms, and connectivity tools can potentially improve analytical and productivity tasks in business as these tools increase in functionality and decrease in cost. Given that definition, I then agree with the premise articulated years ago by Peter Drucker (1967) in "The Manager and the Moron," in which he emphasized that "the computer makes no decisions; it only carries out orders. It's a total moron, and therein lies its strength. It forces us to think, to set the criteria."

Data, even allegedly self-correcting data as in some AI programs, is never the same thing as an answer to a management issue. For example, a 2010 study of companies with 1000 employees or more found that, individually, *each* firm already had more data in its customer relationship management system than was held in the entire Library of Congress (Manyika et al., 2011). The data have just gotten "bigger" since then. As a result, many managers are enthusiastic about the possibilities of digital transformation, yet they lack a clear sense of cause-and-effect in their business, and their people—especially salespeople—get lost in the day-to-day noise.

A poorly executed digital transformation—of which there are many—can *decrease* sales productivity because sales reps obtain even more data, managers review lots of interesting-but-tangential metrics, and much time is diverted from selling and devoted to training about technology, not customers. As always, moreover, the "garbage in, garbage out" rule applies. Use of digital tools is only as good as the data inputs. In many firms, data silos and sales' neglect of CRM (customer relationship management) systems do not inspire confidence in the data used. Many firms need to redesign their processes *before* they can or should use these tools.

But these are managerial, not analog, or digital, issues. The starting point for productive use of digital tools is having good market-relevant questions to ask of the tools, which requires ongoing business acumen.

Is the digital offer typically a complement to or substitute for traditional products?

In my experience, digital offerings generally complement traditional offline, physical products, services, and buying/selling activities. For example, ecommerce has been here for over 30 years. Books.com was selling online while Jeff Bezos was still working on Wall Street. After decades of tax-free sales, ecommerce was just 11.4% of US retail sales in 2019, according to the Department of Commerce (US Census Bureau, 2020).

Further, consumers are discriminating in how they use digital tools for shopping. For example, consumers now spend more time doing online research about cars than they do talking with brick-and-mortar dealers. Yet the vast majority of cars in the US and other countries are still bought in dealerships (less than 1% of the used cars, and less than 5% of new cars, sold in the US in 2018 were online sales). For car buyers, online tools are a complement to, not a substitute for, in-person dealer visits. Research indicates that consumers use independent websites for model comparisons and reviews, manufacturer sites for detailed model information and videos, and dealer websites when they are looking for specific vehicles and information about local inventory (see Cespedes & Hamilton, 2016).

It is an omni-channel world, not a digital-eats-physical world, as evidenced by the strategic decisions made by previously pure-play ecommerce firms. The biggest trend before the pandemic of 2020 was the opening of brick-and-mortar stores by firms like Birchbox, Bonobos, Caspar Sleep, Warby Parker, Wayfair, and—yes, indeed—Amazon.

Will the lockdowns and social distancing caused by the coronavirus change this buying behavior in favor of dramatic gains for digital channels? I do not know—and, despite their tone of prophetic certainty, neither do the many pundits who have predicted this change. But look at what was happening online before the virus.

Social media usage on the major platforms had been essentially flat over the previous four years. In fact, social media usage had *declined* among Americans less than 35 years old, and the only age group using Facebook more were people 55 or older, according to Edison Research (2019). As a marketing medium, online channels were cluttered, subject to increasing expense and diminishing returns, viewed with suspicion by consumers as media attention of foreign hackers raised awareness of cybersecurity issues, and distrusted by many advertisers because of measurement issues on the major platforms and fear of their ads being placed near objectionable content.

Given the ability to block ads and spam, the growing costs of acquiring customers online (which will only become more costly if more firms try to market that way), the experience of "Zoombombing," controls on consumer data by EU regulators and others, and pending restrictions by makers of smartphones and other mobile devices that make targeting of digital ads more difficult—and often impossible—to do, it is unclear how much buying and selling will be done online in the future.

It is also unclear whether social distancing and the closing of brick-and-mortar outlets increased people's eagerness and willingness to transact digitally, or whether it simply demonstrated the limitations of buying, selling, managing, and communicating virtually.

If you are allocating resources—versus talking in a classroom or at a conference about digital transformation—you should treat the technology as a tool and as a complement.

How does digital transformations affect the salesforce?

Sales is a crucial test case for any organizational change, including productive use (or not) of new technologies. Because of the central role of customer acquisition in any for-profit enterprise, changes in selling always have wider organizational implications. So many other decisions and resource commitments in most firms depend upon demand forecasts and sales' ability to meet those forecasts.

Conversely, because the sales force then faces pressures to "make the numbers" monthly, quarterly, and annually, making significant changes in sales activities are risky and difficult. CRM, for example, has been with us for decades. But most companies will tell you that their CRM systems are typically underused by their sales leaders, often ignored by reps, and when reps do input data into CRM, the inputs are inconsistent because different sales-people use different criteria in reporting what is or is not a prospect, deal, probability of closing a sale, or where a prospect is in the pipeline.

As usual in business, the key capabilities are people and management, not technology. Given good management, I see the following as the main areas where, in the next few years, digital tools can have the biggest positive impact on sales activities.

Increasing selling time

The data vary by industry and company, but most salespeople spend the bulk of their time on non-selling activities and much less than 50% of their time interacting with customers and some estimates (Dixon et al., 2011) estimate that, on average, only 26% of salesforce time is spent interacting with customers and prospects. Most companies, therefore, have a big opportunity embedded in their sales models.

Think about the impact in a business if smart use of digital tools can offload other activities and increase selling time by an incremental 10 to 20%. In most businesses, that represents a very significant gain. Further, when improvements lower the total cost of selling activities, prospects that were not profitable enough to target become worth it, increasing your addressable market.

Monitoring online/offline interactions

As a multi-channel combination of online and personal selling becomes the norm in sales models, digital tools become important in measuring and evaluating where, when, and how to deploy online and offline efforts.

For example, time-on-site and page views are often associated with positive engagement, whereas they may result from a confusing or slow-loading site. New-visitor counts are often touted as surrogates for growth, but much of this may simply be driven by the proliferation of new devices and disconnected from actual purchases. In many subscription sales models, a "win" is treated as synonymous with the "last click" when the reality is that the purchase was motivated by a combination of sales and marketing activities throughout the buyer journey. New analytical tools shed light on the types of onboarding that spur initial adoption, the features that retain customers, and the offers that are more and less likely to be effective.

These tools are also important for understanding where in-person selling efforts have the biggest impact and thus where and when your salespeople should have a conversation with customers. In the early stages of many buying processes, customers are not ready to transact, and a better use of resources than a sales call is often content marketing, *if* the company knows its target buyers and establishes the appropriate cadence in its prospecting activities (Cespedes and Heddleston, 2018). At a later stage, ongoing interaction with a knowledgeable and trusted rep is often crucial and online channels, however sophisticated, may not suffice.

The ultimate economic rationale for marketing and sales investments is a "baseline-lift" valuation—that is, a measure of the lift over a baseline of existing sales that is attributable to a specific initiative (Farris, et. al., 2014). That also applies to digital transformation in sales. But you will never know what that lift is (or isn't) unless you can link the investment to the customer-conversion dynamic in your sales model—again, that's a management issue. Further, the lift from a given investment will inevitably alter. You rarely generate three times current revenues and profits by doing three times whatever brought you to your current state, digitally or by offline means.

How do digital products and services affect pricing?

First, let me offer a cautionary note about how *not* to price in this area. If you do not make correct distinctions, you can cut and stumble your way to oblivion. Look at how most newspapers reacted for years to digital competitors: trying to mimic the online firm, but with a higher cost structure, and while giving away their own content online. This was a literal enactment of the old joke about selling below cost but hoping to make it up in volume. Or look at how many retailers responded with self-fulfilling-prophecy actions to ecommerce competitors: reducing headcount in stores, not investing in training sales associates, and generally being oblivious to online/in-person interactions and the impact on sales.

Whether it is for physical or digital products and services—or a hybrid offering, which is increasingly the norm—the issue is linking price and value. For digital offerings, a key is identifying the relevant unit of value as customers experience it, and that is often not intuitive.

For example, for a century movies were released according to a "windowing" model of pricing that captured different buyers' willingness-to-pay. The initial release at the highest price in selected theatres attracted those who valued a movie most highly, followed by release at smaller theatres at lower ticket prices, and then renting (now streaming) the movie for viewing at home weeks later but at lower prices. Studios extracted maximum value from each group without cannibalizing sales to the others. The time frame of windowing shortened in the twenty-first century, but this remained standard industry practice before theatres closed during the pandemic.

The major exception was Netflix, which, even before the crisis, released its movies in theatres on the same day that it made them available for "free" on its streaming platform. Why? Netflix is not in the business of selling individual movies to multiple customers; it sells bundles of movies to individual customers (Smith and Telang, 2019). Consumers assign different values to individual movies in a bundle, but in a large-enough bundle the differences average out. Given tools that accurately predict the average value that subscribers will pay for the bundle, you can set a price slightly below that threshold and extract maximum value from customers.

Similarly, as platform models disseminate more widely, identifying the relevant unit of value becomes essential for pricing digital products or services. Inbound marketing firm HubSpot initially charged a flat monthly subscription fee but then tied its pricing tiers to the number of contacts in a customer's database. This removed a cap on extracting value: as a customer company grew its database, the value of HubSpot's platform at that company increased (it was managing more data), as did its ability to share in that success via performance pricing.

For other platform businesses, the relevant unit is different. Fintech firms typically charge a fee for each transaction processed through their platforms. Usage tends to be episodic, hard to predict, and not, as with HubSpot, part of a marketing or sales cadence at customers. In mobility business models, Zipcar, Maven, and others price on a standard per-hour or per-day rental structure from a choice of supplier-determined pickup and drop-off locations. But Turo, DriveShare, and other person-to-person car-sharing platforms price based on time, distance, auto brand, and other variables that are often part of a bundle with ancillary services.

Identifying the relevant unit for value and pricing also affects how a firm must sell, and to whom. The length of the sales cycle often shortens as you move from an upfront payment to outcome-based performance. But the buyer, relevant value documentation, and the sales conversation also change.

When software firms move from selling licenses to cloud-based subscription pricing, for example, that also means a change in buying. Software licenses are typically purchased as part of capital budgets, whereas annual or monthly subscription payments are usually funded from operating budgets. The former usually means high upfront costs but low annual maintenance costs and unlimited use within the customer organization. Subscription pricing, however, means payments over time and, depending upon the supplier, different usage-, contact-, or features-based packages that sales must learn to justify with operating personnel as well as finance or purchasing people at that customer.

What are obstacles to successful implementation of a digital transformation?

The core barriers will depend on what kind of change is meant. One obstacle is, again, the vague use of "digital" in place of actionable managerial initiatives. Let's consider sales.

Life in a sales organization is filled with short-term deadlines and pressures: sales per quarter, sales per rep, did she or didn't she meet quota. As a sales manager once said to me, "In this job, if you don't survive the short term, you don't need to worry about the long term." In addition, in most companies, salespeople—compared with their colleagues in other functions—have a much higher proportion of their compensation at risk in a variable-pay plan contingent on meeting quota.

Thus, implementation here is highly dependent on minimizing transition and transaction costs. We know from decades of behavioral research that the status-quo bias is a powerful driver of buying behavior in general, and that's also true in sales contexts. The lesson for new-product introductions is to try to minimize behavioral change when introducing an innovation: understand the buyer's usage situation and enhance their productivity within that context while doing your best to minimize required change.

The same lesson applies when implementing digital initiatives in sales organizations: first, ensure that you understand the sales tasks and incentives, help sales people be more productive in performing those tasks and generating outcomes, and then promote the success stories branch by branch.

Second, digital tools are blurring roles and responsibilities between sales and marketing. Traditionally, in most firms, marketing is responsible for generating awareness and then a hand-off to sales. However, on the demand side, digital tools now allow prospects to pursue parallel streams in their search, consideration, and evaluation activities during the buying journey. Meanwhile, on the supply side, digital tools are allowing sales to do many things that, in the past, were in marketing's domain.

A key implementation issue facing many firms in digital transformation is whether their customer-acquisition model allocates responsibilities appropriately across the buying journey, aligning efforts and helping reps know when and how to have a conversation with customers. But providing tools without clarifying those strategic choices and performance management practices only exacerbates redundancy of effort and customer confusion.

So far, most positive examples of digital transformation in sales tend to be software as a service (SaaS) companies using inside sales models. That's not a coincidence. These sales models generate lots of real-time data as prospects click on websites, download content, leave their email addresses, and sign up for online demos; meanwhile, an inside salesforce gives managers more direct, daily control, and influence over their reps' call patterns and other selling activities. It's tougher to do this in outbound field-sales models where salespeople travel and manage more complex buying processes than is generally true in SaaS models. It will take time, learning, and adaptation—again, those are management, not software, issues.

What advice for companies beginning a digital transformation journey?

Journey is indeed the correct term. Integrating any technology or process into an extant business model is a process, not an event or IT project. Further, in business, there is no such thing as "performance" or increased "productivity" in the abstract; there is (or is not) only performance and productivity in your organization with your customers and your products. So context matters greatly here. But I would cite three general types of advice useful in helping to scope-out and prioritize digital investments in a sales organization.

Selling is about buying

The most important thing about selling is, and always has been, the buyer and the buying process—not the particular communication tool(s) used by the seller. It's a company's responsibility to adapt to how people buy; it's not the market's responsibility to adapt to your preferred technology. Make sure you understand how target customers buy in your market today, not yesterday, and how digital transformation—however you define it—can add value to that process and to selling initiatives. Start outside-in with the market, not inside-out with a tool.

For over a half-century, buying has typically been framed in terms of a hierarchy-of-effects model: moving a prospect from Awareness to Interest to Desire to Action. The A-I-D-A formula and its many variants are the basis (often, the unconscious basis) for sales activities and organization in most firms. It's fundamentally an inside-out process and CRM systems are there to provide data about progression (or not) through that company's funnel—the "pipeline" metrics that dominate talk about sales in books, blogs, training seminars, and coaching initiatives.

But research (Cespedes, 2021), and probably some reflection on your own experience, indicates a different buying reality. Rather than moving sequentially through a funnel, buyers now work through parallel activity streams—online and offline, and in multiple sequences. Buying in most categories is now a continuous and dynamic process, not a linear funnel. Understanding where customers are, how they navigate between streams in your market, and how to interact with them appropriately in a given stream is now central to effective selling.

Meanwhile, most sales models are the ad hoc accumulation of years of reactive decisions, usually by different managers pursuing different goals. Even hype about digital transformation may be beneficial if it motivates managers to be more rigorous in examining and updating their go-to-market assumptions and processes. But then move beyond hype to specify the business objectives and where digital can and cannot improve productivity.

Focus on helping the average rep improve

Differences in individual performance are wider in sales than in other functions. In most salesforces, the best salespeople are not just a bit better than the average rep; they are often orders of magnitude better. There *are* sales stars in most companies, and managers must keep them motivated and relevant. Conversely, as the old saying goes, "You hired your problems." The poorest-performing reps are typically poor fits for that sales task or role.

There is an important implication here: small improvements to the abilities and focus of the average performers usually have, in the aggregate, bigger impacts on sales productivity than efforts at either end of that spectrum of performance. Most sales managers know what to do with their stars and their laggards. It's the people in the middle that managers get paid to manage, and that's where digital technologies are likely to have their biggest impact in daily use by most sales organizations.

Beware the silos

Sales is undergoing a sustained data revolution. Among other things, sales managers now receive greater scrutiny from other executives who have access to their data. In my experience, for example, AI and data analytics groups report, more often than not, up through the finance function in firms. Using that data, finance executives ask questions about their companies' big investments in sales, and sales managers must have answers.

The good news is that more data can mean more transparency and continuous-improvement opportunities. The bad news is that most sales managers understand activities that drive the top line in their firms, but not other financial components of selling beyond sales volume. Meanwhile, finance executives and their digital analysts rightly demand value-creation from sales leaders and not only top-line motion, but they are often inexperienced with how buying and selling actually work in their business model. Or, even worse, they assume that the algorithm is the answer, but it's not; it's only a starting point.

In my most recent book, I call this "a dialogue that never happens," and that's how companies get disrupted. Too often, digital transformation becomes an excuse for simply generating and disseminating more data, not information. Data are helpful and necessary, but mute. Good managers know that analysis is not only a search for proximate truth; it's also about motivating the people who must use that information. Salespeople, in particular, will ignore analytics and tools that can't easily be applied where they live: in daily encounters with diverse customers. The data and tools must be placed in context and that requires ongoing interactions.

Frank Cespedes teaches at Harvard Business School. He ran a professional services firm for 12 years, has consulted with companies in many industries, and is a board member of startup firms, established companies, and private equity organizations. He has written for numerous publications and is the author of six books, including *Aligning Strategy and Sales* (Harvard Business Review Press, 2014), cited as "the best sales book of the year" (*Strategy & Business*), "a must read" (*Gartner*), and "perhaps the best sales book ever" (*Forbes*); and *Sales Management That Works: How to Sell in a World That Never Stops Changing* (Harvard Business Review Press, 2021).

References

Cespedes, F. V. 2021. *Sales Management That Works: How to Sell in a World That Never Stops Changing.* Boston, Massachusetts: Harvard Business Review Press.

Cespedes, F.V. & Hamilton, J. 2016. Selling to Consumers Who Do Their Homework Online. *HBR. org,* March 16. Retrieved from https://hbr.org/2016/03/selling-to-customers-who-do-their-homework-online

Cespedes, F.V. & Heddleston, R. Four Ways to Improve Your Content Marketing. *HBR.org,* April 19, 2018.

Dixon, M., Frewer, S., & Kent, A. Are Your Reps Spending Too Much Time in Front of Customers? *HBR.org,* February 8, 2011.

Drucker, P. F. 1967. The Manager and the Moron. *McKinsey Quarterly,* December 1.

Edison Research. 2019, May 30. The Social Habit 2019. Retrieved from www.edisonresearch.com/the-social-habit-2019/

Farris, P., Hanssens, D., Lenskold, J., & Reibstein, D. 2014. Marketing Return on Investment: Seeking Clarity for Concept and Measurement. Marketing Science Institute, Report No. 14–108.

Machlup, F. 1963. Structure and Structural Change: Weaselwords and Jargon, in M. Miller (Ed.), *Essays on Economic Semantics* (pp. 73–75). Englewood Cliffs, NJ: Prentice-Hall.

Manyika, J., Chui, M., Brown, B., Bughin, J., Dobbs, R., Roxburgh, C. & Byers, A. H. 2011, May. *Big Data: The Next Frontier for Innovation, Competition, and Productivity.* McKinsey Global Institute Report, May 2011. New York, NY: McKinsey & Company.

Smith, M. & Telang, R. 2019. Netflix and the Economics of Bundling. *HBR.org*, February 25, 2019.

U.S. Census Bureau. 2020, May 21. 2018 E-Stats Report: Measuring the Electronic Economy, available at www.census.gov/programs-surveys/e-stats.html

Chapter 16

Digital transformation and the role of customer-centric innovation

Interview with the chief value officer, Thales

Andreas Hinterhuber and Stephan M. Liozu

Andreas Hinterhuber and Stephan Liozu discuss several salient aspects of the digital transformation of traditional industrial companies. Stephan Liozu is chief value officer (CVO) of Thales, one of the world's largest aerospace and defense companies that has invested heavily in digitalization. Digital transformation requires new business models that are frequently located in a separate organizational unit. The digital transformation requires a shift from single products to the ecosystem. In order to implement the digital transformation successfully, several issues are key: superior customer insights and true customer orientation; acquisition of new, digital analytics capabilities; partnerships; finally, a careful balance between broad experimentation and a few, narrow bets that deliver results and prove the viability of digitalized solutions.

Andreas Hinterhuber: How do you interpret digital transformation?

Stephan Liozu: Digital transformation means a transformation of the business model first, and then a transformation of technology to support new business models while maintaining existing ones. This notion of dual business models running in parallel is essential to understanding the challenge of successful digital transformation. This interpretation varies from people to people in a large group like ours. Despite a vast communication campaign, people seem to focus on technology and IT first and on business models second.

AH: Can you comment on the process that led to your current digital strategy?

SL: Each year our group conducts a strategic business plan (SBP). Part of this SBP process is asking global business units strategic questions. This process mobilizes key players in teams but focuses the organization on the challenges ahead and some of the most exciting opportunities. Thales has been involved in digital initiatives for many years. The SBP process has begun to frame more and more key digital questions for the past few years. These questions included our role in artificial intelligence, the internet of things (IoT), digital platforms, cybersecurity, and how to extract more value in general from digital innovation. Three years ago the group decided to structure all the group digital initiatives in the form of a digital transformation led by a VP of digital transformation. We were already heavily involved in high tech, so it was a natural transition. Today, we are in full gear and have begun to see some of the exciting innovations deliver results.

AH: Which company is, in your view, best in class in digital transformation? What do you learn from their initiatives?

SL: The key is how you define *best in class*. There's no real success story yet. There are very public failures and many strategic acquisitions. Some sectors have done better than others, like smart building, healthcare, smart farming. I personally track Honeywell, Siemens, Sandvik, Stanley Black & Decker, and Schneider Electric. Many companies are doing transformations, but most of them don't communicate transparent information. It's difficult to know how successful they are.

AH: How will digital technologies enable you to adapt and change the business model?

SL: We need to become better at designing and managing new business models. We are an industrial company, with a legacy of success in recent decades. Now we're asking teams to think differently and to develop new business models and customer value propositions at the heart of the digital transformation. Managing that transition is our number-one challenge. We speak a lot about technologies, but we're less sure about business models apart from those that are well known. We're making progress in the defense market, where we have very good customer intimacy and strong, established relationships. The transition from core to new business models is also not easy. It takes a lot of internal discussion and change management.

AH: Let's address the topic of organizing the digital transformation: many companies separate digital initiatives from the traditional organization. Any comments?

SL: We have a hybrid structure to manage our digital programs. Some of our vice presidents report both to the new digital organization and to the core business structure. That plays the bridging role needed between the two worlds. Then we have dedicated digital factories (France, Canada, and Singapore) that are independent and that manage the digital projects. But the digital innovations remain with the core business units. We have a group of digital champions focused on fostering a strong digital innovation funnel who then feed our three digital factories. There's no perfect design for a digital structure. The best design matches the internal culture and the nature of the markets you serve. Then you have to adapt based on your success rate and the changing market needs.

AH: There's no inherent need to separate digital initiatives organizationally—interesting comment. Do you see digitalization as a complement to or as a substitute for your current product/service offering? A substitute disrupts; a complement provides additional value.

SL: Most of our digital initiatives complement existing or new hardware/equipment opportunities. We've launched a few dedicated startups from the innovation digital process (Soarizon, Citadel, Heropolis, for example). Finally, we've acquired several small digital startups in the areas of AI, analytics, and data to complete our digital platforms (Guavus, for example). We focus on our four core market verticals and engage mostly with existing customers. Top management have made a strong effort to focus our digital efforts on our stronger verticals in order to leverage existing capabilities and relationships. We already have a portfolio of digital products and expect to grow, as a result of the digital transformation, via breakthrough innovation as well as via new business models. This is what our CEO, Patrice Caine, presented at a recent investor conference (see Figure 16.1).

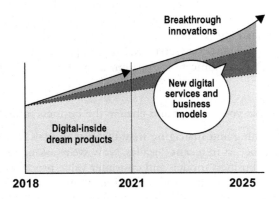

Figure 16.1 Digital innovations as growth drivers, adapted from Caine (2019, p. 14).

AH: What's your view on pricing? What's the role of value-based pricing in the context of digitalization?

SL: We, of course, have a full-time chief value officer (CVO) and have included value coaches in each of the digital factories. These coaches are certified in value-based pricing. We've also developed a value-framing process using the pricing model innovation canvas to frame digital minimum viable products. Most of our digital projects follow this framing process, which includes the steps of value-based pricing, which you know too well, but also essentials of digital pricing. To quantify value, we use the Economic Value Estimation methodology (Nagle & Müller, 2017) and are making progress with this method. It's not an easy process, as we often lack competitive knowledge and customer insights. The value quantification of true differentiation is often where companies really struggle. It takes a lot of iterating and validating with customers in order to produce quantified value propositions for digital innovations. We operate in very complex industries with advanced technologies, and performance differentiators are often hard to evaluate.

AH: How do digital technologies affect branding and communication?

SL: We've conducted a massive branding and brand purpose initiative at the corporate level. We care about the purpose of the firm, and we believe that we're responsible for creating trust and transparency with our digital technologies. We've invested in reliable and explainable AI, for example, and we aim at being socially responsible. This is a message coming from the top, and the initiative is managed by our chief marketing officer.

AH: Let's get to the profit impact. How do you measure the success of the digital transformation? Many companies lack a clear idea of how to measure the impact of marketing initiatives, let alone digital marketing initiatives. A favorite example is Beth Comstock, the chief marketing officer of General Electric in 2008—at that time one of the world's most admired companies, who said—"I would say that we haven't figured it out yet" (Comstock, 2008, p. 1) when asked about the specific approach the company uses to determine the return on marketing expenses. I find this extraordinary.

SL: An extraordinary admission of incompetence. We, of course, measure. We have traditional sales, gross margin, and cash-flow key performance indicators (KPIs). Then we also track traditional software-as-a-service (SaaS) metrics like usage, churn, and customer lifetime value (CLTV). Our strategic department with my help designed the right KPIs to train leadership and finance on how new business models can be different from traditional business models. It's a different model altogether. At Thales, we take impact very seriously, and our focus is on profitable growth while minimizing cash-flow impact. This forces us to stay close to our core market and to leverage our internal capabilities. It's a whole different approach to digital transformation, but the financial markets expect a financial return similar to our core business. So it's a balancing act.

AH: What are the main obstacles to successful implementation of the digital transformation?

SL: There are many barriers to success. I list the top five:

1. Lack of skills and training. We had a developing marketing maturity with our traditional business, and this carried over to the digital business.
2. Tensions between the old and the new can be very hard to manage. There's no easy way to manage the organization structure. But the conflicts can be heated in IT, finance, and product management. Lack of communication, misunderstanding, and lack of trust were a big factor in slowing adoption and progress.
3. Lack of customer-centric innovations at the beginning of the process was a challenge. There are many competing innovations in digital these days. Most of our initial innovations were me-too products. We need to focus more on true customer pains and problems. We're making progress, and we realize the need to focus on customer needs. This is a work in progress.
4. Resistance from finance due to cash-flow impact and pressure to deliver short-term results.
5. Lack of focus: focusing on shiny things is paralyzing. We corrected this in 2018 by focusing on our four core verticals.

AH: Number one is probably the lack of customer-centric innovations, your point 3. As many practitioners, some of whom we interview in this book, and academics point out: technology is a means, not an end. Well said. Let's talk about skills: what capabilities are particularly important in the context of the digital transformation? By contrast, which skills are less important?

SL: What's important is innovative value creation models and a focus on customer pains. We need to be better at discovering and uncovering new and unmet needs leading to digital opportunities. Managing customer value and pricing is essential. A critically important skill is the need to manage ecosystems and to think as a system thinker, not a product thinker. That's a game-changer. Less important skills are cost modeling and process orientation.

AH: Stephan, I appreciate this comment. Managing the digital transformation is all about managing the ecosystem. I fully and totally agree. For which products or services does the digital transformation *not* apply?

SL: This is a hard question to answer because there's no limit to a digital transformation as long as one solves customer pains. It could be in products, services, data, software, etc. There are market segments that might be excluded due to lack of knowledge, but I wouldn't constrain a transformation, especially at the beginning.

AH: What lessons have you learned in the context of digital transformation? What's your advice for a company that has not yet begun and that wishes to learn from your own experience?

SL: Great question: I have strong viewpoints on this.

1. Give the CDO role to an executive with a marketing or commercial background, not to an executive with a technology or IT background. Business, not engineering, should lead digital transformations.
2. Focus, focus, focus: pick your markets carefully and focus your teams on a few successes.
3. Spend more money on the front end of innovation first, and then build a pipeline. Define a strategy followed by the right digital structure. Don't rush to open incubators and factories just because it's trendy.
4. Embrace partnership and co-development to go faster. Fight the not-invented-here syndrome. This is a killer. Buy off-the-shelf solutions and learn from the failures.
5. Invest heavily on up-skilling and reskilling. Also, bring talent from the outside to super boost the development.

AH: The digital transformation is about customer value creation, not technology—well said. What steps have you taken to ensure that the digital transformation increases overall firm profits? In retrospect, what would you have done differently to accelerate the return on investment from digital initiatives?

SL: Focus, focus, focus on a few key innovations with a high level of payback. Digital transformations are not immediately profitable. Don't build digital factories first. Build an innovation pipeline rich in profitable opportunities. You have the option either to outsource the development or to do it internally. Focus on both internal and external digital transformation. You can't do a transformation without investing internally in the right tools.

AH: What comes next? What are future, big topics that CEOs should address?

SL: The big focus is to deliver the impact we've promised to the stock market. We have three years to do it. AI and digital platforms remain hot topics. Finally, CEOs must focus on differentiation in digital. There's a lot of competition, and the focus has to be on winning with disruptive innovations.

AH: What other advice would you give companies that are at the beginning of the digital transformation?

SL: I like to quote Drucker and Chandler. These management gurus have already written the books on transformative innovations. "Culture eats strategy for breakfast"—Drucker. "Structure follows strategy"—Chandler.

AH: Stephan, thank you. The point is: you change the mindset, you change the culture first, then you design a strategy, the organizational question is then almost irrelevant. I appreciate this illuminating exchange of thoughts.

References

Caine, P. 2019. Introduction and progress on medium-term strategic priorities. *Thales 2019 Capital Market Day*. Paris.

Comstock, B. 2008. Q&A with Beth Comstock (GE). *Marketing NPV*, 15 September, 1–2.

Nagle, T. T. & Müller, G. 2017. *The strategy and tactics of pricing: A guide to growing more profitably*. Abingdon: Routledge.

Chapter 17

Digital transformation in luxury industry

Interview with Jean-Christophe Babin, CEO Bulgari

Tiziano Vescovi and Jean-Christophe Babin

The interview "Digital challenges in luxury industry" with Jean-Christophe Babin, CEO of Bulgari Group, analyzes the particularities of digitization in the luxury industry. In this case, it is not a question of digitizing products, but of digitally integrating customer management and the relationship with them, through important effects on communication and branding. It is a question of maintaining the difficult balance between exclusivity and the use of social media, with innovation in languages and targets, but also of promoting a change internally in the skills of the staff in the approach with new generations of luxury customers. Once again, the key issue is the change in corporate culture.

Tiziano Vescovi: How do you interpret "Digital transformation"?

Jean-Christophe Babin: The concept to which we refer in Bulgari is no longer the one of "digital" but of "post-digital", that is, having already ascertained that we are in a mature phase of the development of this theme, we assume that digital is present today and will be present even more in the future and for this reason we are no longer surprised at its existence along the entire value chain.

Therefore, more than "Digital," the key word is "Transformation." I do believe it is the most important one since the strategic choice concerns the "transformation" and specifically if we want to guide or undergo it. In the first case, high risks are assumed, but there are also great advantages to be achieved; in the second, it must be accepted to never fully understand how digital can create new business opportunities.

Tiziano Vescovi: Can you comment on the process that led to the current digital strategy? I.e. scope of the strategic analysis (internal/external), alternative options considered, decision criteria to select the current digital strategy, goals, etc.

Jean-Christophe Babin: Luxury is a sector that is very familiar with innovation and progress, themes strongly present in the current Bulgari digital strategy. Customers in our sector always like to be the first to benefit from a novelty, a new trend, a new experience. For this reason, over the years we have always followed a digital evolution path since the early 2000s by investing in Information Technology, then in the following decade by creating the Omnichannel department and in recent months we are capitalizing on all the experiences that have matured in the company with the creation of an Innovation BU, whose primary aim is to guarantee a long-term competitive advantage based on present and future digital opportunities. The strategy will be based on the open innovation approach because we know that the best insights can come from customers, suppliers, universities, incubators, research centers, etc. and we must be ready to catch them, interpret them and insert them into our Innovation Funnel.

Tiziano Vescovi: Which company is, in your view, best-in-class in digital transformation? What do you learn from their initiatives?

Jean-Christophe Babin: Bulgari revolves around two large polar stars: the product and the customer. We are a company founded on the five senses because for us the luxury experience must completely involve the person who lives it. We always look with great interest to those companies that manage to create a digital experience integrated with the physical one by involving the customer in a network of contact opportunities and endless experience. In recent years, Bulgari has also managed to reach peaks of excellence in clientelling and CRM management, for which we are a reference in the LVMH group (we were the first brand to enter the Genius category in the W&J digital ranking of the innovation index).

Tiziano Vescovi: Digital transformation and business model innovation – How will digital technologies enable you to adapt and change the business model?

Jean-Christophe Babin: Business models are made to last over time, this is their main feature. The new digital technologies are certainly capabilities that allow you to increase touchpoints with the customer, create more efficient and effective pre-sales and after-sales services, broaden the base of potential customers and increase brand awareness. Our current business model is benefiting and will benefit from all these opportunities along the entire value chain. But we are looking for "white space" or those Business Models that can be created only thanks to digital technologies because in this way it is possible to reach customers, market shares and additional revenues exponentially.

Tiziano Vescovi: Organizing the digital transformation – Is this done from within or is there a structure, separate from the traditional business, dealing with digital transformation (i.e. skunk-works)? Any comments on pros and cons of these options?

Jean-Christophe Babin: As mentioned above, I believe that the transformative one is a path, a journey and not a choice between two options made at a certain moment in the company life. Some companies may decide to go all the way; others will decide to stop along the way, while others will undertake to open new tracks. We decided to create an innovation Business Unit because we believe that there are people with great skills and rare resources within us who had shown that they could have a high impact in their specific sectors (e.g. Bulgari Touch: the first leather goods product of luxury in the world, which is at the same time a new communication channel with the end customer or a tool to support the sales force for up-selling, cross-selling, training). It is essential to have a common strategy, clear objectives, career paths, processes, tools and above all to develop and nurture an internal culture that allows obtaining in a stable way those competitive advantages that digital opportunities make available.

Tiziano Vescovi: Digital products – Is the digital offer a complement or substitute to traditional products? What about target customers: new vs existing?

Jean-Christophe Babin: With the current technological development, it is difficult for me to think that a luxury product can be replaced by a digital product. Perhaps the applications that 5G promises will surprise me in this regard. Clearly it must always be clarified first what is meant by a digital product. If we mean a product that alone allows us to provide information to those who interact digitally with it, for example, now all our leather products allow us to do so. For this reason, we focus more on the digital experience of the customer. On one hand, we must guarantee the series of innovative services he expects or is more and more used to get from digital pure players, but, on the other hand, our digital experience must surprise him thanks to the opportunities that digital makes available. Our product will always remain at the center of customer experience because it brings with it the notions of craftsmanship, savoir-faire and knowledge that are rooted in history and that can never be replaced with digital alternatives.

Tiziano Vescovi: Pricing: How do you price digital products and services? What are approaches to understand, measure and quantify customer willingness to pay? Which new revenue streams does the digital transformation enable?

Jean-Christophe Babin: With digital, geographic and temporal dimensions almost lose their meaning. We can also sell in places where our closest store is thousands of miles away. We do this on multiple different digital channels. However, our pricing policy does not vary because it is a way in which we demonstrate the value of our creations to our customers.

Tiziano Vescovi: Communication and branding – How do digital technologies effect branding and communication?

Jean-Christophe Babin: I think that these are the areas in which digital has brought the greatest changes. Today, we are all used to being under the eyes of millions of users 24/24 and we tend to forget that until a few years ago this was not the case. The company used to decide what, how, when and why to communicate. Today the company always communicates even when it decides not to communicate! There are always some threads open on some social network in the world regarding the brand. We are in the era of omni-communication. Only companies that have a brand that has its origins and roots in history and transmits timeless values can cross this turbulent sea without problems. Clearly, we all had to learn new specific languages and ways of communicating for each social media, but this is part of the famous transformation path.

Tiziano Vescovi: Metrics – How to you measure the success of the digital transformation? Do you use different metrics for traditional products than for digital products/initiatives?

Jean-Christophe Babin: I believe that only what is measured is done. Most of existing metrics follow all the business dynamics relating to traditional products. Initially, those of the digital area were made by analogy with the existing ones, but over time ad-hoc metrics have been developed, metrics that are useful for capturing specific phenomena of that world (exposure, awareness, discovery, conversion, retention). We are currently developing a new range of metrics dedicated to innovation that will follow all the milestones of the process from conception to marketing. Our goal is to ensure the best results for the company but also to continuously motivate the resources that work there and retain the best talents.

Tiziano Vescovi: Implementation – What are the main obstacles to a successful implementation of the digital transformation? How did you overcome internal (=other departments, sales manager) or external (=customers, distribution channels) obstacles?

Jean-Christophe Babin: The obstacles to digital transformation are to be found mainly in the corporate culture and therefore in those behaviors that are considered rewarding by the people who work in the company in terms of career and awards. Digital as well as innovation must be recognized as a career path as valid as traditional ones. For this reason, for some years now we have been working to emphasize:

- Change of organizational culture through a clear sponsorship to adopt an open and collaborative mentality;
- The definition of new organizational processes dedicated to innovation and digital;
- The development of new skills for individual employees (entrepreneurship, leadership, networking, scouting).

The other obstacle is external, namely managing to create an external network of collaborators and suppliers able to integrate with our business model. We work in a very particular sector, that of luxury, in which digital recipes from other sectors cannot be automatically, applied although they are considered relevant in sectors such as fashion. Having suppliers who have the business, technological and human knowledge that respond well to our needs is very difficult. For this reason, we believe that the path of open innovation may be the one that will give us the best results in the near future.

Tiziano Vescovi: Capabilities – At the level of the individual marketing/sales manager, what are capabilities that are particularly important in the context of the digital transformation? By contrast, which skills are less important?

Jean-Christophe Babin: Surely the most appreciated capabilities at the moment is ambidexterity, that is the ability to grasp ideas potentially useful for bringing out and exploring new trends while serving the current business, in other words is the ability of a complex organization to be aligned and efficient in its management of today's business demands as well as being adaptive to changes in the environment at the same time.

We know that market insights can emerge everywhere, and we must be able to grasp, organize and evaluate them. Nobody should feel excluded from this process. Digital offers great opportunities in a short space of time, but you have to be good at being ready to take them.

On the contrary, I think specialization is becoming a skill that is losing meaning over time. The sales manager, the product manager, the marketing manager, great specialists who bring great results, often do not have all the tools to read the complexity of the world that is coming with digital. For this reason, today they are supported by the digital sales manager, digital product manager or digital marketing manager. Sometimes creating a duplication of roles and a weighting of the organizational structure. In the long run, this distinction will have to be resolved because it makes no sense to exist.

Tiziano Vescovi: For which products and services does the digital transformation not apply?

Jean-Christophe Babin: There is no rule, it all depends on the ability of the company to integrate digital solutions within the experience offered to the customer. Digital opens up new opportunities for the customer to get excited, know and share, and for the brand to be part of the customer's life in a way that was previously unthinkable. The brand's opportunity and advantage will be in creating a harmony between the physical and digital part and understanding how to exploit new technologies to offer a unique experience.

Tiziano Vescovi: Advice for companies at the beginning of the journey – What lessons have you learnt in the context of the digital transformation? What is your advice for a company that has not yet started and that wishes to learn from your own experience?

Jean-Christophe Babin: It must be clear that digital transformation is not a passing phenomenon, it has accompanied us for the past ten years and will accompany us in the future. For this reason, it cannot be managed by bottom-up initiatives. The ingenious employee, the intuition of a supplier, the enlightened function director are necessary, but not sufficient elements to guarantee a long-term competitive advantage linked to digital technologies. Digital is not a rib of traditional business that is driven by technology only. Digital must be an integral part of the corporate strategy and culture; therefore, it must have a top management representative, it must have a dedicated budget, a strong team and clear, challenging and independent objectives compared to that of traditional business. New working methods dedicated to this area must be developed: design thinking, lean startup, agile.

We have left the moment of improvisation behind us, now the moment of awareness has arrived. The growth and survival of companies passes from knowing how to dominate digital development.

Tiziano Vescovi: Profit impact of the digital transformation – What steps have you taken to ensure that the digital transformation increases overall firm profits? In retrospect: What would you have done differently to accelerate the return on investment from digital initiatives?

Jean-Christophe Babin: I don't like to think with hindsight. The path we have followed is consistent with our strategy. I think the choices made so far have been aligned with the degree of awareness that the company has been able to build and strengthen over time thanks to all the initiatives implemented. Today we have come to create a business unit dedicated to digital products and services as further confirmation that we have taken a further step to accelerate the digital transformation of our brand.

Tiziano Vescovi: Next steps – Once you/your company have progressed on the digital transformation, what comes next? What are future, big topics that CEOs should address?

Jean-Christophe Babin: If I turn my gaze to the future of digital, I think I can only see the tip of the iceberg. In general, we will have to redesign digital experiences with new models that amplify collaboration with our clients. Today digital communication is a one-way experience that with the help of technology will have to turn into true collaboration making our clients active participants. 5G will be a game changer being one of the most powerful enabling technologies that we will have at our disposal. It will revolutionize, on the one hand, the places where we live giving birth to smart cities and on the other hand, it will allow us to experience the digital world in an incredibly immersive way, we start talking about the Internet of Senses. For us who work in luxury and are always looking for immersive customer experiences, knowing that we can digitally delight the customer with all five senses is something that excites me.

Artificial intelligence will definitely become a common tool in most of the big corporations and retailers. We must create new business models were human and AI can easily collaborate, AI should not be feared, on the contrary I see a full power to be unleashed between AI and big data, a sort of co-creation between people and machines.

We will also live in a post-privacy world where privacy issues will be handled globally in a very punctually way, the complexity and importance of the issue of privacy in the future will be placed at the center of our plans and future developments with the development of Attention Economy themes. Just as we guarantee our guests privacy and tranquility in our resorts around the world in the same way we will have to create a digital space made available by Bulgari where customers can enjoy the same sensations.

Finally, yet importantly, sustainability is already and will become more and more an element that will always be present in every Bulgari initiative and will undoubtedly be a further driver of transformation. Bulgari has always been at the forefront in the use of raw materials, and sustainability will become an additional stimulus for our R&D to try innovative and unthinkable ways at the moment.

In general, we have to be able to create a culture and organization that can become an engine for continuous innovation, fostering the unprecedented scale of disruptive technology available today and more tomorrow. We must nurture capabilities and eco-system partnerships able to converge organizations into innovative DNA and digital mindset.

Tiziano Vescovi: In this unexpected situation given by COVID-19, which are the main challenges for a luxury company? How can digitalization help in the relationship with the clients? Which are the new trends in customer behavior connected to digital world?

Jean-Christophe Babin: Every pandemic in history has always given the feeling to those who lived it, that the world would no longer be as before. In the end, history told us something different. I am convinced that, even in the case of COVID, things will be the same way. People are resilient and for this reason they immediately acquire new habits to adapt to the situation they live in, but the desire of sociability, experiences and freedom eventually things will be the same way when the health crisis is over. Certainly, some things will remain attached to us: greater confidence with social networks, e-commerce, online payments, smart working and remote schooling. Many people have been forced by COVID to learn new things; as they have done so now they will feel even freer to be able to enjoy all the potential of the digital world that have previously been denied by their own lifestyle.

Companies, on the other hand, have understood that the domestic customers exist and therefore we must always keep it in mind in the development of domestic markets, relying on domestic customers as much as touristic ones.

Tiziano Vescovi: Other issues – Is there anything else that you think is relevant, or not well understood, in the context of implementing the digital transformation?

Jean-Christophe Babin: The digital transformation must be done in compliance with and respecting the company's DNA, otherwise the transformation fails or the company is damaged. I can consider myself very lucky to be the head of a luxury company, since customers in this sector are willing to buy these products at a price that allows them to bear high research and development costs. The newborn car sector at the end of the nineteenth century was supported by the purchases of a few wealthy customers. Customers who buy luxury products want to stand out and what is more distinctive than being the first to be able to enjoy an innovation? In 2020, digital is what at the end of the nineteenth century was the internal combustion engine, the trigger of a social, industrial and cultural revolution.

Chapter 18

The importance of data in transforming a traditional company to a digital thinking company

Interview with Fabrizio Viacava, chief digital officer of Etro

Francesca Checchinato

Introduction

Etro was founded in 1968 in Milan by Gerolamo Etro. It is an Italian fashion label famous for the Paisley pattern. Etro is a family-run company, where family members have been involved in management and design. In particular, Veronica Etro designed the ready-to-wear line of the women's collection and Kean Etro designed menswear.

The company is present in over 70 countries and its turnover was nearly 300 million euros. Etro operates 150 monobrand stores all over the world.

The company did not embrace technology until 2014 and four years later decided to hire a chief digital officer (CDO), Fabrizio Viacava. Thus, to understand the company's digital transformation Fabrizio was interviewed.

The interview discusses the digital transformation process in Etro and highlights the challenges and success of this company, an example of an Italian family-run company in the luxury fashion market.

Some final remarks summarise the key points emerging from the interview.

Francesca Checchinato: How do you define "digital transformation"?

Fabrizio Viacava: First of all, digital transformation does not deal with technology, it is a change made by people. It is a new framework, a new mindset. Thinking digital is different from thinking in a traditional way. Unfortunately, the latter is no longer effective in our context. To create a digital transformation, firms need digital thinking people, because they are the key to success. It is not a soft vs hard skills issue, both are required. As far as the tools of the digital transformation are concerned, it is important to highlight that they are not just social networks and e-commerce. These are only two of the main features. Digital transformation is more than this. It also means digitalising the performance review and compiling the expense ledger online, not just using Excel or other time-consuming software. The basic assumptions of our digital transformation are to give time back to people, to shorten the value chain, and to make the procedure easier. The point is that behind an easy reality, there is a great complexity: this is digital transformation. Usability is vital.

Moreover, digital transformation is not unique, it depends on the aim, on the business, and on the moment in which the company starts to embrace it. The most important element is the vision – who, where and what the company wants to be in five years or more. To answer these questions, the company needs to think digitally, all the structures must think digitally: from HR to finance, from the supply chain to logistics. My role and my function will disappear in a few years, I hope.

Francesca Checchinato: So, in a few years we will not speak about digital transformation anymore?

Fabrizio Viacava: I hope so. My work is to help Etro to implement the next five-year strategy. If digital exists in this vertical way in five years, I did not do a good job. In a traditional company, the digital transformation office should be not permanent work, it should have a deadline. For a digital company it is different, there is not a digital function, everything is digital. They start their business with the idea of exploiting digital.

Francesca Checchinato: You have a lot of experience in digital transformation; what did you learn from your past experiences?

Fabrizio Viacava: Companies must evolve their digital approach. I worked for Tod's and when I suggested opening Facebook and Twitter accounts it was a revolution; luxury was not consistent with the mass market. Now it is quite normal to have a presence on the main social networks.

Data are important: this is the big deal. I think that one must try to be a pioneer and work on data; developing the strategy on data and learning from data. In this way, measuring the strategy's success is possible. Moreover, digital is an evolving field; managers cannot think "what's done is done". He/she must continuously update their projects and make mistakes. It is necessary to grow. In our culture (Europe and Italy), people want to (and must) avoid mistakes, but on the digital field mistakes happen every day, it is normal. Maybe this is the reason why Silicon Valley is in the US and not in Europe – their culture is more tolerant. If a company does not test and try different solutions, it cannot understand what works and what does not work, it is too complicated.

Francesca Checchinato: Can you tell us something about Etro to understand the context?

Fabrizio Viacava: Etro is a luxury Italian family-run company. For 52 years, the company has performed very well, but a few years ago the general manager and the family understood that the world is changing and if they want to continue to succeed, they have to embrace a digital transformation. Etro is different from the other major fashion luxury companies; their turnover depends on apparel instead of accessories. The producing process is also different: the iconic Paisley pattern, prints and jacquard weave features are the starting points of their collections.

Francesca Checchinato: Can you describe your path to the actual Etro transformation?

Fabrizio Viacava: When the company understood they needed a digital transformation, they decided to create a director role devoted to digitalisation. They needed a CDO because Etro is a traditional company, and it needed to define a new strategy that exploits digital technologies. When I started working for Etro, they created the digital function. Before my arrival, digital was not a function; there were people working on digital in different functions such as retail (for the e-commerce part) or communication, but they did not share a common strategy. We created a function and an organigram that evolves continuously. They decided to create the digital division; I decided how to manage it.

In the beginning, I studied both the internal and external landscape a lot. The company's aim was to enhance the turnover, to implement customer relationship management (CRM) and to plan an effective digital strategy. I interviewed Etro's people to understand the management. I studied what Etro already did and what it was doing – the processes, the business areas, channels and markets, where margin is created and where the potential value could be. I did not suggest changing everything on principle because I needed the people's support; they could not hate me. Digital transformation only works if people are engaged, otherwise even if we have the technology we will fail. To be honest, before starting I had an idea, but these analyses changed my mind.

As far as the external analysis, we studied the competitors and what they were doing.

Starting from these findings, we created a five-year plan, even though I know it is not usual to have a five-year plan; it is an old approach.

Francesca Checchinato: Can you give us some examples about elements that you decided to not implement after your analysis phase?

Fabrizio Viacava: I would have created an e-commerce with a different business model, but it was not possible due to logistics issues and internal processes related to the supply chain. Sometimes you must accept that a company has its logistics and it is not fair to change them. During the first phase of digital transformation, Etro must fill the gap with the market. Nowadays, I think there are two keywords for succeeding: growth and scalability. We must do those things. We have to grow and become scalable faster. Some of our competitors started their digitalisation process eight or nine years ago; we cannot wait anymore. The way companies approach digital is changing and continues to change over and over.

As far as scalability is concerned, it is an important concept that is also applied to data management. If we think about the old model, small retailers know their customers, their preferences and habits. Now, we can replicate the same approach but with thousands or millions of customers, thanks to data. Technology allows us to scale all these data.

Francesca Checchinato: Can you mention any benchmark in the market?

Fabrizio Viacava: Burberry and Gucci for our niche, but they are so different from us. We cannot take them as a real benchmark because of their size and organisational models.

Another benchmark is Zara. It is not exactly in our industry, but it is creating an effective business model. It is important to analyse other industries and the big players, such as Airbnb or Amazon, because our direct competitors are all adopting the same services, applications and so on. Sometimes a company wants to create something just because competitors are doing that and not for a strategic reason. For example, when I arrived, Etro was working on developing an application. I blocked it because it was not a priority and because an application must be different from the website, it must have a specific "reason why", a different value proposition. We will create an Etro app when we can offer to our clients or our VIP clients something more, for example tickets for exclusive events such as Formula 1 or the opera. It is not just creating an app; it is deciding what you want to offer. It is a strategic choice, not a technological problem.

Companies must know who they are and where they want to go. This is the reason why many digital transformations exist; they depend on the company.

Francesca Checchinato: What has changed in the business in the last years due to the digital transformation?

Fabrizio Viacava: The e-commerce area was strengthened. I introduced an important element of our model: fluid stock. We must stop thinking about stocks as silos. There is just one main stock and not the stock of store A or country X. From which country or store we sell the product is just a problem that concerns marginality; consumers should not be involved in this issue.

Wherever a product is based, it is inserted in a platform, an OMS (order management system). Therefore, we know exactly the available number and types of items. The next step is to include the wholesale stocks too. It remains a wholesale property, but it can be used to respond to customers' requests faster than in the past. Our clients are not interested in who the deliverer will be (our boutique, third party boutique, our e-commerce). It is just an internal problem. To maximise our marginality, we have some routing rules that consider both where the item and the client are located. Of course, marginality is the main issue in this system, but we also take into account other elements. If we do not have any other option, we can also ship an item from Japan to Italy. In some cases, we do that to satisfy our customers. The transaction is not convenient, but we can lose marginality today to guarantee it tomorrow.

Another change in the business model was to open the door to the marketplace. We must be there to increase the brand awareness. Just an example: we started collaborating with Farfetch − 92 million clients per month, we cannot even think about losing this opportunity. Suppose we did not sell anything: we should be there anyway to achieve that visibility. This is more important in the Chinese market, where consumers have a completely different purchasing behaviour. Sometimes we think that our model is "the" model, but it is not. Chinese people do not use desktops; they use mobile devices to buy too. In Italy, there are a lot of applications; in China there are just a few, with all of the services included. Entering that market means having a different digital strategy. Opening an e-commerce is not effective and it means losing time as well as resources. We are developing a hard integration with some Chinese marketplaces, but stock is the key element; Etro must guarantee that items will be available.

We will also open an e-commerce in Japan, which is our main retail market, 30% of our direct turnover. This good performance depends on the presence of 48 stores and on the mark up, which is higher than in other countries. Moreover, considering the website traffic, Japan is our second country. Why did we not yet have an e-commerce there? Nobody had an answer because Etro did not have any international digital strategy until few years ago. There were some people that work on international e-commerce but as back office, uploading information without defining a strategy.

Now thanks to our new vision and the availability of data, we have a strategy, and these are questions to ask ourselves − we want to know what is working and how to improve our sales. Moreover, the person involved in the back-office activities now has time to develop strategic activities.

Francesca Checchinato: To sum up: the main business model changes are the fluid stock and the presence in the marketplaces. Any other elements?

Fabrizio Viacava: Yes, omnichannel services. We are working on the "find in store", to check if an item is available in a store. Before the digital transformation, it was not even possible. However, Zara has had it for years. We know that people look for services that they are used to finding on other websites. Then, we are implementing "reserve and collect", "pay and collect" and the already launched "notify me". "Reserve and collect" means that you can book an item seen online because you need to wear it before purchasing it. "Pay and collect" means a customer can buy online but collect in store. It is useful if a customer is travelling. "Notify me" is an alert to inform that a product is available again. These are simple services, but they were impossible to implement with the old structure, before the digitalisation.

Moreover, we changed the categorisation of items to better respond to customer searching behaviour. We used to categorise them by seasons, but consumers search by item categories (jackets, trousers, bags, etc.). Now consumers can find products based on this classification. They do not want waste their time looking season by season. These are simple changes, but they have a great impact on sales. We have also simplified the check-out phase.

Francesca Checchinato: Are products available on the digital channels different from what customers can find in the physical stores?

Fabrizio Viacava: As far as the apparel is concerned, we tried to make the assortment as complete as we can. Instead, concerning home items, it is different; we decided based on data.

We are trying to reduce the number of items, enhancing the assortment depth. Nowadays, we are dealing with 5000 SKU (stock-keeping unit), maybe more. To compare our assortment with a benchmark: Gucci has 900 SKUs. Their turnover is 10 billion; our turnover is 300 million. We need to polarise our offer, to give more visibility to our bestsellers. There are two other reasons for this difference. (1) Apparel provides 70% of our turnover. On average, in the luxury industry, this percentage is close to 30%. (2) Our creative and then productive process is different; it is based on our textile, our Paisley pattern. The decision of the pattern determines every other creative decision, so how we deal with the supply chain is different from the competitors.

We also created some items just for the online channel. Last year, the new communication director decided to insert the invitation for the women's fashion show in a small pochette, created with our texture. Looking at the success and the reaction of the fashion world (PR, influencers, etc.), in three hours we decided to sell them through our e-commerce. We asked the production to produce 200 units. In the end, we sold 392 units, and thus new production was required. The price was 148 euro; it was a tactical decision. No item in our stores has such a low price. Moreover, customers can personalise the pochette by writing their names. This is new for us, even though our competitors started doing it a long time ago. It performed very well. Then, for the new website launch, we tried to do the same with some personalised shirts in a limited edition. It worked very well for the visibility, less so concerning sales.

In the near future, we would like to develop a configurator. It is not a priority, but we are exploring this idea. It is not a front-end problem; it is a process issue. First, we must analyse them because now personalisation is an artisan process, then we can develop a configurator.

Francesca Checchinato: Next steps?

Fabrizio Viacava: As I mentioned before, an e-commerce for the Japanese market. Then, an enlargement of CRM: new functions, new regions. Today, all data are collected and shared among the company, but we want to involve stores because stores are one of the keys of our business. It is important they understand the value of the tool. To collect data from stores, we implemented a digital card. Before the digital transformation, Etro had a paper card, but mistakes are uncountable. Now all the data are certified. The digital card works online too, so all the data are implemented in the CRM; there are no other platforms. Data should be shared with everyone in the company.

We will work on the customer journey creation. For these reasons, we introduced a marketing automation platform. Before my arrival, direct marketing and newsletters were outsourced, both as a platform and people. Now we have a person that works on it; we created an editorial calendar. To create the customer journeys, we introduced a plat-form – marketing cloud by salesforce. One of the main changes is that the marketing cloud is a one-to-one contact platform. Transaction emails, e-commerce related, are sent by marketing cloud too. This will help us to create simple as well as complicated customer journeys. We also want to work on artificial intelligence. Salesforce provided Einstein and we activated this function for our e-commerce and for our CRM too.

We are also working on a digital intelligence project. I want to define it because it is different from the so-called digital intelligence. Usually companies have business intelli-gence to track sales. Our digital intelligence project aims at creating a more evolved data warehouse, where data does not come from sales only but from unstructured data too. Website traffic, social media data, stores visits, etc. The more data, the better. The aim is to use data visualisation to explain the connection between different types of data. The digital team has the leading role, but it is clear that it is a huge problem; all the company's functions are involved. It is a project specifically supported by our team, finance and IT.

Another project to develop during 2020 is digital asset management (DAM/PIM), a big container for all our digital assets. All functions must upload all digital files onto this container, from communication to retail, from store to style office. We want to link all the assets, associating single items with all the related contents. For example, if Etro created an advertising campaign for a jacket, a store window, etc., people will have this information all together, in a bidirectional way. All the company's employees, and thus communica-tion, knows what retail is doing. Data sharing is a key factor in the digital transformation. Product knowledge is available for everyone.

Francesca Checchinato: Can you also measure the effectiveness of all the communication tools?

Fabrizio Viacava: This is an objective for the future, related to our digital intelligence project. DAM just has the aim of sharing information about products or activities. If one needs information about an item (colours, sizes, different models, etc.) or about advertising campaigns, he/she just has to make a query.

Francesca Checchinato: It seems that the digital function takes many marketing decisions. Could you better explain this point to us?

Fabrizio Viacava: I can confirm your impression. In Etro, like in other fashion companies, the marketing function is not a real one like in other industries. The communication is more involved in branding, rather than classical marketing. This is the reason why the digital function develops the editorial calendar and it will create the customer journey. We created the editorial calendar – the communication (function) provides images. Prices are decided by finance and market is defined by the sales function, sometimes in collaboration with retail.

Francesca Checchinato: Another important topic is people. You mentioned the importance of engaging the team and the role of people in transforming the company; how does digital transformation impact on employees?

Fabrizio Viacava: It is important to improve people's capabilities and to motivate people. I have two rules to manage my team. (1) They must think, we are paid to think. Since people are paid to think, they must tell us what they think, share ideas even if they can be wrong. (2) We must be a team. This means that people must defend the team vision and ideas outside the team itself, even if they do not agree with them. This is really hard, but it is necessary to grow as managers and for the company's culture.

Moreover, HR focus a lot on soft skills; those are important but only if the hard ones are taken for granted. Employees must have capabilities to complete their tasks.

Francesca Checchinato: Your team: who works in the digital team and how are you organised?

Fabrizio Viacava: As far as cross function is concerned, my team is composed by a digital architect manager and a digital development manager. The architect designs the digital architecture, defines priorities and the tools to implement them. The development manager ensures that the architect's design is implemented in an appropriate manner.

Then, there is an e-commerce team, where we can also find customer care. There is a CRM team, a person devoted to the online media (and online advertising plans), one person for the analytics and intelligence (who manages SEO and translations, among others), a campaign specialist (that manages the marketing cloud and will help me to create journeys), and two content editors that write content and manage the SEO related to them.

We also work with partners for online advertising. When I arrived, almost everything was managed by internal; that is a big limitation. Companies must involve external experts. We work with the support of some external agencies that help us to optimise campaigns, define better strategies and we also work with maintenance technicians, etc.

Another characteristic of my working approach is that we adopted new tools. We avoid e-mail; there are too many to be effective for our tasks. For example, we introduced Slack and Trello, and different logistics. We also developed Gantt charts but real Gantt charts.

Working through these tools help us to highlight results. It is important to develop and then implement a lot of small projects and do them well. If you just work for a big project and when you deliver it, it fails, the frustration is huge. If you start with some basic steps and then you improve the project month by month, you can see results in a short time too.

Francesca Checchinato: Work 24/7, do you think it is related to digital jobs? Analytics, social media, e–commerce orders never stop.

Fabrizio Viacava: Yes, it depends on the peculiarities of the digital world, but it also depends on passion. And you cannot stop passionate people. This is a positive situation until it becomes stressful. In this latter situation, we must intervene, stressful people do not work well.

Francesca Checchinato: How are you measuring the digital transformation success? Any metrics to highlight?

Fabrizio Viacava: In nine months, we have done a lot of work. Last year, turnover enhanced about +74%. In December, Etro.com was the first store in the world. At the end of 2019, Etro grew in terms of EBITDA, and this result is largely thanks to the digital team. Not because we are the best, but because the company has focalised resources and energies here. Logistics, finance and retail have supported us a lot.

Another metric we monitor is the ROI because we must produce results. But economic results are not the only type of indexes to monitor. We also must look at our database, thus new lead generation is another important metric. This is a concrete result; we can measure leads. No less important is the conversion rate.

It is important to evaluate all metrics together. Sometimes one cannot understand the company's performance looking at just one index. If I look to our conversion rate now, comparing it to the last year I can notice a decrease. Is this negative? No, because our traffic increased 140% and thus orders and sales are increasing.

The culture of analysing data and data-driven decisions are extremely distant to the luxury sector, but when a firm embraces this kind of approach, success follows. Yoox is an example. We (as digital people) have the responsibility to make people aware about the importance of reading data; we must teach how to do that. Last year, we worked a lot in explaining how to use digital data to make decisions.

Francesca Checchinato: Who can access all these data?

Fabrizio Viacava: There are three different levels. A group of people can access the raw data, extract and elaborate them. The family receives a weekly report and a monthly report, and finally board members access an interactive dashboard; they make any query they want.

Francesca Checchinato: At the beginning of the digital transformation, how did people deal with this shift?

Fabrizio Viacava: We had to cope with some resistance which is normal. There are some people that love changes, others that hate them, preferring routines and stable situations. But the GM and the family supported and still support me a lot; they believe in the digital transformation and conveyed this message to all the employees. It was easy to convince people that change is necessary. Some people work only if they receive a request from the chief; if a colleague made the same request, they did not respond properly. It is not fair, but this is the real world.

Another thing that helped me a lot was to speak using metaphors. One of the main mistakes of digital people is to think that everybody understands the digital context. It happens during conferences too. If a person does not work in digital, he/she will not understand anything. It is a language, and a digital transformation officer must be sure that everyone understands it, including other departments. The CDO should try to make things seem less complex than they are. It does not mean to dismiss but to make the situation understandable by people with different backgrounds. We cannot think that all people have the same competences, otherwise misunderstandings start occurring.

Just an example to sum up these concepts. When I proposed this project to the Etro family, I spoke for one hour and a half. They understood the importance of the project, but when I illustrated the required investments, they told me that it was a huge amount of money for a company like Etro. Then, in collaboration with the CFO, we analysed data and compared the investment with a new store opening. In that period, Etro was working on a new opening. In a five-year period our project was less expensive than that. We illustrated data comparing these two investments; we trivialised the investment and the reasoning worked. The family gave me the resources I needed because they really understood the point.

Francesca Checchinato: And what about the distribution chain, the retailers? Any problems during the implementation of the omnichannel approach?

Fabrizio Viacava: To be honest, when we began opening to the marketplaces and e-commerce was no longer an exception, we encountered some resistance. At that point, we decided to adopt a new model and now it works, even though we are already thinking about another evolution of the model. Why did it work? The sales force and retailers were used to receiving incentives when they sold an item through the store. Now they receive an incentive even if they are able to sell online. As I said before, the channel through which the item is sold has no importance for us; it is just a marginality issue. It should be the same for our distribution chain, otherwise they are not motivated to spend time on the project. Therefore, when an item is not available in the store, sellers use the online channel and receive an incentive. In the case of Farfetch, the incentive is related to the time of fulfilling an order because it impacts on our merit score.

Logistics was also complicated because we had to review the logistic process, but they supported us a lot, thus it was not a real problem.

Francesca Checchinato: Looking at your journey: something that you would do in a different way?

Fabrizio Viacava: I would have managed some phases in a different way, especially as far as the implementation is concerned. Some things did not work exactly as we planned both on technical and setting levels. I would revise some elements of our blueprint, which we are now changing. I would have spent more time in analysing some aspects and developing some activities, we run a lot.

Francesca Checchinato: Can you provide us an example?

Fabrizio Viacava: Yes, I can. We introduced more languages because two were not enough in my opinion. The US is our main market and only referring to this market, we must consider three different languages: English, Spanish and Chinese. Our website was available just in one of these three languages and thus we decided to localise all our digital content. For the website, it is not that hard, we just translated content. However, coping with the CRM was complex. We underestimated this complexity and now we are correcting problems to better manage all the languages. Maybe if we thought more about that, we would not have to make repairs now.

We implemented a lot of different projects at the same time. We need to do that because of the gap Etro had with its competitors. Our project is not common in the luxury industry. We cannot implement e-commerce during the first year, CRM during the next year and so on. We have to digitalise the firm quickly. Of course, some of the projects are not perfect, but they work enough to compete in the market. Moreover, to work project by project and not within a system can be risky. Sometimes new implementations in an already created digital system require destroying the past, you cannot just add some functions or applications, you have to develop a new system. Permission and resources to do that are not easy to find.

Francesca Checchinato: Can you give some advice to companies that are at the beginning of the digital transformation process?

Fabrizio Viacava: Do not start immediately, think about the vision. Do not run with excitement to develop exactly what other companies in the same industry are doing. Take the time to define priorities and define where the company has to be in the near future.

Moreover, be curious. Who works in the digital field must be curious and not afraid to try and, eventually, get things wrong.

Finally, look at competitors and other industries' firms.

Conclusion and final remarks

The Etro digital transformation success can be summed up in four main decisions. First, Etro prioritised the availability of data and the ability to analyse data, and instituted a culture of data-driven decisions. This approach enhanced Etro's performance and helped it to define an effective strategy. Before the digital transformation, there was just a draft of the path and the strategy, now starting from data Etro defines the steps to follow. Thanks to this new approach, turnover has grown in a few months.

Second, Etro began data sharing within all the functions. To properly exploit data, they must be shared. Etro has created a Digital Asset Management system to eliminate function barriers on data availability. Thanks to this system, everyone involved in activities related to products has a comprehensive knowledge about all the assets linked to them. These digital assets are shared and are not a single function property. This confirms the importance of creating a systematic approach, where all the functions are involved in the analytics and share common platforms.

Third, Etro improved internal communication. Traditional companies are not used to investing in digital projects so they cannot understand them. The CDO was able to explain the investment, thus the board and the family could compare it with investments they are familiar with, such as a store opening.

Fourth, Etro's digital transformation success is a result of the engagement of the sales force in the project. Online engagement mirrors the offline: the incentive models adopted for the physical transaction worked for the online sales too. If the sales force did not participate in the project, the omnichannel strategy would not properly work.

It is interesting to highlight the need for conducting analyses before starting, but at the same time be quick in transforming. By examining the Etro process, we have learned that a trade-off between time to deeply analyse the context and the need to be on air quickly exists and we can conclude that when the gap with the market is too high, it seems better to act immediately with a comprehensive plan. Logistics, e-commerce, CRM and other digital elements should be planned from the beginning, even if the implementation of some projects need to be postponed and the solution is not perfect.

Part 5

The digital transformation in SMEs: challenges and best practices

Chapter 19

Digital transformation of manufacturing firms

Opportunities and challenges for SMEs

Enrico Pieretto and Andreas Hinterhuber

Digital transformation

The concept of digital transformation

Digital technologies and the internet have fundamentally transformed businesses, consumer habits, products and processes. Products, processes and services have become digitized or have been augmented with digital counterparts and complements and are increasingly interconnected, with great advantages for firms and consumers. New challenges have also emerged for firms, as digital technologies have gained a prominent strategic role. Information technology (IT) strategy has long been treated as a functional-level strategy requiring alignment with the firm's business strategy. Because the business infrastructure has become digital and characterized by increased interconnections, the subordinated role of IT strategy needs to be rethought in conjunction with business strategy; the term *digital business strategy* reflects the fusion between IT strategy and business strategy (Bharadwaj et al., 2013).

Digital technologies have a transformative impact, not only on business strategy but also on the structure of social relationships in the enterprise and in the consumer space (social media and social networking). Products and services integrate digital technologies and are entangled with their underlying IT infrastructures. Digital platforms enable industry disruptions and the rise of new business models. Finally, the increasing price/performance capability of computing, storage, bandwidth and software applications has enabled delivery of new digital technologies through cloud computing. Digital business strategy is defined as "organizational strategy formulated and executed by leveraging digital resources to create differential value" (Bharadwaj et al., 2013).

The increasing diffusion of digital technologies is also changing the producer–consumer relationship (Piccinini et al., 2015). Digital technologies combine information, computing, communications and connectivity technologies (Bharadwaj et al., 2013). Social media and mobile devices are two examples of such technologies that have become part of people's daily lives and whose proliferation has greatly influenced the producer–consumer relationship. The latest developments in digital technologies have empowered consumers and reduced the asymmetry of information that previously characterized the relationship (Labrecque et al., 2013). Consumers can avail themselves of information about attributes and prices of available products and services at their disposal anytime and everywhere; they can exchange information with other consumers, read and watch experts' reviews on the Internet, and contribute to innovations by submitting ideas to firms and generating extensive content on online platforms. From the marketing perspective, this is evidence of a fundamental paradigm shift from customer-centric marketing, where the consumer needs to be strongly influenced by marketing, to customer-driven marketing (Merrilees, 2016), exemplified by the above dynamics.

Firms that recognize and understand these changing dynamics can reap huge benefits by enhancing interactions and exchange of value. Therefore, some of the best-performing firms have adopted a consumer-centric (digital) business strategy. These changing dynamics in society and organizations due to the diffusion of digital technologies are considered drivers of an overarching phenomenon described as digital transformation (Fitzgerald & Kruschwitz, 2013). This definition is not focused on organizations only but takes a wider perspective, considering how the phenomenon of digital transformation is rooted in the changing dynamics of society, to which organizations adapt by leveraging digital technologies through a digital business strategy.

Despite the recent rising research interest in digital transformation in both academia and practice, lack of a common understanding of the concept remains. Morakanyane et al. (2017: 9) conducted a systematic literature review to conceptualize the phenomenon of digital transformation and proposed a new general and inclusive definition: it's "an evolutionary process that leverages digital capabilities and technologies to enable business models, operational processes and customer experiences to create value." Henriette et al. (2016: 3) defined digital transformation as "a disruptive or incremental change process. It starts with the adoption and use of digital technologies, then evolving into an implicit holistic transformation of an organization, or deliberate to pursue value creation."

Both definitions result from literature reviews by the respective authors. The term *evolutionary* captures the nature of a transformative process that develops and unfolds over time. The prominent role of technology, particularly IT, is partly responsible for the evolutionary trait of digital transformation because, on one hand, technologies are subject to evolution; on the other hand, firms learn how to leverage technologies over time, each at their own pace. Digital technologies that characterize the phenomenon of digital transformation are also disruptive because their impact significantly changed how entire industries used to operate. Digital technologies, which enable digital transformation, also require digital capabilities to be built. Digital capabilities are the key skills and capabilities that a company requires to leverage the opportunities enabled by digital technologies. Morakanyane et al. (2017: 9) also define digital capabilities as "technology skills possessed or required by employees, customers and other stakeholders in different areas that can enable the organization to thrive in a digital environment"; this definition considers how both internal and external capabilities influence a firm's capacity to leverage digital technology. A purely conceptual definition of digital transformation is provided by Vial (2019: 121): "a process that aims to improve an entity by triggering significant changes to its properties through combinations of information, computing, communication, and connectivity technologies."

Drivers, impacts and transformed areas

These factors certainly influence the process of digital transformation, but it's important to define the true drivers of digital transformation to distinguish them from impacts. Drivers of digital transformation are attributes that influence and enable the process of digital transformation to take place (Morakanyane et al., 2017). Digital transformation drivers hence include both digital technologies and digital capabilities as well as other internal and external triggers: that is, reasons why organizations engage in digital transformation (Osmundsen et al., 2018). Several studies mention digital transformation drivers: profitability and new revenue growth, customer satisfaction, increased operational efficiency, convenience, increased business agility, increased employee productivity and competitive advantage (Ezeokoli, 2016). Digital technologies and capabilities alone do not enable digital transformation, but they do play a vital role, coupled with factors such as strategy, culture and digitally savvy human capital. Digital technologies and capabilities are the foundation of digital transformation; they create opportunities that organizations leverage to obtain certain benefits.

Digital transformation impacts are the effects that business organizations experience as a result of the transformation process (Morakanyane et al., 2017). These impacts might include improved performance, enhanced relationships, omni-channel access, optimization, automation, cost savings, process improvements and product improvements. While these impacts can be focused on the organization or on the customer, the ultimate impact that organizations pursue through digital transformation is value creation. Higher-level impacts include effects of digital transformation at the industry and society levels, both positive and negative; negative impacts, or undesirable outcomes from digitalization, include cybersecurity and privacy concerns.

The digital transformation process has the potential to effect change throughout the organization, and a holistic perspective should be taken when managing such processes. Throughout the digital transformation literature, some key areas are identified. Westerman et al. (2014) argue that digital transformation takes place in three key areas of the firm – customer experience, operational processes and business models – and that these are the broad areas where companies are building new digital capabilities.

The focus on customer experiences reflects the need to engage consumers (particularly millennials and digital natives) in new ways, provide the highest level of service and establish multiple channels to maximize the exchange of value. Customer experience is a "multidimensional construct focusing on a customer's cognitive, emotional, behavioral, sensorial, and social responses to a firm's offerings during the customer's entire purchase journey" (Lemon & Verhoef, 2016). Customer experience is determined by a complex mix of touchpoints; how firms engage with users by providing immediate, personalized content determines success (Parise et al., 2016). Customer expectations are rising every day as people use their experience with top digital brands like Amazon, Apple and Google as benchmarks for any experience they have in the marketplace: from buying products and services online and in physical stores to interacting with companies before and after a purchase. Satisfying already-high customer expectations requires improved speed and responsiveness; the ability to collect, store and analyze huge amounts of customer data to target the right people at the right time and touchpoint, with the right content, and to provide the personalized top-quality experiences they expect. New digital technologies are changing the rules for customer experience: social media, mobile computing, geolocalization, big data and analytics, augmented reality and virtual reality, AI and machine learning, open new opportunities to interact with customers and to create deeply personalized experiences, both in physical stores and online, also by blending the physical and virtual realms. Companies today have great opportunities to combine existing assets and digital technologies to enhance customer experience. By leveraging the vast amounts of data generated and made available by social networks, mobile platforms and apps, advanced analytics and big data, cloud, and the Internet of Things (SMACIT technologies), companies can enjoy deep knowledge of users to provide customized, high-quality, relevant experiences that continuously improve to address and even anticipate customer needs.

Operations represent the set of activities and processes needed to execute the business model. Transforming internal operations can be a valuable source of competitive advantage that cannot be easily imitated by competitors because it is less apparent from the outside. Digitalizing operational processes can improve many aspects of the organization – efficiency, agility, productivity, safety, the quality of products and services – but ultimately it should impact customer experience positively and create additional value (Verhoef et al., 2019). Hence, enhancing customer experience should always be a priority when the redesign of business processes is concerned and the benchmark of digitalization efforts. The digital transformation of operations will take different shapes depending on the firm's industry, the nature of its operational processes and its capabilities and investment capacity. Service firms, media companies and manufacturing companies will have different opportunities to digitally transform their internal operations. The section "The nine technological pillars of Industry 4.0 and their impact" focuses on digital transformation of manufacturing firms, in which the technologies of Industry 4.0 play a fundamental role.

Stages of digital change

In order to understand how to classify the extent of the efforts of organization towards digital transformation, it's useful to clarify the stages of digital change. Considering the reality of many firms, especially SMEs, the first challenge is related to transitioning from working on paper to working with digital formats. The digitization stage is about converting analog information into digital information to be stored, processed and transmitted by computers. Hence, digitization concerns internal and external documentation processes but doesn't change value-creation activities (Verhoef et al., 2019). Instead, it concerns enabling, improving or transforming business processes (e.g., communication, distribution, customer relationship management) by leveraging IT or digital technologies. Digitalization outcomes are not limited to cost savings; they include more-efficient coordination between processes and, possibly, the creation of additional customer value through enhanced customer experiences. While digitalization might be limited to some functions and processes, digital transformation is a company-wide change involving the transformation of the business model or the creation of new business models; hence, it substantially alters value creation and capture by leveraging opportunities created by digital technology.

Dynamic capabilities, new business models and digital transformation

According to most definitions of digital transformation, the ultimate impact organizations should aim for is business model innovation and the creation of new business models: that is, the development of new ways to create and capture customer value. Because digitalization has increased the speed at which new business models emerge, powered by new ways for firms to interact with customers, business models increasingly have become the unit of analysis, especially in digital contexts. Vial (2019) proposes dynamic capabilities as a theoretical foundation for studying the mechanisms that enable firms to engage with digital transformation to enable strategic renewal. The theory of dynamic capabilities seems to fit the study of digital transformation: dynamic capabilities include "difficult-to-replicate enterprise capabilities required to adapt to changing customer and technological opportunities. They also embrace the enterprise's capacity to shape the eco-system it occupies, develop new products and processes, and design and implement viable business models" (Teece, 2007: 1320). Teece et al. (1997: 515) originally proposed the dynamic-capabilities approach given that the most successful companies globally have demonstrated "timely responsiveness and rapid and flexible product innovation, coupled with the management capability to effectively coordinate and redeploy internal and external competences." The technological developments and societal changes of the last two decades dramatically increased the necessity for organizations to be rapid, flexible and able to orchestrate business model components to respond to an increasingly competitive digital environment, often characterized by disruptive innovations by digital native firms in traditional industries. Teece (2007) collocates dynamic capabilities above the level of operational and ordinary capabilities (routine activities, administration and basic govern-ance). Dynamic capabilities are about the ability to sense and seize opportunities, and to periodically transform the organization to be able to address yet new threats and oppor-tunities (Teece, 2018).

DaSilva and Trkman (2014: 383) clarify the relationship between strategy, dynamic cap-abilities and business models:

> Strategy (a long-term perspective) sets up dynamic capabilities (a medium-term per-spective) which then constrain possible business models (present or short-term perspec-tive) to face either upcoming or existing contingencies. Thus, strategy entails devising dynamic capabilities able to respond to contingencies through the organization's business model. Business models are then bounded by the firm's dynamic capabilities.

Warner and Wäger (2019) proposed a process model to reveal the generic contingency factors that trigger, enable and hinder the building of dynamic capabilities for digital trans-formation, based on interviews with experienced senior executives who lead digital trans-formation projects in incumbent firms. Based on their findings, the authors define digital transformation as (see Figure 19.1):

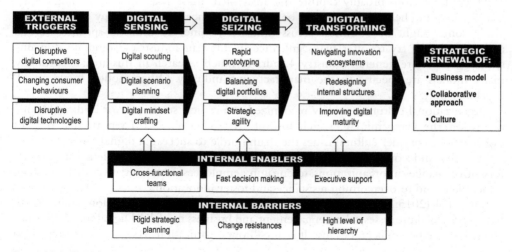

Figure 19.1 Building dynamic capabilities for the digital transformation – a process model.
Source: Warner and Wäger (2019), reprinted with permission, © Elsevier, 2019, all rights reserved.

an ongoing process of using new technologies in everyday organizational life, which recognizes agility as the core mechanism for the strategic renewal of an organization's (1) business model, (2) collaborative approach, and eventually the (3) culture.

Dimensions of digital transformation, readiness and maturity

According to Gurbaxani and Dunkle (2019), digital transformation means the reinvention of the company; this includes its vision and strategy, organizational structure, processes, capabilities and culture. Starting from this vision, the authors identified six dimensions, through research, that led to the development of a framework comprising the factors for successful digital transformation. The objective of the framework is to help executives explore readiness for digital transformation through a checklist, but it also provides a benchmarking tool, through an online survey, to evaluate the company's score against companies that place themselves ahead of competitors in leveraging digital transformation. The six critical enterprise dimensions emerged are:

- **Strategic vision**: existence of a strategic vision and a strategy for executing on the vision.
- **Culture of innovation**: the presence of management practices that encourage innovation.
- **Know-how and intellectual property**: sufficient know-how and intellectual property and how well it is leveraged by the company.
- **Digital capability**: talent available in the company to support digital transformation (strategic and technical expertise and the skills possessed to define and execute digital strategy).
- **Strategic alignment**: a company's ability to make financial investments in digital transformation that correspond to its strategic vision.
- **Technology assets**: the level of the company's use of newer digital technologies (Gurbaxani & Dunkle, 2019).

These dimensions broadly capture the areas firms must assess in approaching digital transformation. The specific items to be evaluated to position a company are many, and while some might be common across industries, some are industry-specific. Several maturity models try to offer firms guidance in different industries, some are domain-specific, others are general. Industry 4.0 drives much research in this area; accordingly, many models are specific to the manufacturing industry. Other models focus on IT capability only.

Organizational culture is widely recognized as a critical success factor in digital transformation, the Capgemini Research Institute (Buvat et al., 2018) reports that 62% of organizations consider cultural issues the main hurdle to successful digital transformation. Developing and communicating a digital culture is necessary to implementing transformative changes in organizations, to engaging employees in the initiatives that are being undertaken, and to overcoming resistance and foster innovation.

Kane et al. (2015:1) define a digitally mature company as "an organization where digital has transformed processes, talent engagement and business models." Digital maturity is not viewed as a static concept, because it is also "how organizations systematically prepare to adapt consistently to ongoing digital change" (Kane et al., 2017: 2); hence, digital maturity concerns an ongoing process of adaptation to an increasingly digital environment. This perspective almost omits technology, underlining how thriving in today's competitive environment requires companies to increasingly focus on people, culture and strategy.

While technology causes digital disruption, traditional companies need to do much more than introduce new technology in their organizations to respond to this threat. Addressing such challenges requires a coordinated, strategic effort initiated by top management and spread across all organizational leaders.

Managing digital transformation

IT strategy and business strategy increasingly need to be aligned in the face of digital innovation (Bharadwaj et al., 2013). The role of executives with respect to the management of digital transformation is debated. Some argue that the chief information officer (CIO) is the ideal executive to lead given the expertise in many dimensions of the digital transformation, but the need to engage other domain leaders is also recognized (Gurbaxani & Dunkle, 2019).

Because the CIO typically is an IT specialist, extending their role to digital innovation, which requires addressing business strategy, might be problematic. Digital transformation requires a holistic, coordinated approach to the promotion of company-wide change; introducing digital technologies is just one aspect of the process. Henriette et al. (2016: 2) note, about the emerging role of the CDO, that "he is dedicated to digital transformation, while having a transverse function of operationalizing digital strategy." Singh and Hess (2017) also note that CDOs are establishing themselves as new executives in companies undergoing digital transformation and point out how their role, responsibilities and capabilities differ from those of CIOs or other CxO positions. The responsibilities associated with digital transformation are complex and challenging for a CEO or other executives to manage, in addition to their original responsibilities. CDOs, unlike CIOs, have neither functional IT responsibility nor profit-and-loss responsibility, and their overall corporate perspective is broader. The CDO must be aware of the full spectrum of opportunities offered by new digital technologies in order to push the company towards digital evolution. The CDO focuses on fostering cross-functional collaboration, mobilizing the whole company across hierarchy levels and stimulating corporate action towards the company's digital transformation. Transformation is the core responsibility of the CDO (Singh & Hess, 2017). Depending on the industry and scope of the digital transformation occurring in the company, the CDO's tasks vary, from customer-experience enhancement through the definition of an omni-channel strategy in retail, to improvement of business operations through implementation of Industry 4.0 solutions in manufacturing.

Singh and Hess (2017) also identified three main roles of CDOs.

- The **entrepreneur role**: explore IT-enabled innovations, establish a digital transformation strategy, help companies innovate with new digital technologies.
- The **digital evangelist role**: inspire people, promote a corporate culture shift, communicate the digital strategy across the whole company to ensure the workforce is engaged, understand the challenges and changes employees go through in the process and manage employee training accordingly.

- The **coordinator role**: coordinate the different functional areas and the many stakeholders of the company affected by the digital transformation, initiate and design the organizational shift from decoupled silos to cross-functional cooperation.

From their analysis of the role of CDOs in six case companies, they also derived five critical skills and competences: IT competency, change management skills, inspiration skills, digital pioneering skills and resilience (Singh & Hess, 2017).

Conclusions on the digital transformation

Changing dynamics in society and organizations due to the diffusion of new technologies drive the overarching phenomenon of digital transformation. Digital disruption (technology and competitors) and changing consumer behaviors force incumbent organizations to activate a continuous, adaptive change process to thrive in the new (digital) competitive environment.

The literature converges in identifying three main areas affected by digital transformation: operational processes, customer experience and business models. Digital maturity is a useful concept for understanding what it means for an organization to be digitally transformed. Maturity is assessed on six key organizational dimensions: strategic vision, culture of innovation, know-how and intellectual property, digital capabilities, strategic alignment and technology assets; this approach also provides a complete view of the levers of digital transformation.

Dynamic capabilities theory is an appropriate approach to explain the phases through which organizations transform. It begins with sensing (identification of opportunities), then seizing (business model definition and redesign and commitment of adequate resources), and eventually transformation (realignment of structure and culture).

A digitally mature, or digitally transformed, organization is characterized by agility that is able to systematically prepare to adapt consistently to ongoing digital change. Digital transformation is more about people, culture and strategy than it is about technology. It requires orchestration; the CDO is an emerging figure in organizations undergoing digital transformation. CDOs are hired as digital transformation specialists; they are dedicated to the orchestration of digital transformation efforts throughout the organization and play three main roles: entrepreneur (explore digital innovations, establish DT strategy), digital evangelist (inspire people, promote a cultural shift, communicate strategy, understand challenges) and coordinator (promote cross-functional cooperation). Figure 19.2 maps the concepts of digital transformation that emerged from the literature on the topic.

The nine technological pillars of Industry 4.0 and their impact

The fourth industrial revolution is based on nine technological pillars:

1. additive manufacturing;
2. autonomous robots;
3. augmented reality;
4. cloud computing;
5. simulation;

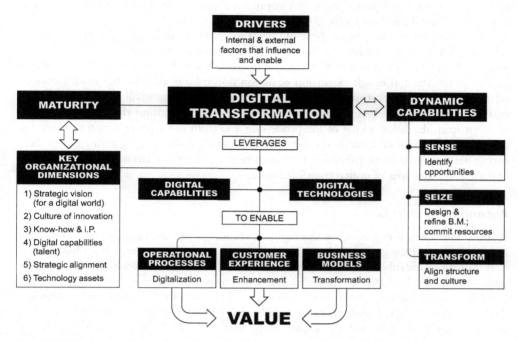

Figure 19.2 Key topics relevant for the digital transformation – own elaboration based on the literature.

6.   Industrial Internet of Things;
7.   big data and analytics;
8.   cybersecurity;
9.   horizontal and vertical system integration (Rüßmann et al., 2015).

Additive manufacturing

Additive manufacturing refers to the production of 3D objects starting from a virtual model. 3D printing has already been adopted by many manufacturing firms, mostly for proto-typing and producing individual components. As technology improves and its advantages become more widely recognized, additive manufacturing will be used to produce small batches of customized products with construction advantages and new design concepts. Research conducted by Make in Italy (2015), in partnership with Fondazione Nord Est, highlighted how 3D printing had become a key element of the competitiveness of traditional sectors of Made in Italy (fashion and home system), with its adoption extending also to small enterprises (2–10M euro in turnover). Interviewed firms indicated areas of greatest benefit (actual or expected) from use of the technology: design and proto-typing, acquisition of the 3D model of existing objects, producing objects with shapes and geometries never possible before, and creation of customized 3D models for clients. Actual benefits resulting from the adoption of 3D printing mentioned included reduced time for designing and prototyping and improved customer involvement (Make in Italy; Fondazione Nord Est; Prometeia, 2015).

The evidence presented shows how 3D printing bears the potential of a transformative digital technology, as it affects several parts of the business model, from improved flexibility and efficiency of internal processes and firm workflow, to the strategic dimension of product innovation, through expansion of design possibilities, and also the manufacturer's relationship with its customers. The possibility of producing small batches of highly customized products, where competitive advantage results not from the scale of production but from the value added of the processing and from the customization possibilities, enables the design of new business models (Bagnoli et al., 2018). Finally, additive manufacturing has a positive impact, through improved client–producer relationships, on customer experience, a key area of digital transformation.

Autonomous robots

Robots have been a key asset for the competitiveness of the manufacturing industry for decades because they can perform heavy, repetitive and dangerous tasks with high precision. Autonomous robots, also called advanced manufacturing solutions, are collaborative interconnected robots that are easily programmed to perform various tasks. New sensors combined with software allow them to perceive their surroundings and avoid collisions. They are also equipped with artificial intelligence (AI), thanks to which they can make decisions autonomously and learn how to perform new tasks by imitation. Thanks to deep learning, they can also perform cognitive tasks: objects recognition, supervision, human resource management. The five qualifying aspects for smart autonomous robots are mobility, sensorial perception, a digital central neural system, energy supply and communication through voice and gestures (Bagnoli et al., 2018).

The term *collaborative* highlights how this technology fundamentally changes the relationship between the worker and the robot: the robot is no longer subordinated but can collaborate with workers and assist them in their jobs, thanks to improved human–machine interfaces. Internal processes affected by autonomous robots include production (more flexibility, higher quality, higher productivity, better workflow due to better communication and coordination), administrative processes, logistics and warehouse management (internal logistics managed by robots and automatic warehouses, autonomous vehicles). Human resources are impacted by the adoption of autonomous collaborative robots: while more high-level competences are required to program, coordinate and solve problems with the new machines, operators enjoy the benefits of improved human–machine interfaces and can better focus on quality control and on developing problem-solving skills since they are relived from repetitive and stressful tasks.

Augmented reality

Augmented reality (AR) is part of a broader category referred to as mixed reality (MR) and is positioned on the reality-virtuality (RV) continuum. The virtuality continuum relates to the mixture of classes of objects presented in any particular display situation, which is augmented by AR applications (Milgram & Kishino, 1994). In AR, real-world environments are enhanced by computer-generated perceptual information, mainly by visual, auditory and haptic sensory modalities. While the concept of AR has been around for decades, the technological developments of the last ten years have matured the technology sufficiently for adoption in industrial settings to support a variety of uses. Although VR is generally preferred for training purposes, AR offers the worker a hands-on experience enriched and assisted by virtual information. Applications of AR include virtual monitoring of production, assistance in complex assembly and maintenance, expert support and quality assurance.

Cloud computing

The National Institute of Standards and Technology (NIST) defined cloud computing as "a model for enabling ubiquitous, convenient, on-demand network access to a shared pool of configurable computing resources (e.g., networks, servers, storage, applications, and services) that can be rapidly provisioned and released with minimal management effort or service provider interaction" (Mell & Grance, 2009).

The NIST also described three service models for cloud computing and four deployment models: Software as a Service (SaaS), Platform as a Service (PaaS), and Infrastructure as a Service (IaaS). SaaS provides the capability to use the provider's applications running on a cloud infrastructure. PaaS provides a web-based platform for applications and services development. IaaS provides a processing infrastructure consisting of virtualized hardware (computing, storage and connectivity resources) where the user can deploy and run arbitrary software. Cloud-computing deployment modes are private, whereby the cloud infrastructure is provisioned for exclusive use by a single organization; community, whereby infrastructure is for exclusive use by a specific community of consumers from organizations with shared concerns; public, provisioned for public use by the general public; and hybrid, in which the cloud infrastructure is a combination of two typologies where, typically, critical data and services are controlled by the private cloud and non-critical information is managed by the public cloud. In manufacturing, cloud computing responds to the needs of the smart factory: big data and analytics solutions, IoT, additive manufacturing and AR can be used to their full potential thanks to cloud computing's ubiquitous connectivity, computing power and flexibility.

Simulation

Simulation, another pillar of Industry 4.0, means producing a model of a situation or process for the purpose of studying and optimizing it. Simulation is powered by advances in sensors, connectivity, software and computing power that characterize the fourth industrial revolution, making it possible for firms to rely on simulation for various purposes that previously required real-world experimentation. In manufacturing, simulation is particularly useful for evaluating the impact of changes in production processes that directly affect the design and development of products; simulation also permits faster product innovation, and virtual prototyping is crucial to realizing small batches of customized products (Bagnoli et al., 2018). Simulation can also be employed to optimize plant and warehouse layouts and to forecast the behavior of complex systems. The most advanced simulation applications are used to map and virtualize every process in a factory, effectively creating its digital twin.

Industrial Internet of Things

The Internet of Things (IoT) is "an information network of physical objects (sensors, machines, cars, buildings, and other items) that allows interaction and cooperation of these objects to reach common goals" (Jeschke et al., 2017: 3). The Industrial Internet of Things (IIoT) is the application of IoT to the industrial environment. The main form of communication enabled by IoT is machine-to-machine, through which devices directly communicate without human intervention. Machine-to-machine communication is expected to contribute to the consolidation of cyber-physical systems (Bagnoli et al., 2018). In production, where most IIoT technology is employed, IoT-enabled machines can monitor and solve potential problems, increasing operational efficiency. On the other hand, connected products equipped with sensors benefit both producers, who can offer additional services thanks to the ability to obtain accurate information about a product's status and usage and offer, for example, predictive maintenance based on remote monitoring of the machine's status; and users, who can benefit from reduced downtimes and improved assistance. IIoT impacts the whole value chain through multidirectional connectivity between processes, components and materials and opens opportunities for new business models and value creation (e.g., servitization business models).

Big data and analytics

The Industry 4.0 paradigm brings about unprecedented needs for data management technologies and solutions. On one hand, businesses have the opportunity to exploit the already vast amount of data they possess though advanced analytics software; on the other hand, the adoption of new IoT-enabled machinery and the production of smart products generates increasing loads of unstructured data that must be collected, analyzed, and translated into information useful for improving processes, products and services, informing decision-making and generating new revenue streams.

Cybersecurity

Industry 4.0 technologies raise interconnectedness to a new level, with cyber-physical systems in the factory; vertical and horizontal integration with partners, customers, suppliers and employees; and vast amounts of data from heterogeneous sources flowing through public and private clouds. While the competitiveness of the manufacturing sector increasingly relies on the new industrial paradigm with IoT and related technologies at its very core, new challenges and concerns about the vulnerability of such interconnected systems emerge.

Data protection and the integrity of networks and systems are crucial for the survival of businesses in the fourth industrial revolution. Given the widespread use of IoT in industrial settings and the large growth it will experience in the coming years, and with 41.6 billion connected IoT devices by 2025, according to an International Data Corporation (2019) forecast, cybersecurity must be a priority. The role of cybersecurity is to protect firms and their stakeholders from potential threats deriving from cyberspace, a complex ecosystem where people, software, services and devices interact through the Internet. Machines and data need to be protected against abuse (intentional or unintentional) and from actions by unauthorized users. Cyberattack typology varies depending on enterprise size: small enterprises are more exposed to external threats, and bigger enterprises, to internal threats (direct attacks by employees) that may undermine business continuity and damage the firm's reputation (Bagnoli et al., 2018).

Horizontal and vertical system integration

To exploit the full potential of Industry 4.0, the pervasive connection and traceability made possible by enabling technologies such as sensors and IoT must involve both the vertical dimension of a firm's internal value chain and the horizontal dimension of the value chain that links the firm with its suppliers and customers. Vertical system integration is meant to realize the concept of smart production systems, such as the smart factory, where all departments and functions are connected under the same integrated system. At the same time, vertical system integration supports horizontal integration through value networks (Bagnoli et al., 2018). Vertical integration supports internal processes thanks to more efficient data flow through the various functions and between smart objects, services and networks. Horizontal integration allows better collaboration with suppliers, who become digitally integrated in the firm's value chain. The relationship with customers also benefits from horizontal integration; the firm is better positioned to meet the needs of customers, who become valuable collaborative partners thanks to the improved interaction between the parties.

The nine technologies that characterize Industry 4.0 enable transformative impacts for firms, especially when more technologies are combined. A combination of two or more technologies is required to achieve certain outcomes or to realize the desired transformative impact; in other cases, one technology is necessary for a system to work properly or warrant the necessary security. Nonetheless, even the adoption of a subset of such technologies can significantly improve operations, enable new business models and improve customer experience.

Human resource management and development in Industry 4.0

It's already been noted that human resource management (HRM) should play a key role in digital transformation processes. Management of human resources as well as development of talent are crucial for manufacturing SMEs and for remaining competitive and achieving Industry 4.0 goals. The effects of Industry 4.0 on the workforce are many. Some have already been mentioned; others include integration of lower-qualified and older personnel, new job profiles, new workplaces, higher technical expertise and employee training required, technology-based trainings and support in failure recognition. On one hand, a new wave of technology requires new highly skilled employees; on the other hand, AR, VR and simulation can assist and guide workers in operations and training, but companies need to learn how to use these new tools. Hence, knowledge and competence challenges call for a strategic and holistic approach to HRM in manufacturing companies. Hecklau et al. (2016) developed a competence model for Industry 4.0 beginning with the identification of its emerging challenges, then they proposed the competences to face those challenges and provided a suitable instrument to visualize the required competences (see Figure 19.3).

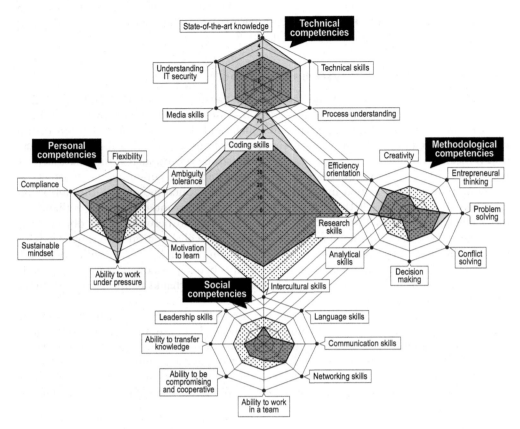

Figure 19.3 Digital transformation competency model.
Source: Hecklau et al. 2016, reprinted with permission, © Elsevier, 2016, all rights reserved.

The competency model can be used for an employee-readiness analysis, as it enables companies to conduct a competence-gap analysis for required competences in Industry 4.0. According to the developers, the tool is designed to assess individual employees, and this assessment should be conducted by an experienced person. The competences need to be weighted according to the department or job profile of the employee being assessed; each job profile requires different advancement levels for each competence. Once the requirements are defined and the competences of an employee evaluated, the model will reveal the existing competence gaps. The tool provides a basis for the HR department to effectively address qualification gaps in Industry 4.0. Suitable training and education strategies defined in advance will be triggered once the model is applied.

Hecklau et al. (2016) also called for further research to focus on the development of specific job profiles. Benešová and Tupa (2017) addressed this issue by defining a set of IT and production job profiles for Industry 4.0. They argue that implementation of Industry 4.0 is a gradual change realized in phases and that, based on the various implementation stages, it would be possible to determine the required jobs. The proposition of a zero phase for SMEs lacking basic information systems for bookkeeping, HR administration, and basic business indicators is consistent with the level 0 of maturity proposed by Mittal et al. (2018) for positioning SMEs that lack prerequisite resources and infrastructures for further Industry 4.0 implementation. In the zero phase, the enterprise will have to implement a suitable advanced information system, which is the entry point of Industry 4.0. The implementation of the zero phase will uncover some problems related to the inadequate computer skills of employees, which are crucial also for subsequent phases of implementation; hence, even reaching the entry point of Industry 4.0 might prove particularly challenging for some manufacturing SMEs. The subsequent transition, from the zero phase to the first phase, is illustrative of the dynamics impacting the workforce. This transition involves the automatization of data collection for the creation of the digital representation of the factory. It requires the mapping of all processes, which demands dedicated professional figures such as process engineers; the increasing volume of data increases demand for cloud services, implying the need for new specialists as well as changing roles among IT employees, as some of the responsibilities switch to the cloud service provider. Once full digitalization is achieved, with the digital representation of the factory in real time (phase 1), the next phases are horizontal integration of processes (phase 2), data analysis and vertical integration (phase 3), and self-controlling manufacture and logistics (phase 4).

The production and IT job profiles identified by Benešová and Tupa (2017), based on the analysis of implementation phases briefly described above, are reported in Table 19.1.

Table 19.1 Production job profiles

Job	Qualification	Skills
Electronics technician	High school education focused on mechanical Practice in the field of handling technology Performing service inspections	Manual skills Ability to learn how to maintain new machines Flexibility; autonomy; responsibility Basic knowledge of electronics, hydraulics Service of the pressure cylinders
Automation technician	High school education in electrical engineering/automatization Practice and experience of machine maintenance and automated lines	Knowledge of safety standards Language skills: English, German, etc. Flexibility; autonomy; responsibility
Production technician	High school education in electrical engineering	Language skills: English, German, etc. Logical thinking Flexibility; autonomy; responsibility Ability and willingness to learn new things; media skills
Manufacturing engineer	Secondary/postgraduate education in electrical engineering	Technical skills Language skills: English, German, etc. Knowledge of technical documentation Ability and willingness to learn new things Organizational skills; cooperation Media skills Communication skills

Source: Benešová and Tupa, 2017

The authors provide evidence of the impact of Industry 4.0 implementation on HR management and development: the enterprise faces aging, digitally unskilled workers; some job positions (the more physically demanding) tend to disappear and the need to hire new specialists to implement new systems emerges; the re-training of employees to work in new, more complex environments is necessary. Finally, especially in the last stages, the difficulty of recruiting programmers and data analysts is likely to be problematic, as these are not widely available in the market.

The problem of shortage of skilled labor force for Industry 4.0 is highlighted also by Matt et al. (2020) as particularly challenging for SMEs competing with large corporations in the fight for skilled workers; in the vision of the smart city of tomorrow, a concept of urban production is proposed with a set of measures to overcome this problem. Through the conceptual paper, the authors acknowledge the importance of highly educated and specialized human resources as crucial to successful Industry 4.0 implementation and the scarcity of skilled labor that will be experienced. The concept of urban production is considered attractive for young, qualified talents due to the high density of higher education in cities.

A qualitative research approach to Italian SMEs in Industry 4.0

A literature review helped identify issues that are particularly relevant for manufacturing firms in Industry 4.0. The perspectives of SMEs are considered in a few articles that address their specific requirements and limitations with respect to Industry 4.0. In general, however, we found a lack of qualitative studies on the perspectives of SMEs that are implementing Industry 4.0 projects. Also, studies of the Italian context, which differs considerably from that of Germany, a more mature market with respect to Industry 4.0, are missing. To address this gap, our next step was to investigate the perspective of real SMEs to understand how they're navigating Industry 4.0. Accordingly, we conducted semi-structured interviews with managers in Italian manufacturing SMEs, located in northeastern Italy, that have undertaken digital transformation projects.

Methodology

A semi-structured interview design was chosen to allow the interviewer to follow topical trajectories in conversations that might deviate from questions prepared in advance. The questions served as an interview guideline and ensured that some comparable qualitative information could be obtained. This approach is considered to allow the different experiences of the interviewees to emerge more freely and represents an opportunity for identifying new perspectives on the topic at hand. Potential candidates were identified among firms that participated in the project by Confindustria Veneto (2019) called "*I Cento luoghi di Industria 4.0*"; managers were reached through LinkedIn via direct message; and telephone interviews were agreed to and conducted subsequently.

Table 19.2 is an overview of the sample of firms and interviewees. Firm size refers to revenue class and number of employees; each interviewee's position is reported along with his experience in the role, expressed in years. Each interviewee was chosen for his relevant

Table 19.2 Overview of firms and interviewees

Firm	Company A	Company B	Company C	Company D
Activity	Machining of metal smallware	Machining of aesthetic metal parts	Screen-printing	Production of packaging machines and transportation systems
Size (employees)	25–50	50–75	10–25	75–100
Revenues (2018)	5–12 million euro	5–12 million euro	1–5 million euro	5–12 million euro
Interviewee	A	B	C	D
Interviewee's position	Managing Director	President	Vice President	IT & Documentation Manager
Experience	11 years	34 years	17 years	4 years

role in the Industry 4.0 implementation project and for a complete understanding and vision of the relevant issues concerning Industry 4.0 and the company.

Firms in the sample

All four firms are SMEs located in Italy. All have a strong European or global projection, and exports are an important, or even their main, source of revenues. Three are suppliers for the automotive sector, serving the most important players in the industry.

Company A

Company A is an engineering firm and specializes in producing a wide variety of rivets and developing fixing solutions. Thanks to the vision of its management, it began a path towards digitalization long before the Industry 4.0 plan was developed by the Italian government. The company spent 70% of its digitalization budget before any incentive was available; this classifies the firm as an early mover, and it's not surprising that it became a case study and an example for manufacturing firms willing to follow a similar path towards digitalization. Company A exported 80% of its production even before going fully digital; this share has remained constant, while revenues have grown significantly after the digitalization of the firm.

Company B

Company B specializes in developing and producing handles, knobs and aluminum profiles for home appliances and automotive. The company is committed to sustainability, and its business culture is based on lean manufacturing principles. Implementations of Industry 4.0, accordingly, built on the lean philosophy, which remains the company's cornerstone even when approaching and introducing Industry 4.0 innovations.

Company C

Company C is a small screen-printing firm and a supplier to global automotive firms. The company adopted lean production in 2011, and then evolved towards Industry 4.0 in 2017, stimulated by contact and collaboration with universities and research centers across Italy and Europe. This company is an interesting example because, as a screen-printing company, its reality is different from that of most companies studied in the literature about Industry 4.0, which addresses mostly metalworking companies.

Company D

Company D designs and produces automatic stretch wrappers for palletized loads and pallet transportation systems. The company is implementing two parallel digitalization paths: one about internal processes, the other about their product. The firm is a supplier of IoT-enabled machines; hence, Industry 4.0 characterizes both the firm's internal process and the smart products it offers.

Results

Discussion of the results follows the main topics covered in the interviews.

Adoption of Industry 4.0 technologies

From the technological point of view, without considering the specific digitalization objectives of each firm, the first steps that manufacturing SMEs must make involve connecting the machines in the factory and upgrading their information systems by adapting them to work together or introducing new software components to fill gaps. These steps are common among firms, as they pose the foundations for further implementations; and this phase is consistent with the level-zero maturity identified in the literature. Interviewee A reported: "The problem was that we had a [digital] information gap from when the requirements for sales orders were entered until the material was taken from the automated warehouse"; after connecting the machines, from the software point of view, the implementation of the MES (manufacturing execution system) filled the gap, allowing the company to gather all the information on the manufacturing process in digital form. Interviewee C reported a similar approach:

> We adopted vertical warehouses and connected most of the machines, both the new ones, already enabled for IoT, and the old ones, by adopting modules based on Arduino developed in collaboration with a local fab-lab, in order to have all the structure connected with the management software.

It's interesting that in both Company A and Company C, old machines were simply upgraded to fit the emerging requirement of connection, refuting the idea that huge investments are always needed to reach the entry point of Industry 4.0; the hurdles vary depending on the infrastructure and the state of IT in the firm. Company C, for example, reduced costs by collaborating with fab-labs to develop the connection modules for its machines.

The digitalization of processes in the factory provides advantages for both internal operations and relationships with external stakeholders. Interviewee D summarized his vision of the integration of data and its advantages:

> The idea is to bring all the data into the data warehouse in an ever more complete way, and then to digitalize the totems in the factory in which the data relating to the order are displayed during the assembly phase; the introduction of barcode scanners will allow the automatic data entry in the MES and a real-time control of the correctness of the assembly operations by the operator; moreover, the progress of the order will be monitored directly by the customer through a dedicated portal.

Digitalization hence increases the transparency of processes and creates value-added information, improving also the relationship with customers. Interviewee A described the great impact of the additional information on which the firm can rely:

> Having all the quality and efficiency measures as well as the causes of machinery stop under control is a great advantage, given the importance of risk management especially in the automotive sector. The customer understands that we apply a logic of continuous improvement and perceives a lower risk in purchasing our products. The additional information we can now provide to customers has allowed as to develop parts for important players in automotive.

Interviewee B talked about how cloud computing is the first enabling technology with which they came into contact: "We needed speed and to be able to collaborate more easily within the company and share knowledge in a more streamlined way but also externally with our suppliers and customers." Cloud computing paved the way to further solutions: the firm implemented software to manage extra-corporate Kanban logics, allowing the sharing of information in real time, interactively, with suppliers, with advantages for speed, synchrony and flexibility. Ubiquitous connection and information also allowed the introduction of a first collaborative robot in a new Industry 4.0 area in the factory; the new station is connected to information systems, which bring extensive real-time information right to the workstation.

Company D not only introduced digitalization of its internal operations but also embedded IoT in the machines it produces. As a supplier of IoT-enabled machines, the firm faced resistance from system integrators it works with, as some were not yet ready to support such technology. On the other hand, while the machines attracted interest because they qualified for fiscal incentives, the firm didn't see the same enthusiasm from customers for the true objective of IoT technology and the advantages it offers. Data-protection concerns also emerged: customers expected the firm to treat data from production with care, storing them in a secure cloud, and deleting them if the relationship was terminated. The firm is committed, nonetheless, to better serving its customers and has clear plans. Interviewee D reported: "The future perspective concerns understanding the services that customers could require, to structure clouding and sensors on the machines to enrich the reports we already offer and to offer, in the future, predictive maintenance." The company already offers tele-assistance on its machines and has the objective of supplying end users with AR glasses to reduce maintenance costs for trivial interventions that can be done by entry-level maintainers guided remotely. Improving customer experience is the prime commitment at company D.

Additive manufacturing adoption is varied: some firms used 3D printing for internal purposes but reported that it is not yet adequate to satisfy customer requirements; others recognize its advantages and were using it to create prototypes but relying on external providers, since limited investment capacity doesn't allow them to introduce it internally and develop the required know-how.

All four companies realized, to some degree, vertical and horizontal integration. The first efforts in SMEs concern enabling connectivity among machines and the integration of software that used to work separately. Digitalization also entails digitalizing documentation and processes previously on paper. Cloud computing solutions also are widespread, addressing various needs: from the flow of information and knowledge within and outside the firm, to data storage for the increasing amount of data coming from heterogeneous sources. The ability to gather data from production thanks to ubiquitous connectivity and the implementation of manufacturing execution system software not only permit improving internal operations but also enable firms to change the value proposition and the business models. While some manufacturing SMEs focus on high-value-added products and mass customization, there is also evidence of the trend towards servitization. Company A transitioned from simply producing what is, in fact, a poor metal product to becoming a specialized competence center for firms that use rivets. It aims to be a market leader in services linked to the product, from design to production, and in providing value-added information about the product itself. Company D, through the development of IoT-enabled machines and software, offers its customers detailed reports on the usage of its machines and is on a path to increase its services for predictive maintenance and AR tools for remote assistance in maintenance interventions.

Adoption paths and how firms come into contact with Industry 4.0 vary. Interviewee C reported that they underwent the implementation of lean production by collaborating with German universities, and this process made them question the entire organization; he also reported that they are active in the world of research:

> We are a firm that is always active from the point of view of innovation, given our positioning on the medium-high range of the market we are stimulated to have contacts with research centers and designers to always search something new. We collaborate with the CERN of Geneva, the CNR of Venice and Bologna and with ENEA with which we collaborate because we can produce in an industrial way that which they experiment.

Concerning the adoption of Industry 4.0, the relationships with industrial associations is also deemed fundamental.

Interviewee A reported that when they began implementing the MES, they were among the first companies, and suppliers in Italy were few; hence, the choice was simple. Now the offer is much wider, and it is also easier to contact not only technology and providers but also the Industry 4.0 support network, including competence centers and digital innovation hubs, which were established in 2016. Company A, among the pioneers, subsequently met with subjects addressing Industry 4.0 and collaborated with them by recounting its own experience to help other firms wiling to follow a similar path.

About the adoption path, Interviewee B said: "We approached Industry 4.0 technologies without making a plan; we got there as needs emerged, following in part the stimuli that new technologies gave us."

Interviewee D mentioned trade magazines, personal curiosity and exhibitions as the main sources that stimulated the introduction of Industry 4.0 solutions in the firm. The most decisive event was a presentation at the IBM Watson IOT Center in Munich, Germany, in which they acknowledged the potential and great expected growth of the IoT sector.

Interviewees agreed that the impact of Industry 4.0 solutions on the work of operators is positive. Interviewee A argued that digitalization substantially simplifies the work of the operator: the new systems allow the operator to receive planning information on a PDA; moreover, the new interfaces guide the operator and assist them through the process of quality control. Interviewee B commented on the introduction of a new collaborative robot:

> The solution introduced has significantly improved the operator's work also in terms of autonomy compared to traditional workstations. Human–machine interaction, through the new digital touch interface, allows the operator to easily solve problems such as micro-stops, tool changes and jams, while, in traditional workstations, many problems require the intervention of specialists. The new collaborative robot has improved various aspects of the operator's work by ensuring a cleaner, more human-centric workplace with better ergonomics.

Interviewee C said that they are in the process of introducing collaborative robots to free the operator from repetitive tasks in order to focus more on value-added activities such as quality control and problem-solving.

Management challenge of Industry 4.0

While acknowledging that the phase of integrating the systems can be technically challenging, as it requires careful analysis and preparation, all interviewees emphasized that Industry 4.0 is much more than a technological challenge.

Many authors claim that the HR department should have a central role in the digital transformation of organizations; SMEs tend to fall short here. Evidence from the interviewed firms shows that if the HR function exists, it's either managed by one person or outsourced. As far as HR development is concerned, the responsibilities tend to be spread among functional leaders. The importance of HR is recognized nonetheless, and each firm has its own way of addressing the challenges related to HR management and development in Industry 4.0.

The main issues and challenges of Industry 4.0 emerging from the interviews were related to the activation of a mindset change, the involvement of people from every department in the process of transformation, the introduction of digital competences and soft skills for Industry 4.0 at all levels, the nurturing of a culture of innovation, and the ability to scan for new opportunities by all departments.

Regarding talent development, Interviewee B noted that they have just one person dedicated to HRM and that they lack the ability to deal strongly with employee career development. On the other hand, concerning talent development, he added: "I believe that in the development of people there must be a widespread and coordinated responsibility within the company, among those who hold leadership roles, in worrying about growing skills in the areas where it's necessary."

Company A introduced a new IT manager to manage the network and solve issues in order to guarantee the systems' operational continuity; they also hired a new management engineer, appointed as CDO. Among other responsibilities, the CDO manages the training of employees; the firm does not have an HR manager. The firm recognized the need to introduce new digital and analytical competences to make sense of the vast amount of data coming from the CPS and to translate it into useful information.

Training in the introductory phase of new technologies is generally done by system providers, either directly to operators or to line managers first, who subsequently train all other employees.

Interviewee A noted: "Training remains a critical step; digital transformation is a big change management operation because people who have been working for 25 years in the same way must be guided and convinced to work in a new way."

Company A introduced many young employees, for whom training on the new systems was easier because they are digital natives; older employees have encountered great difficulties and training has been very long. Interviewee A explained how they managed training with gradual steps, through quick wins. They explained, step-by-step, one operation at a time over the course of a few months to avoid imposing a drastic change on older operators.

The other interviewees stressed the importance of the receptivity of management with respect to Industry 4.0 themes. Beyond technical training, it's important that employees and managers at all levels are engaged in all phases of implementation and that knowledge and a positive disposition towards innovation is spread across all departments for successful digital transformation.

About the initial phase of introducing the concepts of Industry 4.0 in the company, Interviewee D reported,

> Following the meeting with IBM IoT, we perceived the potential of technology and we communicated it to the management; then the idea of involving the whole company started. We tried to translate what we had seen into a language understandable at all levels of the company, we organized an event with the aim of collecting ideas from all people in the company to encourage every employee to contribute ideas, then the best ideas would have been evaluated.

Company D encourages people to get involved, even when it comes to training: "The idea of the company is that people participate in a course because they have curiosity and desire for growth. What we want to communicate is that people must get involved in order to grow."

Interviewee B noted that the adoption of cloud computing reduced the need for IT experts because it shifted the technical burden to the service provider. The company has an IT manager who coordinates the interventions of external specialists in the areas where they are needed from time to time. He also noted:

> We expect the digital skills needed in the company to increase; we do not expect to need highly specialized experts as programmers, it's not what our company is about.

We need people with a higher knowledge of new technologies and the opportunities connected to them than we have today. We have planned for this year that some people within the company must devote part of their time to learning about the technologies available today and understand how we could benefit from implementing them in the future.

Employee age seems to be a relevant factor in Industry 4.0. Company A, for example, had to differentiate training and adapt it to the lower digital literacy of older employees; on the other hand, the average age of employees decreased during digitalization because the company introduced many young, digital-native employees. This was possible because it experienced significant growth. Company C recognized that, in view of future developments of Industry 4.0, it would be ideal to introduce young people into the company. As of this writing, the average employee age is rising because the staff has not grown numerically, and turnover is very limited, as is the possibility of hiring younger employees. Company D is located in an area that is subject to depopulation; Interviewee D noted that this factor might make it more difficult in the future to find skilled workers and also argued that increased collaboration among firms, industrial associations and educational institutions is necessary to realign training with firms' emerging needs.

Digital transformation is more about understanding how to create additional value and improve processes by leveraging new technologies than about technology itself. SMEs understand this and perceive digital transformation as an evolutionary process; accordingly, they continuously scan for new opportunities and nurture a culture of openness to innovation at all levels.

Interviewee D noted: "Having a management that knows about technology and with a good disposition towards innovation is important. We made the difference by involving all departments in the project; this approach favored the creation of a fertile ground on which to introduce the implementations we made."

Interviewee C noted how training and contacts with universities helped them overcome initial skepticism about implementing Industry 4.0 solutions, typical of metalworking companies, in a small screen-printing firm.

Findings

Concerning adoption of Industry 4.0, evidence from our sample SMEs showed that approaches vary considerably, depending on many factors. SMEs typically began by having some processes digitalized, with some documental processes still done on paper; evidence the interviewees provided show that some still are, but firms that have not yet gone fully paperless are in the process of doing so. From the hardware point of view, machines lacked connection; from the software point of view, integration among management software was missing. This typical starting situation of manufacturing SMEs is consistent with what in the literature is called "level zero" of maturity. SMEs hence must go through a "zero phase" in which digitization and connection of machines in the factory are addressed. These are the minimum requirements to begin proper digitalization and the entry point to further Industry 4.0 implementations. All firms had to address software integration issues: the introduction of cloud computing in some firms was accompanied by disposal of some obsolete software requiring high levels of expertise. Management's vision, culture and strategic approach influence the firm's objectives and the scope of digital transformation. The quality of human resources available and the investment capacity influence the degree to which SMEs can build and grow digital competences internally to power digital transformation initiatives and exploit all the opportunities of Industry 4.0. Relying on external providers for some Industry 4.0 technologies prevents some SMEs from creating internal know-how; this choice might limit firms' innovative potential with respect to future, more advanced Industry 4.0 solutions. All firms demonstrated awareness of the importance of developing dynamic organizational capabilities as a prerequisite for sensing and seizing the opportunities of Industry 4.0. This awareness doesn't always translate to a systematic, strong commitment to build dynamic capabilities, although there's evidence that some initiatives are devoted to developing knowledge about the technologies and opportunities of Industry 4.0 among managers.

There's evidence of training for Industry 4.0 soft skills among SMEs that recognize the changing requirements of jobs towards more cross-disciplinary skills, such as communication and problem-solving, as well as the need for Industry 4.0 education to improve the sensing of opportunities at all levels of the firm. SMEs that purposefully create occasions for cross-functional contamination and flatten the organization can engage employees and leverage ideas coming from every level. The promotion of organizational learning and openness to innovation through collaboration with industry associations, higher education, universities and research centers characterizes SMEs that have embarked on digital transformation initiatives; networking with other firms, also outside the firm's value chain, is mentioned as a source to identify areas of improvement.

Two firms developed a holistic plan towards the new paradigm, questioning the entire business model and the value-creation process. These firms radically innovated, moving from simple production towards models based increasingly on the provision of services built around their product. Servitization business models are based on the ability to leverage data from IoT-enabled machines as well as data from the production process and to translate it into value-added information; this outcome is not possible without investing in building analytical competences within the firm.

Continuous improvement of customer experience as well as of collaborative relationships characterizes digitalization initiatives, even in firms that have not achieved a transformative impact at the business-model level (see Table 19.3). IIoT is at the core of Industry 4.0 and is widely adopted; ubiquitous connectivity in the factory and cloud solutions allow SMEs to achieve varying degrees of vertical and horizontal integration. Finally, the lean approach characterizes even companies that are not strictly engineering firms; Industry 4.0 solutions implemented by SMEs tend to build on lean principles towards continuous improvement and a strong customer orientation (see Table 19.4).

Table 19.5 reports the sample firms' adoption of technologies that characterize the Industry 4.0 paradigm. Industrial IoT, vertical and horizontal system integration, big data and analytics and the cloud are employed by all firms and seem to represent the core technologies of Industry 4.0; these alone enable a significant transformative impact from the technological point of view. The table does not describe the degree to which these technologies are leveraged to create additional customer value.

Table 19.6 compares the sample firms against critical enterprise dimensions for strong competitive stance as defined in Gurbaxani and Dunkle (2019). Evaluations are based on limited evidence provided in the interviews. The table indicates how well firms leverage each dimension. While all firms in the sample are well positioned to potentially reap the

Table 19.3 Findings for phases of digital change

Case	A	B	C	D
Digitization (IT infrastructure)	Digitization of documentational processes; connection of machines; coordination existing management software; introduction of MES.	Connection of machines; coordination, integration of software; disposal of obsolete software.	Connection of machines; coordination of existing management software; introduction of MES.	Digitization of documentation; connection of machines; coordination of software, integration of the data warehouse with ERP, PDM, MES, CRM, TIME WORK, BI.
Digitalization (improved processes / customer experience)	Leveraging data from production thanks to IoT enabled machines and MES, together with integration of all software, improves quality and reliability of information, traceability and transparency of all processes. New opportunities to provide additional value to customers emerge.	Cloud computing to improve interfirm agility; cloud solution to manage extra-corporate Kanban logics; horizontal integration with suppliers and customers improve collaborative relationships.	MES enables accuracy in registering timing for each operation and waste percentage; data from MES can be used to provide accurate quotations for customized semi-artisanal products.	Cloud communication with suppliers; production of IoT enabled machines; provision of usage report; tele-assistance; preventive maintenance; predictive maintenance (2020). Proprietary mobile app to verify plant performance. Computer simulation to accelerate processes.
Digital Transformation (BMI and new capabilities built)	Change of the business model from pure production to provision of services linked to the product by leveraging data from the CPS: problem solving; development; engineering; production. New key resources and capabilities: R&D lab; product development; technical departments.	No significant change in business model or development of new capabilities detected.	No significant change in business model or development of new capabilities detected.	Shift towards servitization: provision of detailed usage reports and tele-assistance on installed machines; provision of predictive maintenance under development.

Table 19.4 Findings for Industry 4.0 topics

Cases	A	B	C	D
Responsibilities in digital transformation initiative	CEO and CDO.	CEO; IT manager coordinates interventions of external subjects.	Vice President coordinates digital initiatives.	Opportunities of IoT sensed and introduced to management by IT dept. IT manager has key role in digitalization.
Constraints, challenges and problems encountered	Initial resistance and skepticism among older employees, training of older employees.	Limited investment capacity constrains further implementations and development of internal specialized competences in some areas.	Initial skepticism about implementing Industry 4.0 solutions in a screen-printing firm.	Communicating the value of IoT enabled machines to prospects and customers. Addressing data-protection and cyber-security concerns of customers.
HR management and development initiatives	No HR manager; HR function managed by operations. CDO also manages training for employees.	HR limited to management; HR development responsibility distributed among functional leaders. Training to develop managerial and leadership skills at all levels.	HR function outsourced to external consultant. Courses for all employees on soft skills for Industry 4.0.	Internal HR employee collaborates with functional leaders to analyze hard and soft skills of various positions. Proposes training and promotes networking.
Operator 4.0 impact / vision	Operators equipped with PDA. Improved human-machine interaction. Operators focus on quality control assisted by machines. The job of the operator is simplified.	The collaborative robot relieves the operator from repetitive, heavy tasks. The operator can focus on quality control and autonomous problem solving thanks to the digital HMI.	Implementation of collaborative robot is in progress. Objective is to free the operator from repetitive tasks, to allow focus on value adding activities such as problem solving and QC.	Digitalization of totems in the factory will assist operators in the assembly procedure of machines. AR solution for customers will allow remotely assisted maintenance operations.
Collaborations with external subjects	Industry associations; Politecnico di Milano (the company is a case study for the application of Industry 4.0).	President of the company is also president at t2i, a consortium company dedicated to assist firms in technological transfer and innovation; and at NAT (Network Automotive Triveneto).	German universities (lean); fab-labs (IoT modules); Politecnico di Milano and research centers (CERN; CNR; ENEA).	Industry associations; higher technical institutes; universities (Experior Project with Università Ca' Foscari Venezia).

Table 19.5 Findings for adoption of Industry 4.0 technologies (based on evidence provided in interviews)

Cases	A	B	C	D
Big data and analytics	✓	✓	✓	✓
Autonomous robots (co-bots)		✓	WIP	
Simulation		OUTSOURCED		✓
Horizontal and vertical system integration	✓	✓	✓	✓
Industrial IoT	✓	✓	✓	✓
Cybersecurity				✓
The cloud	✓	✓	✓	✓
Additive manufacturing		OUTSOURCED	✓	
Augmented reality				WIP

Table 19.6 Levers of digital transformation: evaluated by the authors, based on Gurbaxani and Dunkle (2019)

CASES	A	B	C	D
Strategic Vision A clearly defined strategic vision mapped to an understanding of digital needs. Company has a strategy for DT. Senior executive team has a clear understanding of digital technology capabilities and how they will support business objectives. No problem with lack of digital leadership to define strategy.	HIGH	MEDIUM	MEDIUM	HIGH
Culture of Innovation There is a culture of innovation and risk-taking. New ways of thinking and solutions from diverse perspectives are encouraged. Failure while taking a calculated risk is to be learned from; it is not a black mark on one's career. Innovators are rewarded. No problem with cultural resistance.	HIGH	HIGH	HIGH	HIGH
Know-how and IP Increasingly using software to improve operations performance, customer understanding, product know-how, supplier interactions. Sufficient intellectual property assets to implement strategic vision.	HIGH	HIGH	HIGH	HIGH
Digital Capability Availability of digital expertise. Overall, there are necessary visionary/innovative skills within the company to define the right digital strategy. Grades are assigned to individuals based on their level of digital transformation knowledge. Technical talent for innovation is already available in the company. No problem with lack of digital skills to execute strategy.	HIGH	MEDIUM	MEDIUM	HIGH
Strategic Alignment Company willing to fund strategic digital initiatives with uncertain returns. Willingness in the short run to cannibalize existing revenue streams and business models to gain profit in the long run. Collaboration and alignment between M&A, digital and business unit teams. No problem with lack of budget/resources assigned to digital transformation. Investment increase in new forms of software over past three years.	HIGH	MEDIUM	MEDIUM	HIGH
Technology Assets Technology in use: Big data; data mining and analysis/data analytics; mobile technologies; cloud computing; Internet and wireless communications. Sufficient technology assets to implement strategic vision.	HIGH	HIGH	HIGH	HIGH

benefits of digital transformation, firms A and D leveraged strategic vision, digital capability, and strategic alignment better than the others and, by doing so, realized a significant transformation in value creation by changing their business model and/or by developing new internal capabilities.

References

Bagnoli, C., Bravin, A., Massaro, M. & Vignotto, A. 2018. *Business Model 4.0. I modelli di business vincenti per le imprese italiane nella quarta rivoluzione indistriale.* Venice: Edizioni Ca' Foscari.

Benešová, A. & Tupa, J. 2017. Requirements for Education 4.0 and Qualification of People in Industry 4.0. *Procedia Manufacturing,* 11, 2195–2202.

Bharadwaj, A., Sawy, O., Pavlou, P. & Venktraman, N. 2013. Digital Business Strategy: Towards a Next Generation of Insights. *MIS Quarterly,* 37(2), 471–82.

Buvat, J., Solis, B., Crummenerl, C., Aboud, C., Kar, K., El Aoufi, H., Sengupta, A., 2018. *The Digital Culture Challenge: Closing the Employee-Leadership Gap.* Capgemini Digital Transformation Institute.

Confindustria Veneto. 2019. *Industria 4.0 Veneto – I 100 luoghi di Industria 4.0.* [Online] Available at: http://100luoghi.industria40veneto.it/ [Accessed on February 5, 2020].

DaSilva, C. M. & Trkman, P. 2014. Business Model: What It Is and What It Is Not. *Long Range Planning,* 47(6), 379–389.

Ezeokoli, F. E. 2016. Digital Transformation in The Nigerian Construction Industry: The Professional's View. *World Journal of Computer and Application Technology,* 4(3), 23–30.

Fitzgerald, M. & Kruschwitz, N., Bonnet, D. & Welch, M. 2013. Embracing Digital Technology. *MIT Sloan Management Review,* 1–12.

Gurbaxani, V. & Dunkle, D. 2019. Gearing Up for Successful Digital Transformation. *MIS Quarterly Executive,* 18(3), 209–220.

Hecklau, F., Galeitzke, M., Flachs, S. & Kohl, H. 2016. Holistic Approach for Human Resource Management in Industry 4.0. *Procedia CIRP,* 54, 1–6.

Henriette, E., Feki, M. & Boughzala, I., 2016. Digital Transformation Challenges. *MCIS 2016 Proceedings, 33.*

International Data Corporation, 2019. *The Growth in Connected IoT Devices Is Expected to Generate 79.4ZB of Data in 2025, According to a New IDC Forecast.* [Online] Available at: www.idc.com/getdoc.jsp?containerId=prUS45213219 [Accessed on October 16, 2019].

Jeschke, S., Brecher, C., Song, H. & Rawat, D., 2017. *Industrial Internet of Things.* s.l.: Springer International.

Kane, G. C. et al. 2015. Strategy, Not Technology, Drives Digital Transformation. *MIT Sloan Management Review and Deloitte University Press.*

Kane, G. C. et al. 2017. Achieving Digital Maturity. *MIT Sloan Management Review and Deloitte University Press.*

Labrecque, L. I. et al. 2013. Consumer Power: Evolution in the Digital Age. *Journal of Interactive Marketing,* 27(4), 257–269.

Lemon, K. N. & Verhoef, P. C., 2016. Understanding Customer Experience Throughout the Customer Journey. *Journal of Marketing,* 80(6), 69–96.

Make in Italy; Fondazione Nord Est; Prometeia, 2015. *MAKE IN ITALY Il 1° rapporto sull'impatto delle tecnologie digitali nel sistema manifatturiero italiano,* s.l.: s.n.

Matt, D., Orzes, G., Rauch, E. & Dallasega, P. 2020. Urban Production – A Socially Sustainable Factory Concept to Overcome Shortcomings of Quaified Workers in Smart SMEs. *Computers & Industrial Engineering,* 139, 1–10.

Mell, P. & Grance, T. 2009. The NIST Definition of Cloud Computing. *Natl. Inst. Stand. Technol.*, 56(3), 50.

Merrilees, B. 2016. Interactive Brand Experience Pathways to Customer-Brand Engagement and Value Co-creation. *Journal of Product & Brand Management*, 25(5), 402–408.

Milgram, P. & Kishino, F. 1994. A Taxonomy of Mixed Reality Visual Displays. *IEICE Transactions on Information Systems*, E77-D(12).

Mittal, S., Khan, M., Romero, D. & Wuest, T. 2018. A Critical Review of Smart Manufacturing & Industry 4.0 Maturity Models: Implications for Small and Medium-Sized Enterprises (SMEs). *Journal of Manufacturing Systems*, 49, 194–214.

Morakanyane, R., Grace, A. & O'Really, P. 2017. *Conceptualizing Digital Transformation in Business Organizations: A Systematic Review of Literature*. 30th Bled eConference, June 18–21, Bled, Slovenia.

Osmundsen, K., Iden, J. & Bygstad, B. 2018. *Digital Transformation Drivers, Success Factors, and Implications*, s.l.: s.n.

Parise, S., Guinan, P. J. & Kafka, R. 2016. Solving the Crisis of Immediacy: How Digital Technology Can Transform the Customer Experience. *Business Horizons*, 59, 411–420.

Piccinini, E., Gregory, R. W. & Kolbe, L. 2015. Changes in the Producer-Consumer Relationship—Towards Digital Transformation. *In: Wirtschaftsinformatik Conference, Osnabrück, Germany: AIS Electronic Library*, pp. 1634–1648.

Rüßmann, M., Lorenz, M., Gerbert, P., Waldner, M., Justus, J., Engel, P., & Harnisch, M. 2015. *Industry 4.0. The Future of Productiity and Growth in Manufacturing Industries*, s.l.: s.n.

Singh, A. & Hess, T. 2017. How Chief Digital Officers Promote the Digital Transformation of Their Companies. *MIS Quarterly Executive*, 16(1), 1–17.

Teece, D. J. 2007. Explicating Dynamic Capabilities: The Nature and Microfoundations of (Sustainable) Enterprise Performance. *Strategic Management Journal*, 28(13), 1319–1350.

Teece, D. J. 2018. Business Models and Dynamic Capabilities. *Long Range Planning*, 51(1), 40–49.

Teece, D. J., Pisano, G. & Shuen, A. 1997. Dynamic Capabilities and Strategic Management. *Strategic Management Journal*, 18(7), 509–533.

Verhoef, P. C., Broekhuizen, T., Bart, Y., Bhattacharya, A., Dong, J. Q., Fabian, N., & Haenlein, M. 2019. Digital Transformation: A Multidisciplinary Reflection and Research Agenda. *Journal of Business Research*, 122(January), 889–901.

Vial, G. 2019. Understanding Digital Transformation: A Review and Research Agenda. *Journal of Strategic Information Systems*, 28(2), 118–144.

Warner, K. S. R. & Wäger, M. 2019. Building Dynamic Capabilities for Digital Transformation: An Ongoing Process of Strategic Renewal. *Long Range Planning*, 52(3), 326–349.

Westerman, G., Bonnet, D. & McAfee, A. 2014. *Leading Digital: Turning Technology Into Business Transformation*. Boston, MA: Harvard Business Review Press.

Chapter 20

Digital transformation and financial performance

Do digital specialists unlock the profit potential of new digital business models for SMEs?

Nicolai E. Fabian, Thijs Broekhuizen, and Dinh Khoi Nguyen

Introduction

Managing digital transformation is a new strategic imperative for firms to stay relevant in today's dynamic markets. It involves the use of digital technologies to make profound changes to the way value is created and delivered to customers and is appropriated by the firm (Verhoef et al., 2019). Successful digital transformation efforts, such as shown by Amazon, Netflix or Uber, can yield a variety of rewards, while failure can put even market leaders, like Blockbuster, Barns & Noble, and Kodak, on the brink of destruction (Ash, 2020; Lucas & Goh, 2009). Incumbent firms in particular face difficulties to change their existing means of value creation and delivery, as digital transformation involves substantive technology investments with high uncertainty (Karimi & Walter, 2015) and is complex to implement (Libert et al., 2016). Recent literature reviews (Kannan and Li, 2017; Parida et al., 2019; Reis et al., 2018; Verhoef et al., 2020; Vial, 2019) all reveal that there is a dearth of empirical research investigating the impact of digital transformation on firm performance.

The severe implications of digital change have made firms aware that digital technology has the potential to disrupt their business (Kane et al., 2015; Verhoef et al., 2019), and has prioritized the implementation of new digital technologies and digital transformation (Vial, 2019). Yet, many full-scale digital transformation initiatives often fail. Research found that up to 80% of digital transformation efforts fail and that billions of dollars are lost each year due to bad transformation efforts (Tabrizi et al., 2019). One key reason for failure is that firms may find new ways to handle digital technology and create value but fail to appropriate value and realize new money streams from digital transformation (Chanias, Myers, & Hess, 2019).

An apparent gap in the literature is the lack of empirical research on the link between digital transformation and financial performance (Yu et al., 2018). While many IS studies relate digital initiatives to technology projects or the use of technology (Galindo-Martin et al., 2019), few of them link it to business model change (cf. Verhoef et al., 2019). To understand the link with financial performance, we link digital transformation to business model change, as digital transformation explains the changes in value creation and appropriation. By empirically assessing the link between digital transformation and financial performance, we enhance our understanding of how technology can be used to unlock financial gains by adjusting firms' business models.

Another apparent gap is that most studies treat digital transformation as a dichotomous variable that is either present or absent. Similar to digitization and digitalization efforts, digital transformation initiatives vary in their degree of change, and this degree of change may ultimately impact the rate of success (Libert et al., 2016). In our study, we treat digital transformation as a continuum. By measuring the *degree* to which digital technology is used by firms to adjust their existing business model and create new ways of value creation and value appropriation, we can obtain more fine-grained insights in the financial consequences of digital transformation.

Finally, there is an apparent lack of research investigating the conditions under which digital transformation yields positive financial outcomes. Prior studies have highlighted the importance of dedicated digital specialists (e.g. chief digital officers) as a precondition, but have not investigated such boundary conditions. We seek to explore how the presence of digital specialists can shape success or failure of digital transformation (Tabrizi et al., 2019).

In an attempt to increase our understanding of the digital transformation–performance link and address the research gaps, we investigate (1) the impact of digital transformation on firm performance by examining how the firms' degree of digital change in business models affects their financial performance, and (2) by assessing how the presence of dedicated digital specialists within firms may influence this relationship. Using survey data from 189 SMEs, we test our hypotheses and find evidence that digital specialists play a key role in explaining the digital transformation–performance link. The positive link is only visible when digital specialists are present.

We contribute to the literature on digital transformation in several ways. First, we provide empirical evidence for the effect of the degree for digital transformation on firm financial performance for a large set of firms. By taking a business model perspective and making digital transformation comparable across different business models (e.g. platform based, advertising based), we expand the current theoretical discussion on how digital transformation may (fail to) generate financial gains from digital transformation efforts. More importantly, we seek to deepen our understanding of the contingencies of this relationship by underlining the positive performance implications of digital specialists. While case study research highlights the positive performance implications of appointing dedicated digital executives for large incumbents (Singh, Klarner & Hess, 2020), our results suggest that the presence of digital specialists also help SMEs financial performance digitally change their business models more successfully and reduce the potential negative financial impact of digital transformation (Singh et al., 2020). Through their presence, they may not only create digital awareness and help to prioritize digital investments, but they may also strategically envision and smoothen the transformation process, and unlock the value appropriation and thus generate financial gains from digital transformation.

This chapter first presents an overview of existing research on the relationship between digital transformation and firm performance, and how digital specialists may unlock the potential of digital transformation. Based on this review, we create three propositions: one for the main effect of digital transformation on financial firm performance, one for the main effect of digital specialists on firm performance, and one interaction effect between digital transformation and digital specialists. We then conduct our empirical analysis based on a survey, and conclude our chapter based on our findings.

Digital transformation, digital specialists, and digital performance

Digital transformation and business model change

The literature has studied the phenomenon of digital transformation in many different settings (Gupta & Bose, 2020; Quinn et al., 2016), across different industries (Frank et al., 2019; Peppard, Edwards, & Lambert, 2011), and with different technological foci (Brock & von Wangenheim, 2019; Hanelt, Busse, & Kolbe, 2017). Research separates digital transformation from other digitization (i.e., converting analog into digital data) and digitalization (i.e., digitalizing business processes) efforts (Kolloch & Dellermann, 2018; Verhoef et al., 2019), and considers it to be the most pervasive form of digital change, involving significant changes to the company's business model. Conceptually, business models help to explain how firms create and deliver value to customers, and then convert payment received to profit (Teece, 2010). Hence, key to understanding digital transformation is to look at the effectuated change at the business model caused by digital technology. In addition, implementing digital changes to the business model by making use of digital technologies or relying on digital data also serves as a base for competitive advantage and allows firms to compete based on their business model (Teece, 2010; 2018; Verhoef & Bijmolt, 2019). Subsequently, digital transformation requires the implementation of profound changes to the business model. Little is, however, known about the consequences and how digital transformation (i.e., digital changes to the business model) impact firm performance. Only a handful of studies empirically investigate the consequences of digital transformation; yet, these studies do not deal with digital transformation but cover areas such as non–digital business model change, strategy, or other closely related phenomena such as digitization and digitalization (Li, 2018; Chanias et al., 2019).

Based on a review of key empirical studies on digital transformation, we identify three key elements for the digital transformation of business models: business processes, information exchange, market scope, and value system role (see Table 20.1 and Table 20.2). In any of these cases, digital transformation starts with the companies' utilization from a wide array of digital technologies available to the market, including big data, artificial intelligence, Industry 4.0 or internet of things (Loebbecke & Picot, 2015; Ng & Wakenshaw, 2017; Syam & Sharma, 2018; Tortorella & Fettermann, 2017). The use of digital technologies can affect the degree of digital transformation by altering three elements of business model change. First, firms can use digital technology internally to facilitate new ways of value creation and adjust their old business model in favor of a new digital one (Li et al., 2018). For instance, when a local retail shop expands its product offering and becomes an internationally operating online webstore. Second, firms can use digital technologies to collect and exchange information, which in turn can be used to facilitate internal business model transformation (Gupta & Bose, 2020). Information exchange and collaboration can also be used to facilitate relationships with partners (Pagani & Pardo, 2017) like LEGO uses to co–develop new products together with clients, or with suppliers that work closely with retailers for joint storage decisions (Bijmolt et al., 2019). Finally, firms can use digital technologies to seize new business opportunities and enter new markets (Karimi & Walter, 2015, Peppard et al., 2011). For instance, digital technologies can allow firms to connect to new and more price-sensitive customer segments or serve existing customers with new digital products.

Table 20.1 Elements of business model change and digital transformation

Elements of business model change	Mechanism through which the business model is changed	Key reference
Business processes	- Unlock new ways of value creation through changing business processes - Integrate formerly unconnected business areas	Li et al., 2018; Chanias, Myers & Hess, 2019; Li, 2018
Information exchange	- Improve connectivity and information exchange with customers or partners - Collect market information and use big data to improve value creation via innovation	Pagani & Pardo, 2017; Gupta & Bose, 2020
Market scope & value system role	- Introduce new ways of value delivery (e.g., new customer segments, new international markets) - Serve (new) digital products to new and/or existing customers - Take on new (platform) roles within the value system	Karimi & Walter, 2015, Peppard, Edwards & Lambert, 2011

Note that the use of digital technologies can change multiple elements simultaneously. When a physical retailer decides to become a digital platform, it may change its business processes (e.g., using internet technologies to match supply and demand), improve information exchange (establishing new business partnerships, improving pricing strategies based on market data predictions), and expand its market scope and changing its role in the value system (e.g., using e-commerce technologies to become a global platform instead of a local retailer). Hence, the three elements of information collection and exchange, internal changes in value creation, and external delivery of value allow firms to fully benefit from digital transformation.

Digital specialists

Chief executives take on functional roles and responsibilities to improve firm performance (Hambrick & Mason, 1984; Nath & Mahajan, 2008; Peppard et al., 2011). Regarding digital transformation, chief information officers (Grover et al., 1993) and more recently chief digital officers (Tumbas et al., 2018; Singh et al., 2020) play an important role. While the exact responsibilities of such executives may differ (Tumbas et al., 2018), they are generally responsible for the change processes related to digitalization and digital transformation (Hansen & Sia, 2015; Singh & Hess, 2017; Tumbas et al., 2018). As the field and positions are in a constant flux, these executives can be named differently in practice,[1] or, or be fulfilled by existing CEOs to oversee digital transformation. To account for the natural variance in titles, we refer to "digital specialists" and define them as dedicated functional managers who are occupied with and responsible for the firm's utilization of digital technologies and digital transformation (Singh & Hess, 2017; Tumbas et al., 2018).

The effective use of digital technologies will be the hallmark of the digital age. Hence, with a constant flow of new digital technologies, digital specialists have become increasingly important for the firm (Singh & Hess, 2017; Tumbas et al., 2018). In a survey of Fortune 1000 companies, more than two-thirds of the companies report having a digital

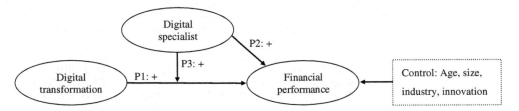

Figure 20.1 Conceptual model.

specialist (a chief digital officer or an equivalent position).[2] The reason for the strong presence of the digital specialists is because the value of digital transformation does not come automatically from the adoption of new digital technologies (Kane et al., 2015). Although digital transformation has an enormous potential to create additional value to the firm and its customers, the returns of digital transformation are highly uncertain and depend on how well firms may utilize technologies to enhance customer value (Warner & Wäger, 2019). The digital specialists facilitate digital transformation through performing various activities that effectively manage and create business values of digital technologies (Hansen & Sia, 2015; Tumbas et al., 2018; Singh et al., 2020). Given the salience of the digital specialists in the literature, we are interested in understanding how digital specialists may facilitate the value appropriation of digital transformation. Figure 20.1 depicts our conceptual model.

Digital transformation, digital specialists, and financial performance

Digital transformation and financial performance

Technology provides the foundation for the introduction of new business models (Lokuge et al., 2019). Yet, its use is not sufficient to realize digital transformation. Only when the technology is used to profoundly change the way a firm earns its money, can we speak of digital transformation. While lower-level digitization or digitalization changes can lead to improvements in operational performance (i.e., when big data are used to make existing operations smarter and/or more efficient), these lower-level changes mainly improve the performance of an existing business model (Verhoef et al., 2019). Digital transformation enables the creation of a new blueprint for value appropriation by reconfiguring asset deployments and business activities (Verhoef et al., 2019). Such change often allows for new revenue streams by offering products and digital services in a different way, by processing customer and market information differently, and by taking on new roles.

Research suggests that there a risk–return trade-off regarding business model innovation: the risks of failure for business model innovation are higher for greater innovativeness but the returns are also higher than incremental business model changes (Lindgart & Ayers, 2014). By analogy, we expect that more pervasive digital transformations have higher risk but are also more profitable than lower levels of digital transformation. Anecdotal evidence provided by conceptual and case-study based papers (Li, 2018) also suggests that more pervasive forms contribute to greater competitiveness and profitability, because such profound changes are more likely to increase customer value, realize economies of scale (e.g., scalability of digital platforms), and make it more difficult for competitors to imitate the business model. The key reason why firms engage in digital transformation is often because existing business models are no longer effective. Digital business models are often much more profitable than non-digital business models. Verhoef et al. (2019) compared the financial performance measures (EBIT and net profits) of a sample of firms and found that firms that run highly digital business models like Facebook, Apple, and Alphabet are much more profitable than firms that run less digital business models offered by companies like IBM, Walmart, and Daimler.

Digital transformation may improve financial performance via realizing cost savings and additional cash streams to create more cost-effective business models (Vial, 2019). A prime example is Domino's Pizza, which used digital transformation to improve customer satisfaction, value creation and financial gains. The company underwent a major transformation and developed an online experience that directly rivals those offered by leading e-commerce brands. With the profound use of digital technology, the company revamped its entire order process and describes itself as "e-commerce company that happens to sell pizza" (Wong, 2018). Domino's managed to gain market share, increase customer satisfaction, drive down costs, and improve profitability.

In sum, firms that engage more strongly in digital transformation are expected, on average, to profit more than those that make less use of digital technologies to transform their business model. While lower levels of digital transformation also allow for efficiency gains, such efforts often face decreasing returns that limit profit growth. The use of digital technologies sparks new opportunities for deeper changes. Firms that engage in digital transformation and, for instance, develop new digital platforms are able to leverage their partners' resources and experience increasing returns to scale (Eisenmann, Parker, & Van Alstyne, 2006). The production and distribution of more digital products are less cost-intensive. Furthermore, products and marketing campaigns can be designed to leverage customers to endorse products, stimulate product virality, and adoption (Aral & Walker, 2011; Van Alstyne et al., 2016). Hence, a greater level of digital transformation is expected to result in greater performance gains from digital technologies. Therefore, we propose the following:

> **Proposition 1:** There is a positive relationship between digital transformation and financial performance.

Roles and impact of digital specialists

While a greater digital business model transformation is expected to enhance financial performance, we expect that this relationship is contingent on the presence of digital specialists who can help overcome barriers, align business model components, and unlock the potential of new technologies. Higher levels of digital transformation are associated with more substantive changes to the organizational logic on how value is created and delivered, and, hence, more easily fail because of the barriers to implement organizational change. Similar to the implementation and adoption of IT artifacts (Strong & Volkoff, 2010; Volkoff, Strong, & Elmes, 2007), digital transformation can raise major barriers such as employee resistance (Svahn, Mathiassen & Lindgren, 2017). Furthermore, existing business models are often deeply embedded in the company and over time become more resistant to change (Christensen, Bartman, & Van Bever, 2016). Especially for traditional (pre-digital) organizations, the implementation of digital business model transformation poses a great threat as existing elements of value creation are under attack, while the future value proposition remains vague (Chanias et al., 2019). Business model components like product or service delivery, internal operations, and key partnerships are subject to change, while new digital technology needs to be integrated simultaneously (Chanias et al., 2019). When not managed properly, digital transformation may result in strategic misalignment between technologies used and the current business model (Dulipovici & Robey, 2013). When managed properly, building blocks are strategically aligned such that the technology strengthens a firm's business value proposition and performance by unleashing positive self-reinforcing processes (Casasdesus-Masanell & Ricart, 2011). Hence, the impact of digital transformation can be greatly enhanced when the transformation is coupled with the creation of effective business models.

To better manage digital transformation, the literature points towards the use of dedicated (digital) executives (Singh & Hess, 2017; Tumbas et al., 2018), who are responsible for major business transformation and integration projects (Matt, Hess, & Benlian, 2015). While tasks and activities vary, digital specialists generally perform three key functional roles (Singh & Hess, 2017). In performing these roles, they can promote digital transformation's impact on value creation and more effectively convert digital transformation efforts to reap financial gains. First, as entrepreneurs, digital specialists search the market for business opportunities based on technological developments and competitor behaviors (Warner & Wäger, 2019; Day & Schoemaker, 2016). These informational resources enable firms to make decision closer to the realities and thus improve performance (Nath & Mahajan, 2008). In addition, digital specialists are able to explore opportunities with digital technologies while concentrating on maintaining technological stability and exploiting value from these technologies (Tumbas et al., 2017; 2018). Thus, digital specialists build digital ambidexterity and balance experimentation and exploitation of new technologies to create novel and replicable business model solutions that help to capture the rents from firm's digital transformation efforts.

Second, digital specialists serve as evangelists to create urgency to respond to digital change and by breaking down internal resistance (Singh & Hess, 2017). This functional role is similar to that of a change agent, but then in the context of digital transformation. Digital specialists play a key role in creating digital awareness to the top management team (Fitzgerald et al., 2014) and disseminating its importance to the rest of the organization. These C-members, especially those in the front line such as sales, could diffuse the digital knowledge to their teams and employees (Singh & Hess, 2017; Singh et al., 2020). Consequently, effective use of technologies in the front line potentially improves revenue.

Additionally, employees might foil the digital transformation attempt as they (1) lack the capability of handling new processes and practices (Singh & Hess, 2017), (2) lack seeing the potential benefits of digital transformation (Svahn et al., 2017), or (3) are accustomed to the existing processes and practices (Polites & Karahanna, 2013). Such resistance increases costs and weakens firms' value capture of digital transformation. Digital specialists may ensure that the technologies are introduced to the organization at an appropriate pace,

Table 20.2 Functional roles and activities of digital specialists

Roles of digital specialists	Activities associated with roles	Value creation of digital transformation (P2)	Value appropriation of digital transformation (P3)	Key reference
Entrepreneur	Technology sensing and opportunity seizing	✓		Warner & Wäger, 2019; Day & Schoemaker, 2016
	Experimentation with technologies to create compatible, replicable digital solutions.		✓	Tumbas et al., 2017; Singh et al., 2020
Evangelist	Reduction of innovation fatigue and changing of incumbent habits		✓	Polites & Karahanna, 2013; Fitzgerald et al., 2014
	Improvement of digital literacy of employees and other stakeholders	✓		Haffke et al., 2016; Singh & Hess, 2017; Singh et al., 2020
	Fostering a digital culture		✓	Singh & Hess, 2017; Kane et al., 2015
Orchestrator	Development of a holistic digital transformation strategy to exploit digital opportunities.	✓	✓	Singh & Hess, 2017; Singh et al., 2020
	Coordination of digital transformation activities to ensure alignment and facilitate a smooth transformation process		✓	Haffke et al., 2016; Singh & Hess, 2017 Tumbas et al., 2017, 2018
	Allocation of necessary resources		✓	Horlacher & Hess, 2016; Sebastian et al., 2017

thus limiting "*innovation fatigue*" or technological anxiety (Fitzgerald et al., 2014; Singh & Hess, 2017). Moreover, they use various strategies to gradually suspend existing system habits and develop new system habits (Polites & Karahanna, 2013) in order to tackle the inertial aspect of resistance and ensure a smooth transition. Specialists also create digital awareness (Fitzgerald et al., 2014) and improve the digital literacy of the employees (Singh & Hess, 2017; Singh et al., 2020). Digital specialists can inform the organizational members of the benefits of new technologies (Svahn et al., 2017). Understanding the value of new technologies builds the innovative and technologically receptive culture that systematically tackle resistance (Kane et al., 2015; Karimi & Walter, 2015).

Digital specialists also play a key role, as orchestrators, in executing the digital strategy. Their presence may help to improve firm performance by establishing a holistic digital transformation strategy and roadmap (Chanias et al., 2019; Singh & Hess, 2017; Singh & Hess, 2020) that guide the development and exploitation of digital technologies (Chanias et al., 2019).

As digital transformation involves a company-wide organizational transformation (Singh & Hess, 2017), successful digital transformation requires the coordination of digital transformation activities and bringing together of separated departments and dispersed knowledge silos (Tumbas et al., 2018; Singh & Hess, 2020). Digital transformation is enhanced when IT functions and other departments are strategically aligned (Tallon & Pinsonneault, 2011), and not just run by the IT department. Digital specialists coordinate the diffusion of digital transformation throughout the organization (Singh & Hess, 2020). Further, they steer the allocation of organizational resources to reap the value of digital transformation. Incumbent firms may need to overcome their technology lock-in and need to divest and seize existing activities that no longer benefit the new (digital) business model (Warner & Wäger, 2019). Digital specialists play an important role as orchestrator by redistributing resources and re-arranging activities, such that their presence helps firms to reap the financial benefits from digital transformation. Based on the arguments on the roles and activities of digital specialists in extant research, we propose that:

Proposition 2: There is a positive relationship between the presence of digital specialists and financial performance.

Proposition 3: The positive relationship between digital transformation and financial performance is moderated by the presence of digital specialists, such that the presence of digital specialists strengthens the positive impact of digital transformation on financial performance.

Methodology

Data

We make use of survey data collected from a larger study[3] to explore the financial implications of digital transformation and the presence of digital specialists. The survey's purpose is to derive insights about innovation activities for small- to medium-sized Dutch enterprises. The sample entails SMEs from a wide range of industries, including services (consultancy, IT), retail, manufacturing (construction), and public services. An online survey was sent out in March 2019 (with two reminders sent after each subsequent week) to roughly 5000 SMEs. In total, 454 respondents participated in the survey, yielding a response rate of close to 10%. After performing list-wise deletion, we obtain a net sample of 189 SMEs. The sampled firms (see Table 20.3) are, on average, 29 years old (M = 29.26; SD = 32.82) and have 38 employees (M = 37.90; SD = 129.50).

Measures

Due to the absence of established measures for digital transformation (Verhoef et al., 2019, Vial, 2019), we developed a new measure for digital transformation using established

Table 20.3 Firm characteristics

	Frequency	Percentage
Age category		
Start-up (< 2 years)	3	1.5
Young (< 5 years)	16	7.9
Adolescent (<10 years)	42	20.8
Mature (> 10 years)	139	68.8
Size category		
Micro (< 10 employees)	106	52.5
Small (10–99 employees)	81	40.1
Medium (100–249 employees)	12	5.9
Large (250 or more employees)	3	1.5
Industry		
Manufacturing	55	27.2
Retail	39	19.3
Public services	84	41.6
Service	20	9.9
Family business		
Yes	95	47
No	107	53
Presence of digital specialist		
Yes	86	42.6
No	115	56.9

Note: The number of observations vary between N=198 and N=202.

guidelines (MacKenzie, Podsakoff & Podsakoff, 2011). We started with a literature review (Verhoef et al., 2019) to lay the conceptual foundation for our digital transformation construct. Subsequently, we used expert interviews with the target group to get a better understanding of digital transformation. The interviews were conducted with executives from a diverse set of Dutch SMEs varying in firm size, number of employees, and industry sector. This was done to get a broad understanding of the execution and consequences of digital transformation, and ensure the content validity of our measure of digital transformation (Rossiter, 2002). After the interviews, we developed an initial set of questions based on our interview findings and insights derived from the literature. We asked a handful of academics to assess the face validity. We also conducted a two-stage Q sorting process to further improve the clarity of our items. After this item generation stage, we performed a pilot test among 101 firms active in the North of the Netherlands. As the test provided initial evidence of unidimensionality (single-factor solution in exploratory factor analysis) and construct reliability (Cronbach's alpha = 0.82), we used the six-item construct in a larger sample for SMEs.

Digital transformation. Based on our scale development, we use six survey items to measure digital transformation (see Table 20.4). The questions are framed towards analyzing the usage of digital technology to make changes to the existing business model of firms. The items cover the three elements of digital business model transformation (i.e., changes to business processes, information exchange, and market scope and value system role).

Table 20.4 Construct measures

Construct	Measurement items	SL
Digital transformation	1. Digital technology has changed the way how we do business	0.858
	2. Digital technology has changed our internal operations	0.777
	3. Digital technology has changed the collaboration with partners	0.825
	4. The knowledge to use digital technology to improve the customer's journey (analyzing and collecting data)	0.815
	5. We have entered new geographical markets and serve new customers with the help of digital technology	0.705
	6. Our role in the value system has changed (e.g. extending markets via online channels, entering new business fields)	0.752
Financial performance	What is the net profit margin of your enterprise?	–

Digital specialist. We opted for a simple measure and asked respondents to indicate the presence of a digital specialist with a simple yes/no answer. In our survey, we presented digital specialists as a person (or a small team) who has digital expertise and is responsible for digital technology and digital transformation processes. Such a definition fits our research context, as SMEs commonly manage digital transformation by individuals or in small teams.

Financial performance. We obtain firm financial performance through a self-reported survey question. Respondents indicate their firms' net profit margin (as a percentage) to measure financial performance.

Control variables. We control for several factors that potentially influence financial performance. To account for firm characteristics that relate to financial performance, we use the natural log of the number of employees to account for firm size (Rai, Patnayakuni, & Seth, 2006), and the natural log of the number of years of existence to account for firm maturity (Karimi & Walter, 2015). We use Dutch industry codes to create four broad categories: retail, services, manufacturing, and public services to control for industry affiliation. Finally, we control for innovativeness using a summative scale of ten items that indicate the implementation of product innovation, process innovation, and organization innovation in the previous three years.

Measurement validation

Prior to using OLS, we use PLS-SEM to validate our latent construct, digital transformation. Digital transformation is internally consistent as the Cronbach's alpha (0.88) and composite reliability (0.9) exceeds the required threshold of 0.7. In addition, convergent validity was established as the average variance extracted (AVE) value of 0.625 surpassed the threshold of 0.5 (Table 20.5). To establish discriminant validity, we find that the square root of the AVE of digital transformation is higher than its correlations with other variables (Fornell and Larcker, 1981), and that all items have low cross-loadings with the other variables (see Table 20.3). Finally, we find that the heterotrait-monotrait (HTMT) ratio of correlation is significantly lower than the most stringent threshold of

Table 20.5 Descriptive statistics and correlations

	Mean	SD	1	2	3	4	5	6
(1) Fin. performance (ln)	2.61	.96	n.a.					
(2) Dig. transformation	2.70	1.00	−.027	.791				
(3) Dig. specialist	.43	.49	−.010	.196**	n.a.			
(4) Size (ln)	1.00	.76	−.475**	.136*	.292**	n.a.		
(5) Age (ln)	1.12	.52	−.252**	−.066	.086	.495**	n.a.	
(6) Innovativeness	3.22	2.20	−.112	.418**	.137*	.236*	.027	n.a.

Notes: * p <.05; ** p < .01; n.a. not applicable. The diagonal displays the square root of the AVE

0.85 (Henseler et al., 2015; Kline, 2011). All tests support the discriminant validity of our variables.

Response and common method bias

Due to the nature of the study, non-response bias can significantly affect the findings. After performing a median split based on respondents' response speed, we find no significant mean difference on digital business model transformation between early and late respondents (t = 0.445; p > 0.10). This test provides some evidence that non-response bias is not a serious problem in our data collection.

Common method bias (CMB) is another common concern of survey research that may bias our results. To alleviate this concern, we used several remedies. First, we applied procedural remedies such as mixed scales (e.g., binary, Likert scale, and self-report net profit margin) and psychological separation (i.e., we insert several unrelated questions in between the independent and dependent variable) in survey design (Podsakoff et al., 2003). In addition, we employed statistical remedies of CMB (Podsakoff et al., 2003). We first use Harman's single-factor test and find that no single factor explains more than 37% of the variance, which is less than the proposed 50% (Podsakoff & Organ, 1986). Second, we employ the stringent marker variable technique using an irrelevant question (i.e., the total share of the firm owned by a single person) in our measurement model. The results of our PLS-SEM model remain similar, suggesting that CMB (if it exists) does not significantly alter our results.

Results

Table 20.6 shows the OLS results that we use for exploring our propositions. We do not find support for proposition 1, as we do not find a positive relationship between digital transformation and firm financial performance (b = .011, p > .05, *in Model 2*). We also do not find support for our proposition 2 delineating the positive influence of the presence of digital specialists and firm performance (b = .21, p > .05). In support of our proposition 3, we find a positive interaction effect of the joint presence of a digital specialist and digital transformation (b= .311, p < .05) on firm financial performance. The absence of a direct effect of digital transformation (as well as of digital specialists), and the presence of an interaction effect suggest that digital specialists do not independently

Table 20.6 OLS regression results

DV: Financial performance	Model 1	Model 2	Model 3
Digital transformation (P1)		.007	−.115
		(.072)	(.093)
Digital specialist (P2)		.208	0.204
		(.129)	(.128)
Digital transformation × digital specialist (P3)			.258*
			(.127)
Size	−.611**	−.680**	−.706**
	(.110)	(.110)	(.110)
Age	−.086	−.088	−.070
	(.163)	(.165)	(.164)
Innovativeness	.17	.11	.14
	(.224)	(.033)	(.033)
Industry dummy	Included	Included	Included
R²	.250	.250	.278
Adj. R²	.225	.221	.241
ΔR²	−	.00	.027*

Note: N=189, * $p < .05$; ** $p < .01$. Unstandardized betas are shown with standard errors in parentheses.

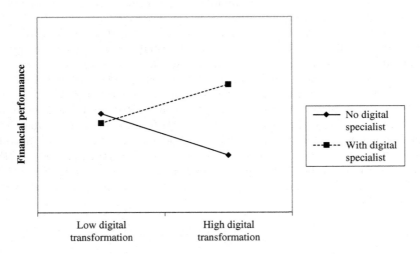

Figure 20.2 Moderation effect of digital specialist on DT – financial performance link.

drive financial performance, but that their presence helps to unlock the potential of digital transformation.

Figure 20.2 facilitates the interpretation of the interaction effect. We find that digital transformation enhances firm financial performance when a digital specialist is present but reduces it when a digital specialist is absent. We concluded that the presence of a digital specialist enhances the firm's ability to capture rents from higher levels of digital transformation.

Discussion

In this study we explored the financial consequences of digital transformation efforts using a large SME sample. Our results indicate that higher levels of digital transformation and digital specialists themselves do not automatically lead to higher financial performance, but that their effectiveness crucially depends on their joint presence. In the absence of digital specialists, firms are not able to capture the rents from increasing levels of digital transformation. Hence, we advance our limited understanding of the conditions under which digital transformation is financially beneficial for companies. Our research yields two important implications.

First, our study contributes to the literature by providing insights into the financial consequences of undergoing digital transformation. To the authors' best knowledge, this is one of the first studies to empirically investigate the impact of digital transformation – measured as the degree to which firms utilize digital technologies to alter their business models – on financial performance for a large sample of firms. Building on the notion that business models function as a blueprint for value creation and appropriation (Teece, 2010), we investigate the financial consequences of business model changes that are caused by digital technology. While accounting for several firm and industry characteristics, we find – similar to earlier smaller-scale studies – that digital transformation does not unequivocally lead to financial improvements. Digital transformation is not a panacea that guarantees performance improvements. On the contrary, our results suggest that the relationship is weak, and that digital transformation only under certain conditions contributes to financial performance. Therefore, our study provides timely needed insights into the nature of digital transformation and replicates the findings of earlier studies that not all transformation initiatives are worth the effort.

Second, we contribute towards an understanding of the contingencies under which digital transformation may be beneficial for firms (Vial, 2019). Our findings point towards the critical role of having dedicated digital specialists to unlock the financial gains from digital transformation. This finding underlines the crucial role of digital specialists as key *enablers*, as they help to secure the rents from digital transformation, yet do not directly influence rents. Digital specialists seem to perform important roles as entrepreneur, evangelist, and orchestrator, and help to overcome inertia and resistance. Existing research has addressed the important roles of such specialists in larger incumbent firms to overcome inertia (path dependency or legacy costs) and break down organizational structures and resistance (Singh, Klarner, & Hess, 2020) in order to alter the course of "large container ships" in response to technological shocks (Eggers & Park, 2018). Our results hint that digital specialists unlock the value of digital transformation for smaller, more agile, and younger firms. As such, we contribute to recent calls to investigate under which circumstances digital transformation pays off for SMEs (Cenamor, Parida, & Wincent, 2019; Quinton et al., 2018). Due to the differences between large and SME firms, our research provides timely insights to better understand the contingency of firm size in digital transformation. Analyzing the digital transformation of smaller firms, we expand the scope of research, and help to contribute to a better understanding of the success drivers of digital transformation (Tabrizi et al., 2019).

Managerial implications

This study provides several relevant and important insights for SMEs and managers. First, our results suggest that managers should be cautious to increase their level of digital transformation, and that they need to carefully assess whether the use of technologies to adjust the business model increases value creation and, more ultimately, value appropriation. By actively measuring and monitoring the financial consequences of digital transformation through business model changes, we show the potential positive impacts of undergoing such transformation. However, our study also shows that digital transformation is not beneficial by itself. Subsequently, we provide a teaser for managers that maybe lower levels of digital change such as digitalization might also be sufficient, and that full-scale digital transformation is not always needed.

Although the hiring of dedicated digital specialists can involve significant costs for SMEs, our results show that these specialists can facilitate the rent capture of digital transformation. Appointing digital specialists seems particularly worthwhile when SMEs intend to make significant changes to the business model using digital technologies. Then these digital specialists may help to improve the success rate of digital business transformation and increase their competitiveness through smoothening the transition process and more effectively appropriating value from the exploitation of digital technologies.

Limitations and future research

Several constraints and limitations arise from the research setting that offer possibilities for future research. First, our study relies on self-reported survey data. While we have taken several steps to alleviate biases, the single source data cannot entirely rule out that common method variance does not affect our results. Furthermore, the cross-sectional data limits the possibility to make causal claims and to account for possible endogeneity (e.g., the appointing of specialists may be related to a firm's aspiring levels of digital transformation, a firm's digital capability, and prior digital performance). Thus, we recommend that future studies collect data from multiple (objective) data sources, as well as to make use of longitudinal or experimental designs to test the causality of our proposed relationships.

Second, while we establish that the presence of digital specialists impacts the digital transformation–performance relationship, we do not know the characteristics of and the modus operandi of the digital specialist. Future research could obtain more fine-grained results by analyzing the digital specialist's characteristics (tenure, educational background, intra- and inter-organizational network) and behaviors (communication strategy, resource allocation decisions), as well as an organization's properties (digital strategy, digital literacy, availability of resources, organizational infrastructure and support).

Finally, although we control for industry effects, we must acknowledge that there is a need for more information on the degree of digitalization and the level of competition within the industry of interest. Therefore, we recommend future research to compare digital transformation not only across but also within industries. In doing so, researchers are also able to identify specific industry-specific contingencies that influence the degree to which digital transformations pay off.

Acknowledgements: This work has benefited from the support and data received from the Northern-Netherlands Innovation Monitor and the Groningen Digital Business Center. We would like to thank them for the support regarding the data collection. In addition, we want to thank Peter C. Verhoef for his guidance as well as sharing valuable insights on earlier drafts of this chapter. Lastly, we want to thank John Q. Dong for his valuable feedback and help in developing the empirical part of this study. The usual disclaimer applies.

Notes

1 KPMG: https://home.kpmg/nl/nl/home/insights/2020/06/why-just-now-the-cdo-is-regularly-taking-place-at-the-board-table.html#:~:text=Job%20title,Officer%20(see%20Table%202).
2 Forbes: www.forbes.com/sites/insights-intelai/2019/05/22/rethinking-the-role-of-chief-data-officer/#5808e5ba1bf9
3 The data used for this book chapter are part of a larger dataset that are also used elsewhere: Fabian et al. (2020). The foci of both studies differ substantially (i.e., explaining firm vs. financial performance), and the overlap of variables used is marginal.

References

Aral, S., & Walker, D. 2011. Creating social contagion through viral product design: A randomized trial of peer influence in networks. *Management Science*, 57(9), 1623–1639.

Ash, A., 2020. The rise and fall of blockbuster. *Business Insider.* Retrieved from www.businessinsider.com/the-rise-and-fall-of-blockbuster-video-streaming-2020–1?international=true&r=US&IR=T

Bijmolt, T. H. A., Broekhuis, M., De Leeuw, S., Hirche, C., Rooderkerk, R. P., Sousa, R., & Zhu, S. X. (2019). Challenges at the marketing–operations interface in omni-channel retail environments. *Journal of Business Research.* https://doi.org/10.1016/j.jbusres.2019.11.034

Brock, J. K.-U., & von Wangenheim, F. 2019. Demystifying AI: What digital transformation leaders can teach you about realistic artificial intelligence. *California Management Review*, 153650421986522.

Casadesus-Masanell, R., & Ricart, J. E. 2011. How to design a winning business model. *Harvard Business Review*, 89(1/2), 100–107.

Cenamor, J., Parida, V., & Wincent, J. 2019. How entrepreneurial SMEs compete through digital platforms: The roles of digital platform capability, network capability and ambidexterity. *Journal of Business Research*, 100(July 2019): 196–206.

Chanias, S., Myers, M. D., & Hess, T. 2019. Digital transformation strategy making in pre-digital organizations: The case of a financial services provider. *Journal of Strategic Information Systems*, 28(1): 17–33.

Christensen, C. M., Bartman, T., & Van Bever, D. 2016. The hard truth about business model innovation. *MIT Sloan Management Review*, 58(1), 31.

Day, G. S., & Schoemaker, P. J. 2016. Adapting to fast-changing markets and technologies. *California Management Review*, 58(4), 59–77.

Dulipovici, A., & Robey, D. 2013. Strategic alignment and misalignment of knowledge management systems: A social representation perspective. *Journal of Management Information Systems*, 29(4), 103–126.

Eggers, J. P., & Park, K. F. 2018. Incumbent adaptation to technological change: The past, present, and future of research on heterogeneous incumbent response. *Academy of Management Annals*, 12(1), 357–389.

Eisenmann, T., Parker, G., & Van Alstyne, M. W. 2006. Strategies for two-sided markets. *Harvard Business Review*, 84(10), 92.

Fabian, N. E., Dong, J. Q., Broekhuizen, T. L. J., & Verhoef, P. C. 2020. Digital transformation: An evolutionary perspective. *Working Paper*.

Fitzgerald, M., Kruschwitz, N., Bonnet, D., & Welch, M. 2014. Embracing digital technology: A new strategic imperative. *MIT Sloan Management Review*, 55(2), 1–12.

Fornell, C. & Larcker D.F. 1981. Structural equation models with unobservable variables and measurement error: Algebra and statistics. *Journal of Marketing Research*, 18(3): 382.

Frank, A. G., Mendes, G. H. S., Ayala, N. F., & Ghezzi, A. 2019. Servitization and Industry 4.0 convergence in the digital transformation of product firms: A business model innovation perspective. *Technological Forecasting and Social Change*, 141 (January): 341–351.

Galindo-Martin, M. A., Castano-Martinez, M. S., Mendez-Picazo, M. T. 2019. Digital transformation, digital dividends and entrepreneurship: A quantitative analysis. *Journal of Business Research*, 101 (December 2018), 522–527.

Grover, V., Jeong, S. R., Kettinger, W. J., & Lee, C. C. 1993. The chief information officer: A study of managerial roles. *Journal of Management Information Systems*, 10(2), 107–130.

Gupta, G., & Bose, I. 2020. Digital transformation in entrepreneurial firms through information exchange with operating environment. *Information & Management*. https://doi.org/10.1016/j.eplepsyres.2019.106192.

Hambrick, D. C., & Mason, P. A. 1984. Upper echelons: The organization as a reflection of its top managers. *Academy of Management Review*, 9(2): 193–206.

Hanelt, A., Busse, S., & Kolbe, L. M. 2017. Driving business transformation toward sustainability: Exploring the impact of supporting IS on the performance contribution of eco-innovations. *Information Systems Journal*, 27(4): 463–502.

Hansen, R., & Sia, S. K. 2015. Hummel's digital transformation toward omnichannel retailing: Key lessons learned. *MIS Quarterly Executive*, 14(2): 51–66.

Henseler, J., Ringle, C. M., & Sarstedt, M. 2015. A new criterion for assessing discriminant validity in variance-based structural equation modeling. *Journal of the Academy of Marketing Science*, 43(1), 115–135.

Kane, G. C., Palmer, D., Philips, N. A., Kiron, D., & Buckley, N. 2015. Strategy, not technology, drives digital transformation. *MIT Sloan Management Review & Deloitte*, (57181): 27.

Kannan, P. K., & Li, H. 2017. Digital marketing: A framework, review and research agenda. *International Journal of Research in Marketing*, 34(1): 22–45.

Karimi, J., & Walter, Z. 2015. The role of dynamic capabilities in responding to digital disruption: A factor-based study of the newspaper industry. *Journal of Management Information Systems*, 32(1): 39–81.

Kline, R. B. 2011. *Principles and practice of structural equation modeling*. New York: Guilford Press.

Kolloch, M., & Dellermann, D. 2018. Digital innovation in the energy industry: The impact of controversies on the evolution of innovation ecosystems. *Technological Forecasting and Social Change*, 136: 254–264.

Li, F. 2018. The digital transformation of business models in the creative industries: A holistic framework and emerging trends. *Technovation*, 135 (October): 66–74.

Li, L., Su, F., Zhang, W., & Mao, J. 2018. Digital transformation by SME entrepreneurs: A capability perspective. *Information Systems Journal*, 28(1): 1129–1157.

Libert, B., Beck, M., & Wind, J. 2016. *The network imperative: How to survive and grow in the age of digital business models*. Boston, MA: Harvard Business Review Press.

Lindgart, Z., & Ayers, M. 2014. Driving growth with business model innovation. Boston Consulting Group. Retrieved from: www.bcg.com/publications/2014/growth-innovation-driving-growth-business-model-innovation

Loebbecke, C., & Picot, A. 2015. Reflections on societal and business model transformation arising from digitization and big data analytics: A research agenda. *Journal of Strategic Information Systems*, 24(3): 149–157.

Lokuge, S., Sedera, D., Grover, V., & Dongming, X. 2019. Organizational readiness for digital innovation: Development and empirical calibration of a construct. *Information & Management*, 56(3): 445–461.

Lucas, H. C., & Goh, J. M. 2009. Disruptive technology: How Kodak missed the digital photography revolution. *Journal of Strategic Information Systems*, 18(1): 46–55.

MacKenzie, S. B., Podsakoff, P. M., & Podsakoff, N. P. 2011. Construct measurement and validation procedures in MIS and behavioral research: Integrating new and existing techniques. *MIS Quarterly*, 35(2): 293.

Matt, C., Hess, T., & Benlian, A. 2015. Digital transformation strategies. *Business and Information Systems Engineering*, 57(5): 339–343.

Nath, P., & Mahajan, V. 2008. Chief marketing officers: A study of their presence in firms' top management teams. *Journal of Marketing*, 72(1): 65–81.

Ng, I. C. L., & Wakenshaw, S. Y. L. 2017. The Internet-of-Things: Review and research directions. *International Journal of Research in Marketing*, 34(1): 3–21.

Pagani, M., & Pardo, C. 2017. The impact of technology on relationships in a business network. *Industrial Marketing Management*, 67, 185–192.

Parida, V., Sjödin, D., & Reim, W. 2019. Reviewing literature on digitalization, business model innovation, and sustainable industry: Past achievements and future promises. *Sustainability*. 11(2): 391.

Peppard, J., Edwards, C., & Lambert, R. 2011. Realizing strategic value through center-edge digital transformation in consumer-centric industries. *MIS Quarterly Executive*, 10(2): 115–117.

Podsakoff, P. M., & Organ, D. W. (1986). Self-reports in organizational research: Problems and prospects. *Journal of Management*, 12(4), 531–544.

Podsakoff, P. M., MacKenzie, S. B., Lee, J. Y., & Podsakoff, N. P. 2003. Common method biases in behavioral research: A critical review of the literature and recommended remedies. *Journal of Applied Psychology*, 88(5): 879–903.

Polites, G. L., & Karahanna, E. 2013. The embeddedness of information systems habits in organizational and individual level routines: Development and disruption. *MIS Quarterly*, 37(1): 221–246.

Quinn, L., Dibb, S., Lyndon, S., Conhoto, A., & Analogbei, M. 2016. Troubled waters: The transformation of marketing in a digital world. *European Journal of Marketing*, 50(2): 2113–2133.

Quinton, S., Canhoto, A., Molinillo, S., Pera, R., & Budhathoki, T. 2018. Conceptualising a digital orientation: Antecedents of supporting SME performance in the digital economy. *Journal of Strategic Marketing*, 26(5): 427–439.

Rai, A., Patnayakuni, R., & Seth, N. 2006. Firm performance impacts of digitally enabled supply chain integration capabilities. *MIS Quarterly*, 30(2): 225–246.

Reis, J. C. G., Amorim, M., Melao, N., & Matos, P. 2018. Digital transformation: A literature review and guidelines for future research. In *World conference on information systems and technologies* (pp. 411–421). Springer, Cham. 10.1007/978-3-319-77703-0_41.

Rossiter, J. R. 2002. The C-OAR-SE procedure for scale development in marketing. *International Journal of Research in Marketing*, 19(4): 305–335.

Singh, A., & Hess, T. 2017. How chief digital officers promote the digital transformation of their companies. *MIS Quarterly Executive*, 16(1): 1–17.

Singh, A., Klarner, P., & Hess, T. 2020. How do chief digital officers pursue digital transformation activities? The role of organization design parameters. *Long Range Planning*, 53(3): 101890.

Strong, D. M., & Volkoff, O. 2010. Understanding organization-enterprise system fit: A path to theorizing the information technology artifact. *MIS Quarterly*, 34(4): 731–756.

Svahn, F., Mathiassen, L., & Lindgren, R. (2017). Embracing digital innovation in incumbent firms: How Volvo cars managed competing concerns. *MIS Quarterly*, 41(1).

Syam, N., & Sharma, A. 2018. Waiting for a sales renaissance in the fourth industrial revolution: Machine learning and artificial intelligence in sales research and practice. *Industrial Marketing Management*, 69: 135–146.

Tabrizi, B., Lam, E., Girard, K., & Irvin, V. 2019. Digital transformation is not about technology. Available at: https://hbr.org/2019/03/digital-transformation-is-not-about-technology.

Tallon, P. P., & Pinsonneault, A. 2011. Competing perspectives on the link between strategic information technology alignment and organizational agility: Insights from a mediation model. *MIS Quarterly*, 35(2), 463–486.

Teece, D. J. 2010. Business models, business strategy and innovation. *Long Range Planning*, 43(2–3): 172–194.

Teece, D. J. 2018. Business models and dynamic capabilities. *Long Range Planning*, 51(1): 40–49.

Tortorella, G. L., & Fettermann, D. 2017. Implementation of industry 4.0 and lean production in Brazilian manufacturing companies. *International Journal of Production Research*, 7543: 1–13.

Tumbas, S., Berente, N., & vom Brocke, J. 2017. Three types of chief digital officers and the reasons organizations adopt the role. *MIS Quarterly Executive*, 16(2), 121–134.

Tumbas, S., Berente, N., & vom Brocke, J. 2018. Digital innovation and institutional entrepreneurship: Chief digital officer perspectives of their emerging role. *Journal of Information Technology*, 33(3): 188–202.

Van Alstyne, M. W., Parker, G. G., & Choudary, S. P. 2016. Pipelines, platforms, and the new rules of strategy. *Harvard Business Review*, 94(4), 54–62.

Verhoef, P., Broekhuizen, T., Bart, Y., Bhattacharya, A., Dong, J. Q., Fabian, N., & Haenlein, M. 2019. Digital transformation: A multidisciplinary reflection and research agenda. *Journal of Business Research*, (September). https://doi.org/10.1016/j.jbusres.2019.09.022.

Verhoef, P. C., & Bijmolt, T. H. A. 2019. Marketing perspectives on digital business models: A framework and overview of the special issue. *International Journal of Research in Marketing*, 36(9): 341–349.

Vial, G. 2019. Understanding digital transformation: A review and a research agenda. *Journal of Strategic Information Systems*, 28(2): 118–144.

Volkoff, O., Strong, D. M., & Elmes, M. B. 2007. Technological embeddedness and organizational change. *Organization Science*, 18(5): 832–848.

Warner, K. S., & Wäger, M. 2019. Building dynamic capabilities for digital transformation: An ongoing process of strategic renewal. *Long Range Planning*, 52(3), 326–349.

Wong, K. 2018. How Domino's transformed into an e-commerce powerhouse whose product is pizza. Retrieved from: www.forbes.com/sites/kylewong/2018/01/26/how-dominos-transformed-into-an-ecommerce-powerhouse-whose-product-is-pizza/#6926840e7f76

Yu, Y., Li, M., Li, X., Zhao, J. L., & Zhao, D. 2018. Effects of entrepreneurship and IT fashion on SMEs' transformation toward cloud service through mediation of trust. *Information & Management*, 55(2): 245–257.

Chapter 21

Supporting pervasive digitization in Italian SMEs through an open innovation process

Cinzia Colapinto, Vladi Finotto and Nunzia Coco

Understanding digital transformation and its implementation

In the last two decades, the emergence of a diverse set of novel and powerful digital technologies, platforms, and infrastructures has transformed both innovation and entrepreneurship in significant ways with broad organizational and policy implications (Nambisan, 2017; Nambisan et al., 2017; Yoo et al., 2010). Indeed, the phrase *digital transformation* has come into wide use in contemporary business media to signify the transformational or disruptive implications of digital technologies for businesses (e.g. Boutetiere et al., 2018), and more broadly, to indicate how existing companies may need to radically transform themselves to succeed in the emerging digital world (e.g., McAfee and Brynjolfsson, 2017; Rogers, 2016; Venkatraman, 2017).

Technological development is often recognized as a prerequisite for deploying innovation efficiently because it enhances inter-organizational collaboration, which better provides access to external resources and new markets, as well as a source of new knowledge, especially when referring to SMEs (Narula, 2004; Nooteboom, 1994).

Digital transformation is an umbrella term that denotes a hodgepodge of technological trajectories, such as additive manufacturing, big data and analytics, collaborative robots, advanced simulation and augmented reality, Industrial Internet of Things (IoT), cybersecurity, and cloud computing (Gerbert et al., 2015). We refer to a bundle of heterogeneous technologies, which are associated with the ability to enable and accelerate the digital connection between products, processes, activities, and firms and that should lead manufacturing towards the so-called fourth industrial revolution, "Industry 4.0."

Despite the recognized benefits of digital transformation, it is unclear how firms handle the adoption of digital technologies, especially SMEs that have significant resource constraints. The investigation on the obstacles focused on single technologies in the extant literature. On the one hand, such fragmentation allows us to identify the mismatch between specific organizational factors and the logic of particular families of technologies. On the other hand, such fragmentation risks slowing down our understanding of digital transformation for two reasons: first, SMEs might not be as granular as assumed by these studies but they might be recalcitrant towards digital as a whole; second, if the assumption that digital technologies break down functional silos in organizations and command integration among different firms in a supply chain and enable end-to-end production and circulation of information, a holistic approach and a systemic view, rather than a local and partial one, might be advisable.

In this fragmented literature, nonetheless, different categories of obstacles are found. Resource constraints are among the most cited factors in the literature. Moreover, given their size, SMEs are often "domain specialists" focusing on the essential operations to develop and manufacture discrete products and lack in general management and infrastructural functions such as IT. As a consequence, they tend to be less aware than larger firms of new trends and transformations and tend to show conservatism behaviors towards new technologies, processes, and ways of doing things. They face mismatches in the search for specialized skills in the labor market, since large and more structured firms signal better, and thus attract, their propensity to invest in new technological advancements (for a synthesis of the literature addressing the gaps, see Coleman et al., 2016).

According to Gabrielli and Balboni (2010), the human capital of SMEs' entrepreneurs and managers play a crucial role (e.g., the more entrepreneurs, managers, and skills are familiar with digital technologies per se, the more they are prone towards their adoption in their organizations). Conversely, not only organizational factors influence the adoption of digital technologies by SMEs. Given their conservative way of thinking and the impossibility of dedicating scarce resources to the experimentation of novel ways of doing things, these firms tend to mimic the practices and strategies of competitors and other firms in the same industry or strategic group (Karjaluoto and Huhtamaki, 2010). Signals given within the organizations, then, are crucial in giving the whole firm momentum and in creating widespread commitment towards the adoption of digital technology and the subsequent change in how firms do business (Bharadwaj and Soni, 2007). Immersed as they are in the operational minutiae of the firm both for human resource scarcity and for their technical and operational experience, entrepreneurs and managers in small firms might be not entirely convinced or able to frame how digital tools might be used to improve their firms' efficiency or efficacy and might be suspicious of the effects of digital transformation.

The open innovation literature has focused on the sharing and flow of knowledge and technological assets across organizational boundaries in pursuit of innovation and entrepreneurship (e.g., Chesbrough, 2003; Dahlander and Gann, 2010; West and Bogers, 2017). Empirical work conceptualizing open innovation in SMEs reports a broad range of approaches to external technological collaborations and a huge variety of partners involved (Brunswicker and Vanhaverbeke, 2015; Nieto and Santamarìa, 2010; Parida et al., 2012). Innovating SMEs extensively rely on external networking to access new relevant knowledge and missing innovation assets, while customer involvement is the most common OI practice for external technology exploitation (van de Vrande et al., 2009). Open innovation practices range from outside-in and inside out to coupled processes that involve co-creation with (mainly) complementary partners. Besides the collaboration, co-creation initiatives span many external actors (West et al., 2014) as universities, government bodies, intermediaries, and citizens. These actors, in innovation processes, are resource integrators, extremely valuable especially for SMEs (Narula, 2004). A similar approach is found in the Triple Helix theory able to interpret the upcoming events of the fourth industrial revolution, which is bringing subversive dimensions for any kind of socio-economic organizations (businesses, universities, or governments) and where the firm is playing a core role (Ryan et al. 2018).

The multi-stakeholder digitization process in Venice

Understanding the reasons behind the delay of SMEs in the adoption of digital technologies is crucial in countries such as Italy. According to the Digital Economy and Society Index (European Commision, 2019), Italy has a long way to go as it ranks 23th out of the 28 EU Members, thus it is still below the EU average. The data about the use of e-commerce and Internet service, recourse to ICT specialists, fast broadband and cloud computing services show low investments in digital technologies. Istat (2019) data confirm that Italian firms are not highly digitized: only 3% have completed the digital transformation. These firms are mainly medium-sized or large firms that account heavily for added value (25%) and workers (13%); however, they cannot sustain the whole national economy.

This chapter presents the findings of a project aiming at supporting the digital transformation of small and medium-sized (with an emphasis on micro-) firms, financed and co-managed by the Chamber of Commerce of Venice (in partnership with Ca' Foscari University of Venice) within the framework of the Italian governmental strategy on digital transformation and Industry 4.0. The purpose of the 2019 project was to introduce and walked through two waves of approximately 40 SMEs towards digital transformation providing mentoring, knowledge, and facilitating networking activities.

The project was designed as an action research because of its distinctive character to address the twin tasks of bringing about change in organizations and in generating robust, actionable knowledge. As such it is an evolving process that is undertaken in a spirit of collaboration and co-inquiry, whereby research is constructed and conducted with members of a social system, rather than on or for them (Coghlan and Shani, 2018; Shani and Pasmore, 1985).

Therefore, an eight months' co-creation process was structured to promote digital transformation, following open innovation practices, inspired by design thinking mainly for three reasons. First, the process is characterized by continuous interactions among different actors, namely the SMEs, five researchers from two different university departments (Computer Science and Management), 20 digital promoters (equally distributed in the two departments/areas and selected by the university researchers), a university foundation (four project managers) and the Chamber of Commerce (three members). This orientation toward co-creation introduces a distinctly social focus and emphasis on collaboration that other methods lacked. The second essential element of design thinking relates to the role of empathy (Patnaik, 2009). Empathy goes beyond mere recognition of the subjectivity of the design domain; virtually all current descriptions of the process emphasize design thinking as human-centered and user-driven as a core value. The third element builds on the design's strong emphasis on the concrete and the visual to emphasize specifically the key role of prototyping. Certainly, prototyping has long been a central feature in fields such as architecture and product development, but design thinking's view of prototyping diverges from the kind of sophisticated 3D prototypes and models traditionally seen in these fields: its function is to drive real-world experimentation in service to learning rather than to display, persuade, or sell; these prototypes act as "playgrounds" rather than "dress rehearsals" (Schrage, 1999).

The initiator was the Chamber of Commerce who opened the first call in December 2018: from an initial group of 45 firms, we got a final sample of 38 firms located in the Venetian area. In terms of firm size, the sample was mainly made up of small firms (49%), while micro-firms accounted for 38% and medium-sized for 13% (2.31 employees on average). Most firms were B2C (42%), and some were simultaneously active on both segments B2C and B2B (21%). Based on a self-assessment questionnaire, each firm positions itself on a continuum from newcomers to champions: our sample firms were mainly divided into two very different groups, namely apprentices (33%) and specialists (47%). In general, they were very keen on investing in digital (95%) and digital training (60%).

Each SME was assigned to a small team of digital promoters (in each team we ensured a mixed knowledge about management and technologies) who fully accessed the digital maturity of the organization. Then, all SMEs were invited to participate in two full days of hands-on training around technology 4.0, and soon after the digital promoters visited the assigned SME to gain a deeper understanding of their situation in terms of digitalization, resources, corporate organization, and strategy. This phase culminated in June 2019 with the selection of five SMEs; in each enterprise we ran a tailor-made co-creation workshop focusing on the main problem encountered. In July, each group suggested an actionable roadmap toward digital transformation indicating the different steps, based on a deeper understanding of each scenario/situation. The full process is presented in Figure 21.1.

Our action research protocol identifies two cycles. The first cycle aimed at diffusing knowledge on new technologies among participant firms and gain insights from them, working on reducing hidden costs. The second cycle, inspired by the design thinking approach, focused on five selected SMEs. The aim was to uncover, generate, and analyze interesting insights emerging mainly from their customers.

In particular, the first cycle generated a collaborative understanding of SMEs' opportunities to adopt 4.0 digital. Therefore, this cycle ended with semi-structured interviews to explore the business model and identify which possible technology could be adopted to develop more fitting strategies.

The second cycle lasted three months and aimed at challenging a technology-driven approach in favor of a strategic decision making one, promoting a human-centered and

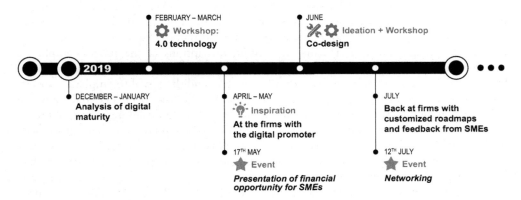

Figure 21.1 Project timeline.

user-driven vision as a core value instead. In other words, the trigger of the process was not the potential of technologies but a focus on the strategy of firms and the interests and perspectives of individual members. After identifying potential strategic evolutions and re-design of firms' business models, technologies were considered and selected as enablers of these transformations.

This phase was inspired by design thinking, and involved the five selected SMEs, to maximize the diversity of contexts in which digital transformation can be implemented. According to Yin's scope of research efficacy (Yin, 2009), case selection was also based on companies' willingness to provide data and detailed information on the crucial innovation project (through which the main strategic change occurred). This choice allowed us to collect and compare grounded empirical information on the challenges faced by the SMEs (Du et al., 2014; van de Vrande et al., 2009).

The research team jointly with the SMEs uncovered, generated, and analyzed interesting insights emerging mainly from the analysis of their customers to complement their previous inner view of their digital maturity. In June 2019, SMEs' customers were interviewed to assess their experience. Then, in July 2019, during a one-day co-creation workshop run by interdisciplinary teams, new concepts for their future roadmaps emerged so that the firms could rethink their business model highlighted by their customers' point of views. Subsequently, all actors prototyped one of the concepts generated during the workshops and discussed how to implement the digital transformation process.

The project data

Throughout the research process, the teams designed and implemented the different phases, protocols, and tools and engaged in collaborative sense-making and sense-giving activities. We follow an abductive approach to data analysis and we constantly moved back and forth between data and theory. Next, we described how the concepts, categories, and their relationships surfaced. Our analysis commenced by reading the empirical material and arranging the events into increasingly coherent narratives. The distinctly evolutionary phases (digital starting point of SMEs, education in Industry 4.0 technology, inspiration phase, ideation through co-design, reflection, and implementation roadmap) afforded constant comparison among the SMEs and the process involvement (see Figure 21.1).

During the grounded sensemaking, the co-authors' conceptual understanding of the walk-through process on digitalization played an integral role in categorization. We reviewed the literature to generate ideas and theories to analyze the initiative's evolution. Leveraging our notes and our observations, we coded the data into categories that clustered into three overarching constructs: governance mechanisms, technology attitude, and organizational structure. Our sources are shown in Table 21.1.

Empirical findings and discussion

Innovation literature suggests that being able to innovate is the result of specific strategic decisions carried out by firms. These decisions are influenced by several variables, including size (De Jong and Vermeulen, 2006), institutional frameworks (Blind et al., 2017), entrepreneurial culture (Shan et al., 2016), the amount of resources dedicated to R&D (Baumann

Table 21.1 Project data

Data source	Data type	Analytical use
Questionnaires	45 self-assessment (SELFI 4.0)	Point of departure – *Gathering data regarding SMEs digital readiness*
	38 In details on digital capabilities and knowledge with the help of digital promoters (ZOOM 4.0)	Collecting key understanding for each contest, richer and more specific on digital technologies than the self-assessment
	Additional demographic data for the 38 SMEs	Detection of differences in sectors and resources
Interviews	Semi-structured interviews (38) – face-to-face with CEOs, each lasting about two hours – 76 hours (audio)	Gaining familiarity with the process of each SMEs and the framing used around digital technology. Collecting qualitative information about each contest and the client firms' aims.
	Internal meetings with the team project team (6). Each meeting lasts 3 hours	Meetings were organized to decide, verify, and gather feedback with the team on data, interpretations, and themes emerging from the authors' coding, by an iterative process.
	Focus on 5 selected interviews transcribed (171 pages) of the SMEs who were selected to obtain the tailor-made co-creation workshop	In-depth qualitative analysis To trace back and triangulate all the process accomplished by each of these SMEs since the beginning.
Reports	38 reports	Triangulation of context information. All the reports were presented to the SMEs and we gather feedback on the solutions indicate
Workshop co-creation outputs	5 co-creation sessions – each tailor-made for the selected SME. Per workshops, we collected the outputs: 4 or 5 concepts per session.	Detection of differences from the report to the WS outputs Triangulation of contest information about selections criteria confirmation and performances assigned by the platform

and Kritikos, 2016), public financing (Szczygielski et al., 2017) and abilities in creating and nurturing networks and collaborations with third parties (Schøtt and Jensen, 2016).

Our research action allows us to verify that the adoption of digital technologies in SMEs has to be guided and external actors must act as "mentors" and gatekeepers for SMEs to benefit both individuals and organizations. As a result, policy interventions should move from a subsidizing logic (i.e. tax deductions for Industry 4.0-related investments or loans) towards a network model (such as the Triple Helix Model) oriented to relations and processes: SMEs should interact more effectively and share information in a process governed by a public agency/actor. Indeed, firm-specific innovation and acceleration processes were developed toward digital transformation based on open innovation strategy and design thinking approach. To conclude, we can identify specific enablers and barriers that accelerate or impede the advancement of the process of adopting digital technologies.

Enablers

Design workshops enabled the observed SMEs to move forward, managing conflicts between opposing frames, typically being held by different generations within firms: younger generations pushing for new business conceptions and strategies collided with older leaders preserving traditional industry recipes and strategic templates. The neutralization of this tension was obtained through the introduction of a different and external point of view, able to ignite a rethinking process.

Design workshops help in "framing" the problems that entrepreneurs were facing in different ways: external interpretive schemes, categories, language, and perspectives are useful in subtracting entrepreneurs from their inertial behavior and cognitive traps due to past experiences, and "tunnel vision," that is the posture according to which an entrepreneur or a manager frames problems relying on previous successful approaches. In particular, design workshops allow them to connect their daily routines with the customer experience, which gives leaders the possibility to move the discussion from features and functionality. The integration of customer experience into their roadmaps allowed them to better understand how to choose technology strategies which will allow the organization to deliver a certain customer experience – an experience that could be designed by capturing core customer and user needs, and identifying appropriate product features and technology choices.

The second enabler was the possibility to "try something" without commitment. Indeed, the possibility to test a concept instead of committing to a specific technology was a way to better understand how to enrich and complement their current service provided for customers and how to evolve it. Prototyping had a huge impact on enlarging the range of possibilities but also in considering a lot of different technologies. Thus, what our research suggests is that in advising SMEs in efforts of digital transformation might benefit from the use of design thinking methods since these allow the firm to focus not on the technology but on systemic and holistic transformations of their ways of doing business. Thus, technologies become more intelligible within a larger frame, rather than unintelligible sets of gadgets and devices whose business function might be obscure or perceived as "not adequate" to the specificities of small and micro-enterprises.

Barriers

We observed that small companies are often trapped in their daily routine, with no resources to spend on innovation activities and no time to embrace digital literacy (inertia). Despite literature, which often suggests that SMEs are more suitable for innovation projects because they can be more flexible in their routines (Christensen et al., 2005), we found that they are stuck in their daily duties and a lot of different tasks are often accomplished by the same resource, which has no time to embrace the knowledge on new technologies. Therefore, the decision-making process on digital transformation, and more in general on innovation activities, relies on the intuitions of a few resources that already play a huge role in the daily activities and have no time to embrace "bigger pictures." Indeed, digital transformation is not only a technological process, and requires personal transformation and managerial cognition renewal as well.

We identify a possible moment of breakthrough for the adoption of new technology: the transaction to new leadership, or better, the facilitation of the emergence of new leaders, and thus new values and perspectives, in the organization. Specifically, in several cases, we noticed that whenever a process of generational handover (father/son, father-in-law/son-in-law) was taking place, the propensity and the interest to acknowledge new technologies and possible networking activities were growing. The newly appointed leaders demonstrated an open attitude to embrace new views into the organization and to consider new ways of working. The intervention of university researchers, external consultants, and personnel of the chamber of commerce, in these cases, was not just beneficial in terms of the specialized content they brought in the process of digital transformation. Rather, these external and recognized actors projected their credibility and legitimacy on the evolutionary trajectories imagined and proposed by young incoming leaders that previously struggled to obtain attention and recognition from the extant leadership in the firm. In other words, open innovation, involving universities and institutions devoted to the support of the digital transformation, works as both a source of ideas and competences but also as a provider of symbolic capital for incoming young firm leaders that should otherwise struggle with a "*liability of youth*" that would feed organizational inertia. Moreover, the attitude toward planning, to fit into their new roles, felt for these appointed leaders as a need to better evaluate their future directions. That is to say that the intervention of universities and institutions requires new firm leaders to upgrade their skills in planning and presenting their intended courses of actions, since the language of these institutional actors puts a prize on structured means-ends chains that might then be assessed through rigorous and objective measures (such as budget, expected results and returns on the investment, and feasibility). In other words, open innovation involving firms, university researchers, and institutional personnel contributes more than just providing ideas or skills: it creates a shared language, a set of categories and ways of thinking, talking, and presenting that provide the innovation process with structure. However, during the moment of transaction, we also witness many communication conflicts between the two generations, therefore sometimes we notice an even bigger difficulty to move forward.

The main takeaway of our experience is that firms need to be supported and mentored in the development of adaptive capabilities for continuous experimentation aimed at testing traditional assumptions lying at the core of their way of doing business. Fast and low-cost prototyping as a way to devise and strategize according to different future scenarios will be fundamental in preparing firms to be adaptable enough to deploy digital technologies when the imagined changes materialize.

This project reveals a peculiar manifestation of the effects of the triple helix model. While aimed at solving discrete needs and demands – specifically, the need to understand and then deploy specific digital technologies in SMEs – the collaboration among institutions, universities, and firms generated "byproducts" that are, ultimately, crucial in digital transformation. One of these byproducts is the creation of informal relations among small firms. Thick networks of relation among SMEs, we suggest, are not solely conduits for the collaborative development of discrete solutions through the pooling of ideas and knowledge. They rather produce a sense of identity and cohesion that helps these firms to first and foremost make sense of the challenges and to develop a sense of community that enables the subsequent search for solutions.

A final remark for further research is to consider the strategy transition from a focus on products to one that emphasizes services or rethinks a firms' offering in terms of bundles of products and services. As often stated, digital technologies, with their potential to generate, accumulate, and store data at every step of the value chain, allow firms to provide clients with sophisticated bundles of products and services, especially when they move towards smart products. The transition for SMEs is easier said than done; packing products with services, requires firms to abandon a focus on products and production processes, while triggering design efforts and the conception of ecosystems of products and services with the immersion and in-depth understanding of customer experiences. Anchoring strategic planning in customer experiences still generates fierce resistance in these firms. This is an area where open innovation approaches as those illustrated in this chapter might be beneficial.

References

Baumann, J., and Kritikos, A. S. 2016. The link between R&D, innovation and productivity: Are micro firms different? *Research Policy*. https://doi.org/10.1016/j.respol.2016.03.008.

Bharadwaj, P. N. and Soni, R. G. (2007). E-commerce usage and perception of e-commerce issues among small firms: Results and implications from an empirical study. *Journal of Small Business Management*, 45(4), 501–521.

Blind, K., Petersen, S. S., and Riillo, C. A. F. 2017. The impact of standards and regulation on innovation in uncertain markets. *Research Policy*. https://doi.org/10.1016/j.respol.2016.11.003.

Boutetière, H., Montagner, A., and Reich, A. 2018. Unlocking success in digital transformations. *McKinsey & Company*. www.mckinsey.com/business-functions/organization/our-insights/unlocking-success-in-digital-transformations.%0A.

Brunswicker, S., and Vanhaverbeke, W. 2015. Open innovation in small and medium-sized enterprises (SMEs): External knowledge sourcing strategies and internal organizational facilitators. *Journal of Small Business Management*, 53(4), 1241–1263.

Chesbrough, H. W. 2003. The new imperative for creating and profiting from technology. *Harvard Business Publishing*. https://doi.org/10.1111/j.1467-8691.2008.00502.x.

Christensen, J. F., Olesen, M. H., and Kjær, J. S. 2005. The industrial dynamics of open innovation: Evidence from the transformation of consumer electronics. *Research Policy*, 34(10), 1533–1549.

Coghlan, D., and Shani, A. B. R. 2018. *Conducting action research for business and management students*. London: Sage.

Coleman, S., Göb, R., Manco, G., Pievatolo, A., Tort-Martorell, X., and Reis, M. S. 2016. How can SMEs benefit from big data? Challenges and a path forward. *Quality and Reliability Engineering International*, 32(6), 2151–2164.

Dahlander, L., and Gann, D. M. 2010. How open is innovation? *Research Policy*. https://doi.org/10.1016/j.respol.2010.01.013.

De Jong, P. J., and Vermeulen, P. A. M. 2006. Determinants of product innovation in small firms: A comparison across industries. *International Small Business Journal*. https://doi.org/10.1177/0266242606069268.

Du, J., Leten, B., and Vanhaverbeke, W. 2014. Managing open innovation projects with science-based and market-based partners. *Research Policy*. https://doi.org/10.1016/j.respol.2013.12.008.

European Commission. 2019. Digital Economy and Society Index (DESI). https://ec.europa.eu/newsroom/dae/document.cfm?doc_id=59897

Gabrielli, V. and Balboni, B. (2010). SME practice towards integrated marketing communications. *Marketing Intelligence and Planning*, 28(3), 275–290.

Gerbert, P., Rüßmann, M., Lorenz, M., Waldner, M., Justus, J., et al. 2015. Industry 4.0: The future of productivity and growth in manufacturing industries. *Boston Consulting*. https://doi.org/10.1007/s12599-014-0334-4.

Istat. 2019. CITTADINI, IMPRESE E ICT. www.istat.it/it/files/2019/01/Report-ICT-cittadini-e-imprese_2018_PC.pdf

Karjaluoto, H., and Huhtamäki, M. (2010). The role of electronic channels in micro-sized brick and-mortar firms. *Journal of Small Business and Entrepreneurship*, 23(1), 17–38.

McAfee, A., and Brynjolfsson, E. 2017. *Machine, platform, crowd: Harnessing our digital future*. New York: WW Norton & Company.

Nambisan, S. 2017. Digital entrepreneurship: Toward a Digital technology perspective of entrepreneurship. *Entrepreneurship: Theory and Practice*. https://doi.org/10.1111/etap.12254.

Nambisan, S., Lyytinen, K., Majchrzak, A., and Song, M. 2017. Digital innovation management: Reinventing innovation management research in a digital world. *MIS Quarterly*, 41(1), 223–238.

Narula, R. 2004. R&D collaboration by SMEs: New opportunities and limitations in the face of globalisation. *Technovation*. https://doi.org/10.1016/S0166-4972(02)00045-7.

Nieto, M. J., and Santamaría, L. 2010. Technological collaboration: Bridging the innovation gap between small and large firms. *Journal of Small Business Management*. https://doi.org/10.1111/j.1540-627X.2009.00286.x.

Nooteboom, B. 1994. Innovation and diffusion in small firms: Theory and evidence. *Small Business Economics*. https://doi.org/10.1007/BF01065137.

Parida, V., Westerberg, M., and Frishammar, J. 2012. Inbound open innovation activities in high-tech SMEs: The impact on innovation performance. *Journal of Small Business Management*. https://doi.org/10.1111/j.1540-627X.2012.00354.x.

Patnaik, D. 2009. Wired to care. *Pearson Education, Inc*. https://doi.org/10.1017/CBO9781107415324.004.

Rogers, D. L. 2016. *The digital transformation playbook: Rethink your business for the digital age*. New York: Columbia University Press.

Ryan, P., Geoghegan, W., and Hilliard, R. 2018. The microfoundations of firms' explorative innovation capabilities within the triple helix framework. *Technovation*. https://doi.org/10.1016/j.technovation.2018.02.016.

Schøtt, T., and Jensen, K. W. 2016. Firms' innovation benefiting from networking and institutional support: A global analysis of national and firm effects. *Research Policy*. https://doi.org/10.1016/j.respol.2016.03.006.

Schrage, M. 1999. *Serious play: How the world's best companies simulate to innovate*. Boston: Harvard Business Press.

Shan, P., Song, M., and Ju, X. 2016. Entrepreneurial orientation and performance: Is innovation speed a missing link? *Journal of Business Research*, 69(2), 683–690.

Shani, A. B., and Pasmore, W. A. 1985. Organization inquiry: Towards a new model of the action research process. *Contemporary Organization Development: Current Thinking and Applications*. Glenview, IL: Scott Foresman, 438–448.

Szczygielski, K., Grabowski, W., Pamukcu, M. T., and Tandogan, V. S. 2017. Does government support for private innovation matter? Firm-level evidence from two catching-up countries. *Research Policy*. https://doi.org/10.1016/j.respol.2016.10.009.

van de Vrande, V., de Jong, J. P. J., Vanhaverbeke, W., and de Rochemont, M. 2009. Open innovation in SMEs: Trends, motives and management challenges. *Technovation*. https://doi.org/10.1016/j.technovation.2008.10.001.

Venkatraman, V. 2017. *The digital matrix: new rules for business transformation through technology*. Vancouver: Greystone Books.

West, J., and Bogers, M. 2017. Open innovation: Current status and research opportunities. *Innovation: Management, Policy and Practice*. https://doi.org/10.1080/14479338.2016.1258995.

West, J., Salter, A., Vanhaverbeke, W., and Chesbrough, H. 2014. Open innovation: The next decade. *Research Policy*, 43(5), 805–811.

Yin, R. K. 2009. Case Study Research: Design and Methods. *Essential guide to qualitative methods in organizational research*. https://doi.org/10.1097/FCH.0b013e31822dda9e.

Yoo, Y., Henfridsson, O., and Lyytinen, K. 2010. The new organizing logic of digital innovation: An agenda for information systems research. *Information Systems Research*. https://doi.org/10.1287/isre.1100.0322.

Part 6

Conclusion

Chapter 22

Our roadmap to digital transformation

Francesca Checchinato, Andreas Hinterhuber, and Tiziano Vescovi

Based on the literature about digital transformation and on the cases presented in this book, we can conclude that it is more correct to refer to digital transformations than digital transformation. DT is an evolution that depends on the company's approach to a new way of working, where digital technologies are just enablers of a firm transformation. This transformation is driven by people that adopt technologies, but there isn't "a best way" to transform because there are various internal and external constraints (factors) to consider. We agree with Tekic and Koroteev (Chapter 7): "digital transformation is a multifaceted phenomenon in that it has different aspects/implications for different companies."

Thus, digital transformation depends on the company (its size, its organization, its relationships with customers and partners), on the industry, and on the moment in which the company decides to start the process and it seems a "Holy Grail: it is not so easy to find and not so easy to capture" (Michael Nilles, Chief Digital Officer, Henkel, Chapter 8).

The main questions are: "Is digital transformation necessary? Can a firm survive without transforming its business?" The answer is not easy and, of course, depends on contingencies that determine the specific path to follow, but adopting digital technologies seems to be the only chance to survive in the middle-long period. The depth of the transformation is the challenge, because the deeper the transformation, the higher must be the market system's involvement. In the B2B sectors, transformation affects the relationships between suppliers and their clients. If leading companies of a sector start transforming their business, this transformation will have an impact on the whole supply chain. Their suppliers need to embrace the digital transformation if they want to continue to work for them. Suppliers that fit the transformation will survive; the others will be replaced by new-born digital companies.

In the B2C market, digital transformation is in some way pulled by consumers. They need information, emotional contents, customized products, and 24/7 services, etc. Companies must develop strategies to cope with this new landscape and they need digital technologies to automatically respond to these requests. As for B2B, companies that evolve and provide a marketing mix that fits consumers requirements will survive; the others will be replaced. Even if the product is unique and consumers recognize its quality, in a few years its digital part (i.e. customization, pre-sale services, prototypes …) will be part of the so-called "expected product" and thus it will be required.

Therefore, companies can decide if they want to have a proactive role or a passive role. In other words: to just accept the situation and implement the digital transformation accordingly or to anticipate competitors, suppliers, and clients, and act to implement the digital transformation, enhancing the chance to set the rules on their environment.

Therefore, if DT is necessary, it is important to define the most effective way to move on the road of the digital transformation success.

First of all, before starting the process, three main questions must be answered: why transform, what to transform and how to transform. Concerning the first one, it is important to define the reason why this process should start. Digital transformation is often triggered by changing customer behaviors and expectation, by changing competitors' landscape, or by internal drivers, like improving internal processes, reducing labor, producing, or distribution costs (Osmundsen, Iden and Bygstad, 2018; Rojers, 2018).

We can delineate four phases that drive the future of digital transformation, explaining some of the challenges and solutions about "how" to implement a company's digitalization (Figure 22.1).

An internal and external analysis is the first step to define how to transform a company, the processes that require deep changes and the priority, the objectives to achieve in the short term but looking at the bigger plan of the entire transformation. Time is needed. Companies face a trade-off between acting fast and planning carefully. Often, a chief digital officer is hired, usually a person coming from other businesses, who needs to study the company, to have a deep knowledge of the company. As we have learnt from the Etro case study, this step requires time, and all the company must be involved in the analysis.

Then a "technologies starting package" has to be implemented, in order to begin approaching the digital opportunities. The technologies to include in the package depend on the digital transformation main aims, but they are usually related to data analysis platforms. In fact, one of the main aims is usually to improve customer satisfaction and thus profits, but to effectively understand consumers insights, companies need to collect

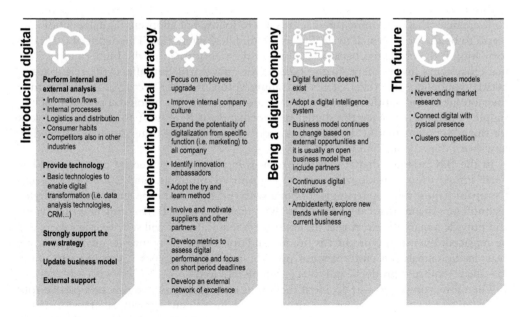

Figure 22.1 The four phases of digital transformation.

and then analyze structured, semi-structured, and unstructured data: they have to create new platforms (Erevelles, Fukawa and Swayne, 2016) that allow them to understand their customers.

Thus, digital transformation requires a new approach and new capabilities, it brings people out of their comfort zone. In this phase the support from the CEO and the advisory board is crucial for motivating employees and making them aware that the transformation is coming, there are no reasons to refuse it. Business models start changing based on the possibilities that technology offers. This change should be just the first one, because in the new era, business models must evolve constantly, to respond to new dynamic competitive scenarios. Finally, to introduce digitalization, experts or governmental support is often needed, especially for SMEs that require both capabilities and resources and because they usually display a "cognitive inertia that prevents them from fully engaging with digital platforms" (Cenamor, Parida and Wincent, 2019). An important, but yet not fully explored, trigger of the digital transformation of traditional companies are internal starts-ups, as the interview with Jörg Hellwig, CDO of Lanxess (Chapter 13) illustrates.

Once the digital transformation starts working, firms must evolve to implement the strategy and benefit form DT. They enter phase 2. The focus is the people; they must be supported in this new situation as well as be inspired with a new corporate culture, based on digital principles (data-driven decisions, dynamism, short time schedules …). HR functions should create a specific plan for their employees, who are facing new tasks and routines. Most of the time they are asked to work on higher-level tasks, where analytics and strategic capabilities as well as an entrepreneurial mindset are required. Companies should provide specific learning projects to develop these new skills. This supports Wedel and Kannan (2016), who stated that a successful development and implementation of marketing analytics in firms require not only a system able to collect different types of data but also analytics education and training. As we learned from Unilever case study (Chapter 5), long learning education, planned by the company for all the employees involved in the digital transformation, is one of the options to solve the competencies problem. They develop upskill and reskill programs to maintain the market employability and guarantee that people working in Unilever are able to cope with analytics without barriers. Moreover, training avoids people fighting against the DT, just from fear. Not just training how to use technology but also coaching. People must learn that in the digital systems they must purpose projects and solutions, even if they could fail; they do not have to be afraid to try. Trial and error is a common method of problem-solving in digital companies and employees need to be confident with this new method. As highlighted by Folonari (Chapter 11), European Head of Business Strategy of Akqa, finding people that really believe in the DT, the so-called ambassadors, helps the implementation success. Building teams and innovations around them can help with improving the process.

In this phase, the DT is no longer linked to a specific function – the function that usually triggers the adoption (usually marketing or sales) – but all the company's functions start embracing the technologies to achieve results. This is an imperative step to exploit DT, otherwise benefits are just partial, and efforts encompass benefits. A poorly executed DT is worse than a non-DT, because people spend time in creating reports or collecting data instead of using that data to take decisions to drive the overall company strategy. Likewise, to exploit the DT opportunities partners must be motivated to embrace this change and adopt the company's technologies system. How can a CRM work if retailers are not integrated? How can an e-commerce work if logistics are not integrated? As highlighted by Viacava, CDT of Etro (Chapter 18) this issue is crucial to implement the digital transformation: wholesalers and retailers need to understand this change; they must receive benefits to embrace the system.

Moreover, since results help people to understand the benefit of DT, companies should develop tailored metrics to assess digital performance. They should focus on short period deadlines, too. Metrics are not just the traditional one but they must follow the new business dynamics. During this phase, most companies are developing a new range of metrics that will follow all the milestones of the digital process. The interview (Chapter 8) with Michael Nilles, CDO of Henkel, highlights key metrics: digital efficiency/productivity, digital growth (net of cannibalization), and digital growth.

Then we arrive at the third phase: being a digital company. What does it mean? First, digital function or a chief of digital transformation does not exist anymore because all the company (and partners) are digitalized and they are adopting an integrated business intelligence system where all the processes can be managed and controlled. Metrics do not focus on digital, but they focus on the overall performance. Being digital implies implementing new business models – as Stefan Stroh, CDO of Deutsche Bahn emphasizes (Chapter 3). The mantra must be "never stop learning" and "never stop evolving." A digital company must change their business models based on the market. If there is any evolution on the market concerning technologies or consumer patterns, then they must adapt their business models to fit the new landscape. Organizational ambidexterity, exploring new trends while serving current business, is the strategic paradigm of digital enterprises. Companies go through a dual transformation by exploiting current competencies and enabling innovative capabilities through strategic transformation. Evidence shows that it is positively associated with higher performances (Mitra, Gaur and Giacosa, 2019).

What's next? There are some timid signs of a turnaround, to connect physical spaces to digital companies. The case of Amazon's stores is an example. Nonetheless, it is easier for a digital company to be successful in a traditional context than for a traditional company to be successful in a digital landscape. The reason is not only related to the company's history, but to both the culture and the processes. Even if they start competing with more traditional companies or companies in their transformation route, digital enterprises base their decision on an intelligent system, and they are used to changing their business models to find the best way to create value continuously. They are also used to analyzing markets and finding opportunities based on never-ending market(ing) research, to keep their market knowledge updated. Finally, since digital transformation engages a market system and not just a single company, competition is shifting from competition between firms to competition between ecosystems with a set of challenges that will need to be further explored.

References

Cenamor, J., Parida, V. and Wincent, J. (2019) How entrepreneurial SMEs compete through digital platforms: The roles of digital platform capability, network capability and ambidexterity, *Journal of Business Research*. Elsevier, 100(December), 196–206. doi: 10.1016/j.jbusres.2019.03.035.

Erevelles, S., Fukawa, N. and Swayne, L. (2016) Big data consumer analytics and the transformation of marketing, *Journal of Business Research*. Elsevier, 69(2), 897–904. doi: 10.1016/J.JBUSRES.2015.07.001.

Mitra, A., Gaur, S. S. and Giacosa, E. (2019) Combining organizational change management and organizational ambidexterity using data transformation, *Management Decision*, 57(8), 2069–2091. doi: 10.1108/MD-07-2018-0841.

Osmundsen, K., Iden, J. and Bygstad, B (2018). Digital transformation: Drivers, success factors, and implications. *MCIS*, September.

Rojers, J. P. (2018). Digital transformation, business model innovation and efficiency in content industries: A review. *The International Technology Management Review*, 7(1), 59–70.

Wedel, M. and Kannan, P. K. (2016) Marketing analytics for data-rich environments, *Journal of Marketing*, 80(6), 97–121. doi: 10.1509/jm.15.0413.

Index

Printed in the United States
by Baker & Taylor Publisher Services